STORAGE

AGRICULTURAL
EDUCATION
IN A
TECHNICAL
SOCIETY

AGRICULTURAL EDUCATION IN A TECHNICAL SOCIETY

An Annotated Bibliography of Resources

MARY RUTH BROWN   EUGENIE LAIR MOSS
KARIN DRUDGE BRIGHT

Foreword by
HOWARD SIDNEY

AMERICAN LIBRARY ASSOCIATION
Chicago 1973

Cover photograph courtesy of
USDA–Soil Conservation Service

Library of Congress Cataloging in Publication Data

Brown, Mary Ruth, 1925–
  Agricultural education in a technical society.

  1. Agriculture—Bibliography.   I. Moss, Eugenie
Lair, 1921–   joint author.   II. Bright, Karin Drudge,
1947–   joint author.   III. Title.
Z5071.B78   016.63   72-7501
ISBN 0-8389-0128-X

International Standard Book Number 0-8389-0128-X (1973)

Library of Congress Catalog Card Number 72-7501

Copyright © 1973 by the American Library Association
All rights reserved. No part of this publication may be reproduced
in any form without permission in writing from the publisher,
except by a reviewer who may quote brief passages in a review.

Printed in the United States of America

To our husbands, who
patiently suffered neglect
during the writing of this book.

CONTENTS

FOREWORD	ix
PREFACE	xi

A: BOOKS	1
Basic Background Books	2
General Works	2
Chemistry and Physics	4
Mathematics and Statistics	6
Economics	8
Agriculture in General	10
Biology	12
Genetics and Cytology	14
Microbiology	16
Parasitology	18
Ecology	19
Plant Sciences	20
General Works	20
Taxonomy	25
Anatomy, Physiology, and Morphology	29
Plant Breeding	33
Plant Diseases and Control	33
Weeds and Weed Control	36
Field and Forage Crop Production	37
General Horticulture	42
Ornamental Horticulture	49
Mycology	56
Forestry and Forest Products	57
Animal Sciences	63
General Zoology	63
Animal Husbandry	68
Poultry Husbandry	78
Feeds and Feeding	79
Veterinary Medicine	80
Entomology	83
Apiculture	86
Pests and Pest Control	86
Zoos and Zoo Management	88
Fish and Marine Resources	88
Physical Sciences	90
General Works	90
Agricultural Engineering	94
Soils and Fertilizers	99
Agricultural Chemistry and Physics	103
Climatology and Meteorology	104
Food Sciences	105
General Works	105
Food Composition and Analysis	106
Food Products	110
Food Processing and Preservation	114
Food Microbiology and Toxicology	116

Food Quality Control	118
Food Distribution and Trade	119
Nutrition	120
Food Supply	121
Natural Resources	122
General Conservation	122
Wildlife Management	126
Parks and Recreation	128
Social Sciences	129
Agricultural Economics and Agribusiness	129
Rural Sociology	136
B: PERIODICALS	138
Indexing and Abstracting Services	139
General Periodicals	140
C: PAMPHLETS AND OTHER INEXPENSIVE MATERIALS	155
D: FILMSTRIPS, TRANSPARENCIES, AND OTHER AUDIOVISUAL AIDS	158
E: GOVERNMENT DOCUMENTS	164
Catalogs, Indexes, and Other Guides	165
Monographs	167
Periodicals and Other Serials	170
F: GUIDES TO SOCIETIES, ORGANIZATIONS, AND INDUSTRIES CONCERNED WITH AGRICULTURE AND RELATED FIELDS	180
PUBLISHERS DIRECTORY	188
AUTHOR INDEX	193
TITLE INDEX	203
SUBJECT INDEX	218

FOREWORD

Agricultural Education in a Technical Society is the first publication prepared specifically for educators working in secondary and postsecondary agricultural education. The objective is to assist educators, administrators, and librarians in vocational technical education in agriculture.

Agriculture is a growth industry in terms of production services and the use of modern technologies. Our standard of living is dependent upon continued research, modern production techniques, and the use of the best-known procedures and methods in the operation of agricultural businesses.

The land-grant colleges have achieved a remarkable record in agricultural research and education that has completely revolutionized the agricultural industry. The implementation of this knowledge on the farm and in the factories has freed the mass of our people from the drudgery of manual labor. The largest segment of people now employed in agriculture and related occupations need skills and technical knowledge.

It is apparent that our greatest need in the forseeable future is for postsecondary agricultural technical programs of less than a baccalaureate degree where agriculturalists can acquire the skills and technical know-how for successful performance on the job. Agricultural educators must therefore have the latest and best information readily available to teach the application of skills and technologies to students preparing for agricultural careers.

Since 1966 the number of students enrolled in postsecondary programs in agriculture has tripled. Teachers in community colleges, technical institutes, and postsecondary institutions are finding it difficult to locate teaching materials for agricultural education at the technical level.

Agricultural Education in a Technical Society is the first annotated bibliography of resources compiled to assist teachers in agricultural education. This bibliography is extremely useful since all known resources are in one publication. The authors are to be complimented for their careful selection, complete listing of current materials, and inclusion of publications of proven value.

All resources in the bibliography have been carefully reviewed and compiled by knowledgable educators and librarians with expertise in agricultural education. This bibliography meets an urgent need for a growing industry and for the increasing number of students and teachers in agriculture. It is an invaluable publication for educators in secondary and postsecondary education in agriculture that will aid in improving the quality of instruction for students and will make teaching more satisfying for faculty.

My congratulations to the authors for producing a unique and excellent publication. This contribution to agricultural education will be appreciated and used by many to make life better for many people throughout the world.

HOWARD SIDNEY
Chairman
Division of Agriculture and Natural Resources
Agricultural and Technical College
State University of New York
Cobleskill, New York

PREFACE

As a result of the increased emphasis on technical education, new schools and institutes are rapidly being established all over the United States. The success of these schools is seriously handicapped by the lack of dependable guides for selecting and building collections of print and nonprint materials; that is, materials specifically designed so as to aid in implementing and sustaining ongoing and projected curricula.

Therefore, it is the objective of this work to bring together in one volume annotated lists of books, periodicals, government documents, sources of pamphlets, audiovisual aids, information on societies, industries, and other organizations concerned with agriculture and related fields.

The levels of material in this book are designed primarily for use in junior colleges and technical schools, although it is felt that much of the material included could also be useful for vocational classes at the secondary level.

Norman K. Hoover's *Handbook of Agricultural Occupations* (Danville, Ill.: Interstate, 1969) and *Agriculture, Forestry, and Oceanographic Technicians,* edited by Howard Sidney (Garden City, N.Y.; Doubleday, n.d.), were used as reference sources for determining the subfields of agriculture to be included. A broad selection of areas covered in the book include forestry, food processing, crop production, gardening, ornamental horticulture, livestock production, veterinary medicine, fisheries, marketing, farm equipment, and agribusiness.

The following were used as guides for determining criteria for inclusion of titles: the U.S. Office of Education two-year post high school curriculum guides, specific subject bibliographies, book reviews, and the recommendations of teachers and other knowledgable persons. It was necessary in some cases to include books on a more advanced level either because few sources were available on the subject at the specified level, or because they were included in the U.S. Office of Education curriculum guides.

An attempt was made to include only current materials; however, some older books were included for their outstanding value, or because more current materials were not available on the subject. Only materials in the English language believed to be in print up to January 1, 1971 are included. A few books published outside the United States were included because their value was not limited to any geographic or climatic region. Textbooks, reference books, and supplementary reading materials are listed. Local and state materials were omitted unless they were found to have relevance to other geographic areas. Symposia were included only when of special value to subject areas.

The annotations used tend to reflect the opinions of reviewers, who are experts in the topic on which the book is written, and the opinions of classroom teachers. Items to be considered for first purchase are starred. The opinions of reviewers, the authors' familiarity with the book's use by classroom teachers, and recommendations of University of Kentucky College of Agriculture subject specialists were used as criteria for starring.

Acknowledgment is given to the University of Kentucky Libraries and to the Veterans Memorial Public Library of Bismarck, North Dakota for the use of their resources. The authors wish to express their appreciation for the sustained cooperation and advice of the University of Kentucky College of Agriculture faculty and to the staff of the University of Kentucky Agriculture Library for their support throughout the development of this project. Special thanks are extended to Mrs. Betty Jo Howard who typed the manuscript, Miss Gail Ruffner who served as proofreader, and to Miss Mary Glee Buck who gave valuable assistance in many phases of this book's preparation.

MARY RUTH BROWN
Head Librarian
University of Kentucky
Agriculture Library
Lexington, Kentucky

EUGENIE LAIR MOSS
Reference Librarian
Veterans Memorial
Public Library
Bismarck, North Dakota

KARIN DRUDGE BRIGHT
Librarian
Horry-Georgetown
Technical Education Center
Conway, South Carolina

A

Books

The profusion of book materials published in the agricultural sciences made it necessary to establish definite policies for selection and arrangement of titles annotated in this section. Books are arranged alphabetically by author under broad subject headings. When there is more than one title by the same author, the jointly authored titles follow those by the single author. In cases where the subtitle is unnecessary for understanding the book's content, it is omitted. Prices, listed for each entry, are quoted from the 1970 *Books in print* but they are subject to change at any time. Publishers are indicated by an abbreviation in each entry, and their addresses are listed alphabetically by abbreviation in the Publishers Directory. Inclusion of materials in this section was based on the following criteria:

1. Books are included only when there are favorable reviews written on them, when recommended by faculty members and other knowledgable persons, or when listed in vocational/technical agriculture curriculum guides and general bibliographies for basic vocational/technical collections.

2. Titles having less than 100 pages are not considered to be books.

3. General reference materials are excluded because they are covered in so many other sources (such as Winchell's *Guide to reference books*).

4. Titles are included only when known to be in print.

5. In some areas, such as floriculture, where the amount of literature available is enormous, only a representative list of titles can be given in order to maintain a balanced as well as a comprehensive bibliography. The basic background books are a mere sampling of some of the best materials available. In other fields where there is a dearth of books published, titles may not necessarily be of highest quality but are listed because there are no better ones. Other titles not included in this section may be found in *Cumulative book index, Books in print,* and other similar bibliographies.

6. Highly expensive books which are of questionable value for basic agricultural vocational/technical collections are excluded.

7. U.S. and state government published books are omitted here. The most important ones are annotated in section E, Government Publications, and others may be found in *Monthly catalog of U.S. government publications* and *Monthly checklist of state publications*.

8. With one or two exceptions, no books are included which have copyright dates later than 1970. (Maintenance of an up-to-date collection is essential, and the librarian, administrator, and instructor should constantly read reviews of books in current periodicals, brochures, and publishers' announcements to familiarize themselves with newly published works.)

BASIC BACKGROUND BOOKS

General Works

A1 **Barzun, Jacques,** and **Henry F. Graff.** The modern researcher. rev. ed. 1970. Harcourt. 430p. $8.50.

A step-by-step guide written for the technical school student and the 4-year college student as well as for the researcher. Gives helpful directions on how to use the library, how to take notes efficiently, how to construct clear sentences, and how to write a well-structured paper.

A2 **Bates, Robert L.,** and **Walter C. Sweet.** Geology: an introduction. 1966. Heath. 367p. illus. $7.95.

Presents physical and historical geology in a nontechnical manner for those who desire only a basic introduction, which is equivalent to a 1-semester course. Covers all of the major geologic processes, theories, and materials.

A3 **Beal, George M., Joe M. Bohlen,** and **J. Neal Raudabaugh.** Leadership and dynamic group action. 1962. Iowa State. 365p. illus. $6; paper, $2.50.

A classroom text, written in casual style, on group dynamics and leadership techniques. Discusses the fundamentals of group operation and mechanics and the relationship of the individual member to the organization. Provides an explanation of how to evaluate group effectiveness and leadership of members and how to improve these qualities to achieve proper intragroup relations and activity.

A4 **Brown, Leland.** Communicating facts and ideas in business. 1961. Prentice-Hall. 388p. illus. $9.50.

First, sets forth the fundamental principles of skillful communication; then explains the importance of communicative ability and the technique of logical and definitive thinking; and finally, shows how these skills may be applied to problem solving and everyday conversation.

A5 **CRC Handbook of Laboratory Safety.** Ed. by Norman V. Steere. Chemical Rubber. 568p. illus. $22.50.

A handy source of information on laboratory safety. Chapters contributed by specialists. In addition to general treatment of the subject, specific features such as first aid, legal liability, hazards, and protective equipment are discussed. A good book for aspiring laboratory scientists and technicians and a good supplementary handbook for general laboratory courses.

A6* **Eastman, Edward R.** How to speak and write for rural audiences. 1960. Interstate. 164p. illus. $4.95.

Because the author is experienced as an editor of an agricultural periodical, he writes a highly authoritative, helpful book. Written from the do-it-yourself point of view for the student desiring basic knowledge of how to write and speak. Offers suggestions on how to acquire writing and conversational skills based on the fundamental principles of penmanship and speech.

A7 **Evans, Ralph M.** An introduction to color. 1948. Wiley. 340p. illus. $16.

A presentation of the basic scientific principles of color from the physical, physiological, and psychological view. Not only discusses such topics as color systems, color measurement, and the effects of color on behavior, etc., but also describes its application and importance in art and technology.

A8 **Gilman, William.** Science: USA. 1965. Viking. 499p. $7.95.

A description of the development of science and technology in the last 20 years and an evaluation of their current position in the American political, social, and economic world. A serious subject presented in journalistic style with touches of humor. Good background reading for students interested in pure or applied science careers.

A9 **Grantz, Gerald J.** Home book of taxidermy and tanning. 1969. Stackpole. 160p. illus. $7.95.

How-to book on taxidermy and tanning with illustrations which explain the processes discussed. Information on equipment, techniques, mounting of large and small animals, cleaning and processing of skins and furs, and preparation of them for collection or study. Good for students learning tanning skills.

A10 **Hay, Robert D.** Written communications for business administrators. 1965. Holt. 487p. illus. $9.50.

Explains how to compile information, arrange it logically, and present it clearly in the formal report. Also deals with particular problems such as how to handle or write the letter of refusal, the acceptance letter, the ordinary request, and the special request.

A11* **Hiscox, Gardner D.,** ed. Henley's twentieth century book of formulas, processes, and trade secrets. new rev. and enl. ed. by T. O'Conor Sloane. 1957, reprint 1968. Books. 867p. illus. $5.

A unique reference because it provides thousands of chemical recipes, formulas, and processes, which have been gathered from scattered sources, in 1 compact volume. Contains formulas for products ranging from wines to cosmetics. Includes a glossary of chemical terms, a directory of sources for chemicals and supplies, a laboratory procedures outline, and a quick reference index.

A12* **Karr, Harrison M.** Now you're talking. 2d ed. 1968. Interstate. 230p. illus. $4.95.

Devoted to the improvement of oral communicative ability. Identifies problems with talking in various situations ranging from group discussions to formal speeches. Offers suggestions on, and examples of, how to maintain vocabulary, voice, and self-control as well as how to prepare and organize a speech.

A13 **Lastrucci, Carlo L.** The scientific approach: basic principles of the scientific method. rev. ed. 1967. Schenkman. 272p. illus. $5.95.

Explains in nontechnical language what science is and how scientific theory can be used to analyze and solve problems. Coordinates the 2 related aspects of science, its philosophy and its methods.

A14 **Lawrence, Nelda R.** Writing communications in business. 1964. Prentice-Hall. 145p. $4.95.

Enumerates the basic requirements of effective writing in workbook format through which examples of good and poor business letters are presented for comparison. Types of business communication discussed are letters, press notices, summaries, reports, minutes, and memos.

A15 **Leidy, W. Philip.** A popular guide to government publications. 3d ed. 1968. Columbia. 365p. $12.

An annotated list of popular U.S. government publications, organized alphabetically by subject. Includes price and order number for ordering from the Superintendent of Documents.

A16 **Mason, Ralph E.,** and **Peter G. Haines.** Cooperative occupational education and work experience in the curriculum. 1965. Interstate. 525p. illus. $7.50.

Designed to assist educators, administrators, and others in planning and developing vocational/technical programs in business, industry, and agriculture. Text covers 4 broad areas: new needs arising in work; curriculum patterns as related to the work environment; fundamentals of the cooperative education program; and fulfilling needs in specialized fields of occupation.

A17 **Mather, Kirtley F.** The earth beneath us. 1964. Random. 320p. illus. $17.95.

Written by a college geology instructor with over 40 years of experience. Thus, an authoritative text though simple and written for the layman. A pictorial account of the physical development of the land. Awarded the Thomas A. Edison tenth award for excellence in literature for young people.

A18* **Mills, Gordon H.,** and **John A. Walter.** Technical writing. rev. ed. 1962. Holt. 434p. illus. $7.95.

A guide to the principles and mechanics of good technical writing. Offers instruction in surmounting the common problems in technical composition which include writing definitions, descriptions, explanations and analyses, and organizing reports.

A19   Moore, Arthur D. Invention, discovery, and creativity. 1969. Doubleday. 178p. illus. $4.95.

Provides numerous suggestions and experiments designed to help the young person understand original thinking, creativity, and inventiveness and to aid him in developing these qualities in himself. An appendix tells methods of patenting inventions, as well as the greatest inventors in American history.

A20   Myers, Louis M. Guide to American English. 4th ed. 1968. Prentice-Hall. 499p. illus. paper, $4.95.

Rated as one of the best texts for college freshman English. Covers grammar, research paper information, use of dictionaries, use of the library, making outlines, logic for argumentation and debate, correction symbols, and exercises.

A21   Opdycke, John B. Harper's English grammar. rev. and ed. by Stewart Benedict. 1966. Harper. 286p. $5.95.

Intended as a student text or reference work on English grammar. Recommended by high school teachers and other knowledgable people for students and laymen interested in the fundamentals of sentence structure and word usage and their application in writing.

A22* Robert, Henry M. Robert's rules of order, newly revised. new and enl. ed. by Sarah Corbin Robert. 1970. Scott, Foresman. 594p. $5.95.

The standard code of parliamentary procedure and the basic guide to leading the business meeting for societies, conventions, and other organizations. Although primarily a reference manual, this handbook is also useful for instructors who may select topics from it for courses on any desired level.

A23   Shelton, John S. Geology illustrated. 1966. Freeman. 446p. illus. $10.50.

An introductory text for nonmajors who need to acquire the basic vocabulary of geology and a familiarity with this earth science. Clearly written, beginning with simple descriptions and ideas, and gradually leading the reader to more complicated points. More than 350 figures, mainly aerial and ground photographs or sketches, illustrate the points made in the book.

A24* Sutherland, Sidney S. When you preside. 4th ed. 1969. Interstate. 190p. illus. $4.95.

An instructional text designed to teach leaders the details of group mechanics and organization. The author is broadly experienced in leading conferences, workshops, etc., and thus his theories and ideas have been soundly tested first in his personal work. Has appeal for any audience but will best serve the readers on the junior college level and above.

A25* Technician Education Yearbook. 1963/64– . Prakken. $10.

Contains facts about U.S. technical education, case studies of outstanding programs, occupational information, and other advice of essential value to instructors. Includes the business-oriented, agricultural, engineering, and the medical technologies.

A26   Thorstensen, Thomas C. Practical leather technology. 1969. Van Nostrand. 272p. illus. $17.50.

A descriptive introduction to leather processing. Illustrations, including flow charts and line drawings, are used to help clarify technical descriptions of processes. Economic aspects of the industry are examined, current research is analyzed, and future applications are conjectured in addition to the discussion of leather technology.

A27   Wells, Walter. Communications in business: a guide to the effective writing of letters, reports, and memoranda. 1968. Wadsworth. 428p. $8.95.

Presents the basic concepts and principles of business letter, report, and memo writing. Arranged by function, these types of business communication are discussed in terms of style, structure, and purpose.

Chemistry and Physics

A28   Aitchison, Gordon J. General physics. 1970. Chapman. 522p. illus. $7.20.

A highly readable book which describes physics as a substantial, practical science. Intended for first year college students who probably will not pursue their study of the subject beyond 1 course. The student is encouraged to test his understanding of concepts and principles in well-planned problems and exercises.

A29   Baker, Jeffrey J., and Garland E. Allen. Matter, energy, and life: an introduction for biology students. 1970. Addison-Wesley. 225p. illus. paper, $3.25.

An introductory text which presents basic chemistry from the biological point of view. Features discussions of chemical principles such as energy conversion, chemical bonding, and molecular structure and activity. Offers supplemental reading lists, illustrations, and an easy-to-use index.

A30* **CRC Handbook of Chemistry and Physics: A Ready-Reference Book of Chemical and Physical Data.** 51st ed. 1970. Chemical Rubber. 2400p. $24.95.

A universally known, quick reference book basic to any educational institution collection. Contains a vast amount of critically analyzed data and information in chemistry, physics, and mathematics. Revised almost annually with each edition incorporating new tables and new material.

A31 **Campaigne, Ernest.** Elementary organic chemistry. 1961. Prentice-Hall. 312p. illus. $9.95.

A well-organized text intended for nonscience majors who need a background course in organic chemistry. Clearly written, with technicalities kept to a minimum.

A32 **Conn, Eric E.,** and **Paul K. Stumpf.** Outlines of biochemistry. 2d ed. 1966. Wiley. 468p. illus. $10.95.

An introductory textbook which offers a clear, authoritative discussion of the fundamental concepts of modern biochemistry. May serve as a valuable reference for teachers or as a classroom text for students having some background in biology, chemistry, mathematics, and physics.

A33 **Fairley, James L.,** and **Gordon L. Kilgour.** Essentials of biological chemistry. 2d ed. 1966. Van Nostrand. 314p. $9.

A basic biochemistry work emphasizing the general principles rather than detailed information on specific organisms. Written for students with varying educational backgrounds and with a wide range of interests, this text is easily adapted to the needs of technical school students who must have a biochemical background to work successfully in the agricultural industries.

A34 **Fuller, A. J. Baden.** Microwaves. 1969. Pergamon. 289p. $7.50.

Written for the undergraduate student interested in elementary microwave theory, this text is partially devoted to microwave techniques and partly to transmission line and waveguide theory. For most technical school students, its value is limited because much of the material is too advanced for even a 2-year college level of training. It will, however, serve the student who desires a full explanation of the subject, including discussions on transmission lines, electromagnetic fields, conducting media, devices, and measurements.

A35 **Hoffman, Katherine B.** Chemistry for the applied sciences. 1963. Prentice-Hall. 429p. illus. $9.95.

Serves the high school or college student who has had courses in mathematics and algebra. A comprehensive presentation of the important aspects of the science from the general history, chemical theory, and methods and the application of them, to the chemical properties and nature of matter.

A36 **Jevons, Frederick R.** The biochemical approach to life. 2d ed. 1968. Basic. 226p. illus. $5.95.

A provocative introduction to biochemistry with few technicalities to mar its clarity. Uses biochemistry to explain life and its processes and assumes only a high school background in biology and chemistry.

A37 **Karplus, Robert.** Introductory physics; a model approach. 1969. Benjamin. 498p. illus. $13.75.

An introductory textbook covering the general physical concepts and principles. Intended for nonmajors with very little scientific background. At the end of each chapter, several problems with solutions are given. Author manages to include a large amount of material but detracts from the great quantity of it by writing in an informal style.

A38* **Mallette, M. Frank, Paul M. Althouse,** and **Carl O. Clagett.** Biochemistry of plants and animals. 1960. Wiley. 552p. illus. $10.95.

An excellent source for the agricultural science student interested in learning plant and animal biochemistry. Text is divided into 3 major sections: general biochemistry, plant biochemistry, and animal biochemistry. Assumes basic background in chemistry in students using this textbook.

A39* **Merck index: an encyclopedia of chemicals and drugs.** 8th ed. Paul G. Stecher, ed. 1968. Merck. 1713p. illus. $15.

A standard work for students, laboratory technicians and researchers, physicians, and chemists, containing 9,500 entries on drugs and chemicals. This encyclopedia is cross-indexed by chemical, proprietary, generic or nonproprietary name. Monographic entries range from 2 lines to ½ page in length and give such information as chemical formula, molecular weight, history, origin, and distinguishing properties of the chemical or drug.

A40 **Pauling, Linus.** College chemistry. 3d ed. 1964. Freeman. 832p. illus. $9.25.

A carefully organized and clearly written textbook for an introductory course in general chemistry. Covers a broad range of topics including modern chemistry and a chapter on the 34 basic principles. The third edition of the author's *General chemistry,* published in 1970 by Freeman, is also worthy of consideration.

A41 **Quagliano, James V.,** and **L. M. Vallarino.** Chemistry. 3d ed. 1969. Prentice-Hall. 844p. illus. $9.95.

Written for the beginning student in terms he can understand. A comprehensive presentation of the fundamental theories, methods, and applications of chemistry. A valuable reference or textbook which may be used on the high school or college level.

A42 **Rogers, Eric M.** Physics for the inquiring mind. 1960. Princeton. 778p. illus. $12.50.

A complete survey of the methods, nature, and philosophy of physical science. Concepts and fundamental principles of physics from matter and energy to atomic and nuclear physics are presented in basic language for the nonscientist.

A43 **Stock, Ralph,** and **Cedric B. F. Rice.** Chromatographic methods. 2d ed. 1967. Van Nostrand. 256p. illus. $6.75; paper, $3.50.

Outline of chromatographic techniques presently practiced. Suitable as either a beginning text or a comprehensive review. Greater emphasis is placed on paper and gas chromatography than on the less widely used methods.

A44 **White, Emil.** Chemical background for the biological sciences. 2d ed. 1970. Prentice-Hall. 152p. illus. $6.95; paper, $3.25.

Chemistry book for the biology and biochemistry student and others beginning a study of this science. Presents information in concise, outline format and demonstrates possible applications of chemistry to biology. Treats inorganic and organic chemistry, emphasizing the essential aspects of each.

Mathematics and Statistics

A45 **Alder, Henry L.,** and **Edward B. Roessler.** Introduction to probability and statistics. 4th ed. 1968. Freeman. 333p. illus. $7.50.

An elementary text on probability and statistics for students who have had high school algebra. Examples and exercises are provided, and answers are given for some. Topics of discussion include random sampling, chi square, descriptive statistics, probability and nonparametric statistics, hypothesis derivation and testing, and variation analysis. Shows application of these mathematical principles in the agricultural, medical, geological, and psychological sciences.

A46 **Benice, Daniel D.** Introduction to computers and data processing. 1970. Prentice-Hall. 370p. illus. $8.50.

A true introduction to the computer which assumes that readers have no background in this field. Proceeds from a general description of the computer's history to information on hardware and software and specific subjects such as flow charting, coding forms, and input-output systems. Offers a chapter on each of the 5 computer languages.

A47 **Curry, Othel J.,** and **John E. Pearson.** Basic mathematics for business analysis. rev. ed. 1970. Irwin. 290p. illus. $10.

Invaluable reference on the mathematical concepts and techniques most frequently used in the business world. Among the features provided, the outline of the arithmetic relationships and the fundamentals of algebra, and the discussion on the mathematical principles basic to marketing are two of the most valuable.

A48 **Diamond, Solomon.** The world of probability: statistics in science. 1964. Basic. 193p. $5.95.

A broad survey of statistics and its application to science requiring only a knowledge of high school algebra for background. Concrete examples are used to demonstrate the importance of probability theory and statistical methods in modern science. Should be useful as a supplementary text for students taking a beginning course in statistics.

A49* **Donahue, Warren,** and **Everett Robertson.** Foundations of technical mathematics. 1970. Wiley. 478p. illus. $9.95.

Presents basic mathematics used by technicians for solving practical problems. Covers fundamental algebra, applied geometry, right-triangle trigonometry, graphing, and slide rule usage. Can be used for self-study and is an excellent foundation upon which to base more advanced work.

A50 **Ferrar, William L.** Mathematics for science. 1965. Oxford. 328p. illus. $4.95.

Intended for use by students who must develop sound mathematical background. Assumes little prior mathematical knowledge. First part of book considers precalculus subjects such as trigonometry and analytic geometry. Second part deals specifically with calculus. Appropriate for a technical school library as a supplementary text or a reference for instructors.

A51*   Finney, David J. Introduction to statistical science in agriculture. 2d ed. 1962. International Pubn. Ser. 216p. illus. $8.25.

Supplies a nontechnical explanation of statistics and the uses of mathematics in agriculture. Numerous examples and illustrations clarify points made in the text. Intended for the non-mathematically-minded reader, it uses hardly any algebra. Deals mainly with fundamental mathematical principles.

A52   Freund, John E. Statistics; a first course. 1970. Prentice-Hall. 340p. illus. $8.95.

A well-developed survey of statistics requiring little prior mathematical knowledge in the reader. Contains exercises designed to teach problem solving and to test and improve understanding of statistical principles. Explains mathematics in simplified terminology and depends on words, not proofs, for exposition.

A53   Hull, T. E. Introduction to computing. 1966. Prentice-Hall. 212p. illus. $10.60.

A comprehensive account of data processing and computers. First concentrates on the computer's general characteristics and then deals with more specific topics such as debugging, flow charts, algorithms, numerical techniques, and computer languages. Devotes much attention to Fortran, one of the most widely used languages. Gives advice on the use of computers in problem solving. Intended to serve senior high school level and above.

A54   Huntsberger, David V. Elements of statistical inference. 2d ed. 1967. Allyn. 398p. illus. $12.35.

Invaluable for those seeking a sound background in statistical concepts. Treats only those elements which may be applied to several different fields. Emphasizes the methods instead of the actual application of them, and keeps the textual style at a very elementary, but at the same time deep, level.

A55   Klaf, A. Albert. Arithmetic refresher for practical men. 1964. Dover. 438p. illus. paper, $2.50.

A review of mathematics written for the practical man. New ideas are introduced through a question and answer format. Interest, business and finance, ratio and proportion, averages, graphs and charts, powers and roots, and logarithms are some of the topics covered. A number of mathematical short cuts are given to aid in solving common problems.

A56   Lytel, Allan. Handbook of algebraic and trigonometric functions. 1964. Sams. 160p. paper, $3.50.

A thorough representation of basic mathematical tables and formulas. Arranged in 2 parts, the first containing algebra, geometry, trigonometry, and calculus formulas and the second consisting of mathematical tables and conversions. Intended for students at the high school and junior college level.

A57*   McGee, Roger V. Mathematics in agriculture. 2d ed. 1954. Prentice-Hall. 208p. illus. $9.95.

Treats subjects such as mathematical operations, percentage, measurement equations, ratio and proportion, averages, graphs, and the utilization of these to answer specific problems. Intended to serve agriculturists who need a basic knowledge of practical arithmetic. Provides tables of measurements and examples of realistic situations to teach readers how to apply mathematics in the agricultural world.

A58   Mack, Cornelius. Essentials of statistics for scientists and technologists. 1967. Plenum. 174p. illus. $5.95.

Presents briefly the fundamental statistical concepts and methods from the scientific and technological points of view. Assumes little prior mathematical background in the reader. Technical terminology is explained when used, and examples and problems are given to illustrate statistical theories and procedures. Appendix offers an adequate number of useful tables.

A59   Mendenhall, William. Introduction to probability and statistics. 2d ed. 1967. Wadsworth. 393p. illus. $13.25.

A thorough examination of probability and statistics beginning with introductory math and proceeding to an examination of descriptive procedures, probability laws, distributions, and other topics. A highly valuable text for the non-mathematician because the author relies more on verbal discussion than on mathematics for his explanations.

A60*   Slade, Samuel, Louis Margolis, and **John G. Boyce.** Mathematics for technical and vocational schools. 5th ed. 1968. Wiley. 515p. illus. $7.95.

The current edition of a source which has been frequently used since 1922. Covers the specialized application of mathematics in technical and trade professions. Gives instruction on topics ranging from elementary arithmetic to more complex subjects such as logarithms, algebra, and geometry. Contains numerous practical problems and questions for self-testing.

A61 Snedecor, George W., and **William G. Cochran.** Statistical methods. 6th ed. 1967. Iowa State. 593p. illus. $9.95.

Useful as a text or reference guide to the utilization of statistics in experimental work. Problems and answers are given for practice. Chapters begin with simplified explanations and end with more technical formulas and methods of application. Text has been used in the agricultural and biological sciences for many years.

Economics

A62 Albers, Henry H. Principles of management: a modern approach. 3d ed. 1969. Wiley. 702p. illus. $10.50.

Communication, motivation, and decision-making are some of the topics covered in this survey of managerial techniques. Presents principles which can be applied on any level from president to foreman and in any type of work from advertising and marketing to personnel.

A63 Austin, Charles F. Management's self-inflicted wounds; a formula for executive self-analysis. 1966. Holt. 320p. illus. $9.25.

An unusual, highly readable book. First, it identifies the wrong way of doing things, then examines the results of such actions, and finally, attempts to indicate the right way to handle situations. Over seventy examples are given of improper behavior. Written in casual language with anecdotal illustrations. Highly beneficial for any student or administrator who is about to assume or who is active in a managerial role.

A64 Bean, Louis H. The art of forecasting. 1969. Random. 196p. illus. $10.

Explains the simple methods of forecasting stock market fluctuations, business cycles and trends, political action and voting behavior, and weather and crop fluctuations. Written by a well-known economist who is famous for his forecasting ability.

A65 Boeckh, Everard H. Boeckh building valuation manual. v.1, Residential and agricultural. 1968. Amer. Appraisal. 288p. $38.

Only the first volume of this 3-volume set applies to agriculture. It is devoted to estimating building costs for residential and agricultural construction. Valuable for teaching estimating.

A66 Calhoon, Richard P. Personnel management and supervision. 1967. Appleton. 423p. $3.95.

Attempts to relate management theory with actual practice. Emphasizes the supervisory role and deals mainly with aspects of personnel management. Considers top management and staff personnel management.

A67 Crissy, William J., and **Robert M. Kaplan.** Salesmanship: the personal force in marketing. 1969. Wiley. 366p. illus. $8.95.

A salesmanship textbook with a balanced blend of the principles behind selling and the practical application of these concepts. Views the salesman as a marketer.

A68 Emery, David A. The compleat manager: combining the humanistic and scientific approaches to the management job. 1970. McGraw-Hill. 201p. illus. $8.95.

An invaluable book for the manager, his fellow workers, and the organization which he manages because it identifies the basic problems of the director and the directed and then gives possible solutions. Concentrates on teaching the manager techniques of self-analysis as well as methods of improving his faults and developing his strengths.

A69 Hebard, Edna L., and **Gerald S. Meisel.** Principles of real estate law. rev. ed. 1967. Schenkman. 523p. illus. $11.25.

A thorough, authoritative, and modern analysis of real estate law. Using a nontechnical approach, this textbook covers all phases of the field from ownership rights to real estate brokerage.

A70 Holloway, Robert J., and **Robert S. Hancock.** Marketing in a changing environment. 1968. Wiley. 498p. illus. $9.95.

A basic textbook. Discusses the major factors which affect marketing today as well as the different phases of the system which are involved. Theoretically these factors are the fundamental principles upon which all marketing trends are founded.

A71 Johnson, Herbert W. Creative selling. 1966. South-Western. 367p. $4.80.

Intended for those interested in or studying for a sales career. Deals with opportunities and compensations in selling, personal preparation for selling, understanding consumer behavior, and selling concepts and techniques.

A72* Kelley, Pearce C., Kenneth Lawyer, and **Clifford M. Baumback.** How to organize and operate a small business. 4th ed. 1968. Prentice-Hall. 624p. illus. $9.95.

Presents the most important factors in organizing and managing a small business. Business fields discussed include manufacturing, wholesale, retail, and services. Location, growth opportunities, financing, personnel management, and small business assistance agencies are some of the topics treated.

A73 **Kerby, Joe K.** Essentials of marketing management. 1970. South-Western. 696p. illus. $9.50.

Discusses the dynamics of marketing, a multi-faceted, ever-changing field, in terms of fundamental principles instead of details and explanations of transient, present-day marketing techniques. Encourages the reader to formulate his own idea of marketing, a way of thinking which should stand the test of time.

A74 **Lasser, Jacob K.** How to run a small business. 3d ed. rev. and enl. 1963. McGraw-Hill. 475p. $8.95.

A handbook written in popular style to serve the small businessman. A useful guide which includes information on financing, insurance, accounting, taxes, management, promotion, and other business activities.

A75* **McMichael, Stanley L.** McMichael's appraising manual. 4th ed. 1951. Prentice-Hall. 731p. illus. $8.95.

A detailed description of appraisal methods with many supporting illustrations and tables. Covers appraisal of 3 types of property—farm, business, and residential. A very worthwhile handbook.

A76 **Mason, Ralph E.,** and **Patricia M. Rath.** Marketing and distribution. 1968. McGraw-Hill. 566p. illus. $6.96.

Topics considered in this concisely worded, well-illustrated text include products, marketing, distribution, promotion and selling, and customer service. Considerably valuable in the phases of agribusiness which deal with agricultural product marketing, product processing, and agricultural supplies. Presents an enormous amount of factual information and provides a full section on farm product marketing. A good reference for post-secondary students who need a background in economics and marketing.

A77 **Pederson, Carlton A.,** and **Milton D. Wright.** Salesmanship principles and methods. 4th ed. 1966. Irwin. 756p. illus. $13.35.

Discusses sales skills which have been proven effective. Text is slanted toward service-oriented salesmanship, and this involves putting the customers' interests first and acting as an expert and counselor for them. An invaluable source for the beginner.

A78 **Riggs, James L.** Production systems: planning, analysis, and control. 1970. Wiley. 604p. illus. $9.95.

An extensive survey covering production management and systems. Intended for the student and simple enough for the college sophomore. Technical mathematics and methods are replaced by a balance of easy-to-understand verbal explanations and analytical theories in this text. Contains helpful aids such as summaries, notes in the margin, references, examples, questions, and problems.

A79 **Robinson, O. Preston, Christine H. Robinson,** and **George H. Zeiss.** Successful retail salesmanship. 3d ed. 1961. Prentice-Hall. 467p. illus. $9.95.

Gives thorough account of salesmen, salesmanship, and retail trade. Concentrates on how to make sales, how to please the consumer, and how to make sure he is completely content with the items purchased. Tips on consumer preferences and behavior are included.

A80* **Semenow, Robert W.** Questions and answers on real estate. 6th ed. 1966. Prentice-Hall. 681p. illus. $8.95.

Discusses the fundamentals of the real estate business and especially emphasizes the law and its application in this field. A frequently used reference tool which not only provides questions and answers for students taking the real estate examinations but also includes definitions of terms and information helpful in daily problem solving. Most of the principles reviewed in this guide are applicable in all the states.

A81 **Tousley, Rayburn D., Eugene Clark,** and **Fred E. Clark.** Principles of marketing. 1962. Macmillan. 716p. illus. $9.95.

Standard marketing text for general courses which consists of details on the sizes, locations, ownership, controls, types of operation, and merchandise lines of retailing. Also presents the basics of consumer behavior which are so important in salesmanship and marketing of products.

A82 **Walters, Charles G.,** and **Gordon W. Paul.** Consumer behavior: an integrated framework. 1970. Irwin. 548p. illus. $15.35.

A complete presentation of the major principles and the details of consumer behavior. Although the text contains some review material on marketing, it is basically written for the undergraduate or graduate student who has had training in business economics and marketing. Treats the consumer from the individual as well as the group point of view.

A83 Wingate, John W., and **Carroll A. Nolan.** Fundamentals of selling. 9th ed. 1969. South-Western. 595p. illus. $5.88.

A comprehensive survey of salesmen and salesmanship which includes information on retailing and the profession, selling. Concentrates on fundamental selling theories and techniques and their uses in any type of sales situation.

A84 Zacher, Robert V. Advertising techniques and management. rev. ed. 1967. Irwin. 666p. illus. $9.25.

The broad scope of this reference on advertising extends beyond the application of advertising in marketing to its use as a communicative medium in any field or situation. Presents an introduction to advertising for beginning students which develops in them an appreciation for the importance of this tool in effective product distribution.

Agriculture in General

A85 Beuscher, Jacob H. Law and the farmer. 3d ed. 1960. Springer. 406p. illus. $6.

This book, copyrighted in 1960 and reprinted in 1965, is concerned with the importance of law in farming. Numerous illustrations and examples are given in nontechnical language to explain the complexities of farm law. Although not completely up-to-date, it should orient students to the common legal problems in farm ownership and maintenance. (Not many books on agricultural law are available, but most colleges of agriculture publish bulletins and other similar materials which explain laws of importance to farmers within the particular state.)

A86* Binkley, Harold, and **Carsie Hammonds.** Experience programs for learning vocations in agriculture. 1970. Interstate. 604p. illus. $7.95.

A well-structured and concisely written reference on work-study programs for vocational/technical agriculture students. The book begins with a definition of vocational agriculture and an explanation of the need for directed experience as a part of the preparation for farm or off-farm agricultural work. Next, it deals with programs in farming occupations and then, finally, describes experience programs in nonfarm agricultural jobs.

A87 Boalch, Donald H., ed. Current agricultural serials: a world list of serials in agriculture and related subjects (excluding forestry and fisheries) current in 1964. v.1 1965, v.2 1967. International Association of Agricultural Librarians and Documentalists (order from PUDOC). 2v. $12.50.

Volume 1 contains a current (as of 1964) list by title of more than 13,000 serials in agriculture and kindred subjects. Volume 2 contains a subject and geographical index. New or changed titles are reported in the *Quarterly bulletin* of the IAALD. Annual subscription to the *Quarterly bulletin* is $13.50 for libraries, institutions, and associations and $6 for individuals. Subscriptions should be sent to H. E. Thrupp, Secretary/Treasurer, Tropical Products Institute, 56-62 Gray's Inn Rd., London, WCIX8LU, England.

A88 Brandsberg, George. The two sides in NFO's battle. 1964. Iowa State. 301p. $4.95.

A clear, objective presentation of the National Farmers Organization written by a newspaper reporter who based his information on observation, interviews, and current writing. He describes the social and economic conditions responsible for the founding of this controversial farm organization and gives an account of its goals and activities, its opposition, and its ideals.

A89 Courtenay, Philip P. Plantation agriculture. 1965. Praeger. 208p. illus. $6.

One of the few introductory texts on plantation agriculture. Contains authoritative facts about plantations, tropical crops, and their distribution, but is slanted geographically in its presentation. A useful reference for tropical agriculture, agricultural geography, and agricultural economics courses.

A90 Crampton, John A. The National Farmers Union: ideology of a pressure group. 1965. Nebraska. 251p. illus. $5.50.

This intellectual approach to a controversial subject gives an objective yet commiserate picture of the Farmers' Educational and Cooperative Union of America. The author examines the organization's ideology, discusses its history and organizational problems, studies its educational and lobbying tactics, etc. Controversial statements are thoroughly documented.

A91 Hannah, Harold W., and **Norman G. P. Krausz.** Law and court decisions on agriculture (with particular reference to Illinois). 1968. Stipes. 465p. paper, $12.50.

Two agricultural law professors are responsible for this compilation of judicial decisions. The farmer and his land are the subject of most cases in this volume. For readers with no legal background, the style and material has been simplified. Valuable for providing the reader with a brief background in legal history and philosophy and a familiarity with common law regulations.

A92* Higbee, Edward. American agriculture: geography, resources, conservation. 1958. Wiley. 394p. illus. $8.95.

A complete examination of American agriculture covering all crops and agricultural regions of the United States. Deals with the fundamental resources of agriculture including soil, water, and plants, and presents a study of certain farms whose techniques and systems are known to foster the constructive utilization and preservation of these resources. A good reference for background information for all agricultural sciences students.

A93 Hill, Johnson D., and **Walter E. Stuermann.** Roots in the soil; an introduction to philosophy of agriculture. 1964. Philosophical Lib. 162p. $4.75.

A brief review of agriculture covering the time period from the dawn of civilization to the present written from the historical and philosophical points of view. Intended for the layman rather than the professional philosopher, theologian, or economist. The authors claim that there is an agricultural philosophy and that it is based on agricultural policies formulated by the government.

A94* Hoover, Norman K. Handbook of agricultural occupations. 2d ed. 1969. Interstate. 385p. illus. $6.75.

A vocational guidance handbook broadly outlining occupations in agriculture and related areas. Gives detailed directions on educational requirements and considers possibilities for promotion. Provides occupational briefs for more than 50 common nonfarm jobs. The appendix contains a guide to institutions which offer training in the agricultural occupations covered in the text.

A95 Krebs, Alfred H. Agriculture in our lives. 2d ed. 1964. Interstate. 696p. illus. $5.95.

Acquaints the reader with the fundamental principles of agricultural science. Livestock, crops, plant diseases, animal diseases, weeds, insects, marketing, and modern agricultural techniques are among the topics covered. Intended to help provide young people with the understanding of agriculture needed for educated career choices and for basic appreciation of the science.

A96* Midwest Farm Handbook. 7th ed. 1969. Iowa State. 505p. illus. $6.95.

A frequently used reference manual for students, farm owners, and agricultural extension workers. Covers various aspects of farm problems and advice for solving them based on statistics and other practical details. Topics included are livestock marketing, grain marketing, seed and soil treatment, dairy cattle management, feed changes, pest control, fertilizers, farming techniques, and vegetable growing.

A97 Mighell, Ronald L. American agriculture: its structure and place in the economy. 1955. Wiley. 187p. illus. $6.50.

Presents the economic aspects of American agriculture as the Census Bureau reported them in the 1950s. Condenses and organizes the statistics and other details compiled by the bureau, and through them, explains the role of agriculture in the national economy of the midcentury. Portrays farms and farm life in simplified manner and thus serves as a good introductory or supplementary text for agricultural history, agricultural economics, and general agriculture courses.

A98 Miller, Howard L., and **Ralph J. Woodin.** AGDEX: a system for classifying, indexing, and filing agricultural publications. rev. ed. 1969. Amer. Voc. Assoc. 1v., unpaged. $4.

A step-by-step explanation of a standard indexing-filing system for agricultural materials. Teachers, librarians, and administrators using this system will find that it facilitates organization and retrieval of all kinds of agricultural literature including technical and professional information. A highly valuable classification scheme because of its comprehensiveness.

A99 Mosher, Arthur T. Getting agriculture moving: essentials for development and modernization. 1966. Praeger. 191p. illus. $6.50; paper, $2.50.

Consists of 3 parts: a description of (1) the fundamental principles basic to agriculture; (2) the necessary factors for development in agriculture; and (3) the causes for promotion of this development (such as land improvement). Stresses the interrelationships between these productive factors. Invaluable for those technicians in agricultural experience programs as well as for other workers in backward areas.

A100 Ochse, J. J., and others. Tropical and subtropical agriculture. 1961. Macmillan. 2v. illus. $35.

A universally useful treatise on tropical and subtropical agricultural crops and techniques for propagation. An excellent reference for horticulture students, this text begins with common problems such as crop improvement methods, climate, and soil quality, and then proceeds to detailed descriptions of production and botanical characteristics of crops. An authoritatively documented book.

A101* Sidney, Howard, ed. Agricultural, forestry, and oceanographic technicians. 1969. Doubleday. 344p. illus. $11.95.

Acquaints the high school and the 2-year college student with the job opportunities available in agriculture. Detailed descriptions are given for each occupation listed, and jobs in agriculture, fisheries, forestry, conservation of natural resources, and marine biology are included. Articles on the occupations are written by specialists, and related jobs are grouped within each particular field. An invaluable source of career information which also provides a list of schools that have programs in agriculture, forestry, and oceanography for technical training.

A102 Special English: Agriculture 1, 2, 3. v.1 *Soils,* v.2 *Field crops,* v.3 *Horticulture, farm, forestry, and livestock.* 1966-67. Macmillan. 3v. paper, $2 each v.

An excellent guide for classroom use or self-study. Teaches the proper usage of the technical language of agricultural science and explains terminology through illustrations, practice exercises, dialogues, readings, and drills. Suitable for the student or anyone working in or interested in agriculture. Sets of tapes are available for purchase with the volumes.

A103* Winburhe, John N., ed. A dictionary of agricultural and allied terminology. 1962. Michigan. 905p. $15.

An agricultural dictionary for the specialist as well as the student. Gives a broad survey of the agricultural terminology, new and old, which is found in the literature and which is frequently used in verbal communication. Contains thousands of entries and includes names as well as general terms.

A104 Wrigley, Gordon. Tropical agriculture: the development of production. rev. ed. 1969. Praeger. 376p. illus. $13.50.

An essentially elementary approach to tropical crop production which covers crop protection, cattle management, pasture cultivation, crop ecology, and crop culture. Text is illustrated by line drawings, photographs, tables, and charts. Concentrates on the conditions which affect tropical agriculture and on the preservation of fertile soil when crop production is increased. Presents much valuable information on meeting the food needs of the world through tropical crop propagation.

Biology

A105 Austin, Colin R. Fertilization. 1965. Prentice-Hall. 145p. illus. $4.95; paper, $2.95.

Intended for the student or worker interested in fertilization. Written in a concise, easy-to-understand style, this work not only serves as an excellent introduction to reproductive physiology but also it reports the latest developments in related fields for more advanced readers. Useful for comparative study because chapters begin by describing general concepts and then give detailed examples in several different organisms.

A106 Beaver, William C., and George B. Noland. General biology: the science of biology. 8th ed. 1970. Mosby. 546p. illus. $10.50.

A highly comprehensive text designed as an introduction for advanced secondary school and college readers. The fundamental principles of biology are thoroughly outlined, and specific topics discussed are plants, viruses, human biology, animal biology, the continuity of life, and organisms and environment. Material has been completely updated, and the volume contains a good bibliography.

A107 Bernal, John D. The origin of life. 1967. Universe. 345p. illus. $12.50.

Written from the author's point of view, this treatise sets forth Bernal's theories on the beginnings of life. His approach to this subject is based on the premise that if life made itself once, it can be done again. Such a subjective account is apropos for this topic because the origin of life is for the most part an unsolved problem of science. As a result, the text should be read with an open mind.

A108 Curtis, Helena. Biology. 1968. Worth. 854p. illus. $11.50; Nat. Hist Pr. trade ed. $15.

An introduction to biological science which emphasizes animal biology. Written in clear, nontechnical style for beginning college biology students, and includes helpful diagrams, illustrations, and short explanatory essays that supplement the text. Useful for reference or course instruction, it gives detailed information on genetics, ecology and evolution, cells, behavior, organisms, and development.

A109 DeLaubenfels, David J. A geography of plants and animals. 1970. Brown. 133p. illus. paper, $1.95.

Authored by a geographer who is concerned with the variation in distribution of the plants and animals in the biosphere and who believes that the major problems of ecology are caused by the environmental differences in various regions. Discusses populations, communities, biota, ecosystems, and formations as factors causing these differences.

A110 Dodson, Edward O. Evolution: process and product. rev. ed. 1960. Van Nostrand. 352p. illus. $9.50.

Suitable for a college textbook or simply for a guide to evolution for interested readers. Approaches the subject from 2 angles: (1) evolution by natural selection; and (2) actual manifestations of phylogeny. Not only discusses the step-by-step progression from the most primitive to the most complex forms of life but also explains the causes for change and the quantitative results.

A111 Edmondson, Walles T., ed. Fresh-water biology. 2d ed. 1959. Wiley. 1248p. illus. $39.95.

A standard reference for identifying plants and animals found in the inland waters of North America. Includes fungi and bacteria, but not aquatic vertebrates or parasites which are only internal. Revised edition of the classic book of the same title by Ward and Whipple, but with a change of scope. It no longer contains general information on the ecology and habits of aquatic flora and fauna and so is not useful as a textbook of limnology or general biology.

A112* Gray, Peter, ed. The encyclopedia of the biological sciences. 2d ed. 1970. Van Nostrand. 1027p. $24.95.

An encyclopedia serving nonbiologically specialized readers and teachers who need a complete, detailed reference to biological and related subjects. Gives authoritative descriptions of terms and concepts which go far beyond mere dictionary-type definitions. Articles are written by scientists, and they include comparisons between various aspects of the biological sciences and those of other allied fields.

A113* Henderson, Isabelle F., and **William D. Henderson.** A dictionary of biological terms. 8th ed. by John H. Kenneth. 1963. Van Nostrand. 640p. $12.50.

The first 7 editions of this standard dictionary of English origin were published under a different title, *A dictionary of scientific terms.* Definitions are included for the biological, botanical, and zoological sciences and also for genetics, cytology, physiology, anatomy, embryology, and other allied fields.

A114 Keeton, William T. Elements of biological science. 1969. Norton. 582p. illus. $7.95.

Examines the major elements of biology in a manner which is informative, but not necessarily comprehensive. This survey covers such topics as the origin and character of life, elementary chemistry, Darwin's theory of evolution, and the cell. The author describes Darwin's evolution hypothesis as the underlying principle which unifies biology. Keeton also presents helpful discussions on problems of organisms and possible solutions to them.

A115 Klots, Elsie B. New field book of freshwater life. 1966. Putnam. 398p. illus. $4.95.

A completely indexed identification manual containing many animal (both vertebrate and invertebrate) and plant organisms found in the fresh waters of America. Additional features are a chapter on collection and preservation of specimens, an identification chart for insect nymphs, an introduction to aquatic ecology, and a glossary of ecology words.

A116 Knudsen, Jens W. Biological techniques; collecting, preserving, and illustrating plants and animals. 1966. Harper. 525p. illus. $13.95.

Deals with all types of collection and preservation methods including taxidermy. Covers every kind of organism from bacteria and fungi to plants and mammals, and explains ways of obtaining specimens such as poisoning, electrocution, and shooting. Designed for the reader interested in developing skills for this type of work in biology or museums.

A117 Nason, Alvin. Essentials of modern biology. 1968. Wiley. 508p. illus. $9.95.

Provides up-to-date coverage of biology for readers with little prior knowledge of the subject. A more brief account than given by its parent volume, *Textbook of modern biology,* it shows the interrelationships between chemistry and physics and biology, and then discusses these in terms of molecules and cells, lower plants and animals, and, finally, more complex organisms.

A118 **Newbigin, Marion I.** Plant and animal geography. 3d ed. 1964. Dutton. 298p. illus. $5.75.

The author, a British biologist, is here concerned with animal and plant communities, their distribution, and the factors responsible for their vitality and dispersal. The approach to the geography of living organisms in this book is very broad, and it is comparable to that used in many general works on ecology.

A119 **Pimentel, Richard A.** Natural history. 1963. Van Nostrand. 436p. illus. $10.

Information given in this volume consists of facts about ecology, geology, and the taxonomy of plants and animals. Learning aids which supplement the text include line drawings and excellent reference lists for every chapter. Written simply and accurately, this reference may be easily read by the layman and student alike.

A120 **Telfer, William,** and **Donald Kennedy.** The biology of organisms. 1965. Wiley. 374p. illus. $8.95.

A companion volume to Stern's *Biology of cells,* intended to acquaint the college student with modern biology. Devoted to the biology of the organism which the authors believe is a subject lost in the present emphasis on cellular and molecular biology and population biology. Authors write in a very clear, concise style.

A121* **Weisz, Paul B.** Elements of biology. 3d ed. 1969. McGraw-Hill. 486p. illus. $9.20.

Taken from Weisz's larger book, *The science of biology,* and simplified in order to make the content available to many readers with no basic scientific knowledge. Concentrates on the fundamentals of biology and eliminates much technical detail. Style of writing is clear, and the illustrations and the suggested reading lists with each chapter are good.

Genetics and Cytology

A122 **Afzelius, Björn.** Anatomy of the cell. 1966. Chicago. 127p. illus. $6.50.

Comprehensive, current treatment of cellular anatomy consisting of chapters on cell membranes, mitochondria, ribosomes, nuclei, lysosomes, and microbodies. An illustrated, well-organized text which is especially recommended for supplementary reading in introductory biology courses.

A123 **Asimov, Isaac.** The genetic code. 1963. Grossman. New Amer. Lib. 187p. illus. $4.50; paper, 75¢

Presents the dynamics of genetics and explains its hereditary function and the cell's biochemistry. Written by a popular science author and biochemist. Deals extensively with the influence of DNA on heredity.

A124 **Auerbach, Charlotte.** Genetics in the atomic age. rev. ed. 1965. Oxford Book. 111p. illus. $3.

A fairly simplified account of genetics and the effects of radiation on heredity. Technicalities are kept to a minimum in order to provide the beginning biology students and the interested nonspecialists with a background in basic genetics.

A125 **Bonner, David M.** Heredity. 2d ed. 1964. Prentice-Hall. 112p. illus. $4.95.

Acquaints the reader with the fundamental processes of heredity and with the interrelationships between physical, biological, and chemical research in genetics. Gives specific attention to gene chemistry and the parts of the cell which influence genetics.

A126 **Borek, Ernest.** The code of life. 1965. Columbia. 226p. illus. $5.95.

Invaluable for its coverage of basic genetics and general molecular biology. Begins by discussing the gene and nucleic acids and then proceeds to the research done which reveals the correlation between nucleic acid structure and function. An excellently informative work.

A127* **Brewbaker, James L.** Agricultural genetics. 1964. Prentice-Hall. 156p. illus. $4.95; paper, $3.75.

A genetics text specifically adapted to agriculture. Demonstrates how principles of heredity are important in animal husbandry and other agricultural fields through discussion of variations in populations, mutations, parasitism and symbiosis, breeding programs and prevention of variations, and genetic progress through selection. Assumes a prior knowledge of basic genetics and statistics in the reader.

A128 **Cook, Stanton A.** Reproduction, heredity, and sexuality. 1964. Wadsworth. 117p. illus. paper, $2.95.

A broad survey of the major aspects of reproduction with information on Mendelism, mitosis and meiosis, polyploidy, crossing over and linkage, chromosome structure, nature and structure of the gene as it corresponds with protein synthesis, gene action in development, mutation, and many other topics. Intended for a background text. Contains a wealth of facts and is well written and adequately illustrated.

A129 Darlington, Cyril D., and L. F. LaCour. The handling of chromosomes. 5th ed. 1970. Hafner. 264p. illus. $9.95.

A simple guide to laboratory study of chromosomes for botany or zoology teachers and medical or agricultural laboratory workers. Gives step-by-step directions on how to prepare chromosomes so they are visible, dissection of certain materials such as salivary glands, and how to make special drawings or microphotographs.

A130* Gardner, Eldon J. Principles of genetics. 3d ed. 1968. Wiley. 518p. illus. $10.95.

A standard work that covers modern genetics generally and systematically, beginning with discussions of the simple principles and ideas and gradually progressing to the more complex theories and facts. Divided into 3 sections: (1) fundamental genetics; (2) character and function of genetic material; and (3) heredity and evolution of populations. Text is entirely rewritten and contains new illustrations, examples, and problems as well as modern ideas and recent discoveries about the gene. Essentially a simple work introducing elementary biological terminology and genetic principles.

A131 Goldstein, Philip. Genetics is easy. 4th ed. rev. and enl. 1967. Lantern. 295p. illus. $6.

Suitable for high school and college students just beginning a study of heredity. Examines the main principles of genetics and explains them in highly accurate style, simplified language, and through clear illustrations.

A132 Pfeiffer, John. The cell. 1964. Time. 200p. illus. $7.20.

Covers cell genetics, physiology, morphology, and function in attractive format and easily understood style. Designed to serve all ages, it is most valuable for its unique photographs and illustrations, which provide an education in themselves.

A133 Rieger, Rigomar, Arnd Michaelis, and Melvin M. Green, eds. A glossary of genetics and cytogenetics. 3d ed. 1968. Springer-Verlag. 506p. illus. $16.50.

Not only provides the definitions for genetic and cytogenetic terms included but also describes them in the context of real situations where they are used and further clarifies their meaning in explanatory diagrams. Indispensable for the student, teacher, and researcher in these fields.

A134 Stern, Herbert, and David L. Nanney. The biology of cells. 1965. Wiley. 548p. illus. $9.95.

Designed for the freshman biology student who has had no specialized instruction in the field. May be read with its companion, *The biology of organisms,* by Telfer and Kennedy, for a complete introduction to basic biology. Topics examined are basic genetics and cytology, fundamental cell chemistry and physics, maintenance of cellular order and specificity, and regulation of cell behavior. Throughout the text, emphasis is placed on the organism as a whole, and the biochemistry, morphology, and physiology of various types of organisms are presented within this perspective.

A135* Swanson, Carl P. The cell. 3d ed. 1969. Prentice-Hall. 150p. illus. $5.95; paper, $2.50.

A completely revised version that has replaced the frequently used 1960 and 1964 editions. Not only portrays the basics of cell behavior, function, and structure but also examines the many recent discoveries in cell science. Concisely and authoritatively written in a style appealing to the student and instructor alike.

A136 White, Philip R. Cultivation of animal and plant cells. 2d ed. 1963. Ronald. 288p. illus. $9.

Acquaints the reader with the basic principles and methods of tissue culture, from obtaining and preparing the material to isolating and establishing the culture. For the student who has had no experience in laboratory work or cultivation of plant and animal cells.

A137 Wilson, George B., and John H. Morrison. Cytology. 2d ed. 1966. Van Nostrand. 319p. illus. $8.95.

Written on a level understandable by secondary school and college students. Gives only the highlights of cytological science and expects the reader to supplement the general information given with materials from the chapter reference lists.

A138* Winchester, Albert M. Genetics: a survey of the principles of heredity. 3d ed. 1966. Houghton Mifflin. 504p. illus. $9.50.

An informative, widely known college genetics text which provides one of the most elementary approaches to heredity and genetics. Instructs the novice in the major concepts of genetics, the chemical basis of heredity, the biochemical pattern of gene activity, the genetic code, chemical regulation of enzyme manufacture by virus transfer in bacteria, and DNA duplication and synthesis. Helpful examples of animals and plants are used.

Microbiology

A139 **Andrewes, Christopher H.** The natural history of viruses. 1967. Norton. 237p. illus. $10.

Examines human, animal, and plant viruses in relation to their spread, ecology, latency, classification, congenital properties, cancerous tendencies, and evolution. Although all 3 types are sufficiently covered, more emphasis is placed on the animal viruses. The general characteristics of viral organisms are extensively discussed.

A140 **Barron, Arthur L.** Using the microscope. 3d ed. 1965. Chapman. 257p. illus. $5.75.

Part 1 consists of information on the parts of the microscope, simple optical concepts, evaluation of equipment, illumination, proper use of the microscope, measurements, and special instruments and their utilization, and part 2 covers photomicrography. An excellent guide for the knowledgable microscope user as well as for the beginning student. Appendixes contain directions for mounting slides and other practical instructions and facts.

A141 **Brock, Thomas D.** Principles of microbial ecology. 1966. Prentice-Hall. 306p. illus. $8.50.

A unique treatise because it is one of the few volumes written on the subject in any language. Intended audience includes microbiologists, soil specialists, sanitary engineers, and students interested in the field. Topics considered are population ecology, dispersal, microbial ecosystems, interactions between microbial populations, etc. Deals with qualitative highlights instead of quantitative details.

A142 **Buchanan, Robert E., John G. Holt,** and **Erwin F. Lessel,** eds. Index Bergeyana. 1966. Williams & Wilkins. 1472p. illus. $25.

A companion volume to *Bergey's manual of determinative bacteriology* and a standard reference on the nomenclature of bacteria. Tells who first proposed the name of each taxon listed and when, and gives citation for original published description of each. Includes and denotes names of bacteria not accepted as valid as well as officially accepted names. Also gives source from which each taxon was isolated and thus acts as a partial host index for bacteria.

A143 **Carpenter, Philip L.** Microbiology. 2d ed. 1967. Saunders. 476p. illus. $7.50.

The author writes this broad survey of microbiology in a concise, well-organized manner. This book is concerned primarily with bacterial biology, microorganisms, roles and ecological relationships of microscopic life forms, and the dependency of microscopic pathogens on their hosts.

A144 **Curtis, Helena.** The viruses. 1965. Doubleday. 228p. illus. $5.95; Nat. Hist. Pr. paper, $1.45.

Spans the entire field of general virology and presents the historically important facts as well as those of recent significance. Describes several of the men responsible for the development of virology and for progress in virological research. Satisfactory reading for the layman and the high school or college student.

A145 **Frobisher, Martin.** Fundamentals of microbiology. 8th ed. 1968. Saunders. 620p. illus. $9.

For the advanced high school or college reader who has had biology and chemistry. Classified as one of the better fundamental microbiology textbooks, it is a very popular work for reference and course study. Supplies details on microorganismic biology, nomenclature, and other related topics, and discusses practical applications of microbiology in therapy, industry, etc.

A146 **Gebhardt, Louis P.,** and **Dean A. Anderson.** Microbiology. 3d ed. 1965. Mosby. 488p. illus. $9.25.

A thorough approach to the elements of microbiology. This is designed for the student who has little knowledge of the biological or physical sciences. It is written in clear, concise language, and it has an accompanying laboratory guide. The authors' treatment of sanitary and industrial microbiology is exceptionally good.

A147 **Gillies, Robert R.,** and **Thomas C. Dodds.** Bacteriology illustrated. 2d ed. 1968. Williams & Wilkins. 163p. illus. $10.

A collection of color photographs in atlas format which provides the student with a guide for identification and differentiation of bacterial cultures and colonies in the laboratory. Contains many plates of stained bacteria as seen through the microscope. This work consists of 3 parts: (1) instructions on techniques of staining, cultivation, and classification; (2) essays on 16 bacterial genera which describe the structure, staining nature, usefulness and biochemical test results, and the animal vaccination procedures for each; and (3) diagnostic skills.

A148 Hall, Cecil E. Introduction to electron microscopy. 2d ed. 1966. McGraw-Hill. 397p. illus. $17.50.

One of the best texts on electron microscopy, it gives sufficiently extensive coverage of the subject for a single semester course. Omits discussion of the research value and the commercially available models of this microscope and dwells on the skills essential for electron microscope use and the fundamental principles basic to this science.

A149 Harrigan, W. F., and Margaret E. McCance. Laboratory methods in microbiology. 1966. Academic. 362p. illus. $13.50.

Presents step-by-step directions for developing laboratory methods used in microbiological evaluations. A good manual for the microbiology laboratory worker interested in quality control and product improvement as well as for the student. Written in terse language with interpretations of the methodologies described. More complex systems for microbiological analysis may be devised from these simplified procedural outlines.

A150 Jones, Ruth M. Basic microscopic technics. 1966. Chicago. 334p. illus. $6.50.

Contains not only the standard staining and specimen preparation information found in most introductory microscopy texts but also descriptions of the elements of microscope science, the living organisms, embryo preparation, and drawing. Appendixes give equipment sources, reagents, supplies, stains, and formulas. Emphasizes animal rather than plant study.

A151 Kay, Desmond H., ed. Techniques for electron microscopy. 2d ed. 1967. Blackwell Scientific. 576p. illus. $10.08.

An authoritative manual written by 10 specialists for the prospective electron microscopist and the professional. Invaluable for practical information and instruction in experimental methods.

A152 Lawrence, Carl A., and Seymour S. Block, eds. Disinfection, sterilization, and preservation. 1968. Lea & Febiger. 808p. illus. $30.

A detailed account of purification and preservation techniques which examines modes of action, testing methods, antiseptics and disinfectants, surgical and hospital disinfection, antimicrobial preservatives, and chemical and physical sterilization. Provides a basic idea of the means for fighting microbial contamination.

A153 Luria, Salvador E., and James E. Darnell, Jr. General virology. 2d ed. 1967. Wiley. 512p. illus. $14.

A composite survey of biological, biochemical, and molecular virology. The unifying theme of this work is the relationship of animal, plant, and bacterial viruses, and the audience served is the novice and the researcher in the field. Discusses results of research conducted since 1953 and thus supplements research reported in the first edition.

A154* Needham, George H. The microscope: a practical guide. 1968. Thomas. 115p. illus. $6.50.

An indispensable book covering all aspects of microscopy except photomicrography. Useful for reference and for a short course text. In some areas, such as phase contrast microscopy, the author uses a postintroductory approach. Describes the specific parts of the microscope and the function, capacity, and actual use of the instrument.

A155 Pelczar, Michael J., and Roger D. Reid. Microbiology. 2d ed. 1965. McGraw-Hill. 662p. illus. $10.95.

Well-developed chapters on molds, yeasts, viruses, protozoa, algae, and rickettsiae are included in this text for college students having a background in general biology. Emphasizes bacteria and gives a balanced description of principles and applications. Technical details are kept to a minimum, and those given are accurate and directly related to topics described.

A156 Salle, Anthony J. Fundamental principles of bacteriology. 6th ed. 1967. McGraw-Hill. 822p. illus. $12.95.

This volume features an up-to-date account of the basic bacteriological concepts and phenomena, and it illustrates the major points made in the text with thoughtfully chosen examples and diagrams. Has been used for almost 30 years as a reference and text in this field. A well-organized, pleasantly readable work.

A157 Sigel, Mola M., and Ann R. Beasley. Viruses, cells, and hosts. 1963. Holt. 175p. illus. paper, $1.88.

Appropriate for all levels from seventh grade to adult. Chapters provide facts about the morphology and arrangement of cells, structure of virus particles, virological research methods, virus-cell relationships, and the tendency for viruses to produce cancer. Style of writing is simple in the beginning, then becomes more advanced toward the end.

A158 Smith, Alice L. Principles of microbiology. 6th ed. 1969. Mosby. 669p. illus. $9.75.

A book which deals with 3 main areas of consideration: an introduction to the subject and the tools used to study it; a review of the specific kinds of organisms; and the diagnosis and control of microbiologically caused human disease. Suitable for reading by a beginning microbiology student on the college level.

A159* Society of American Bacteriologists. Bergey's manual of determinative bacteriology, by Robert S. Breed [and others]. 7th ed. 1957. Williams & Wilkins. 1094p. $15.

A standard guide to bacterial nomenclature with a description of each type according to class, species, and subspecies. Identification keys are provided. This is a companion to *Index Bergeyana,* which has the host and habitat index.

A160 Stanier, Roger Y., and others. The microbial world. 3d ed. 1970. Prentice-Hall. 873p. illus. $15.95.

Supplies an introduction for beginners in the field and outlines the basic biological characteristics of microorganisms. Useful to the student, food technologist, public health official, and the sanitary engineer.

A161 Stanley, Wendell M., and Evans G. Valens. Viruses and the nature of life. 1961. Dutton. 224p. illus. $4.95; paper, $1.95.

The senior author of this highly authoritative work was awarded the Nobel Prize for chemistry for his work in virology. The book covers the properties and behavior of viruses, their structure, proteins, and nucleic acids. Admirably illustrated with excellent, carefully chosen photographs. Presents essential principles in largely nontechnical style.

A162* Stehli, George J. The microscope and how to use it. Tr. by William A. Vorderwinkler. 1960, reprint 1970. Dover. 157p. illus. paper, $1.50.

Full of details on the parts of the microscope and how to adjust them, what kind of instrument to use, what to look for in certain specimens, where to find them, how to do blood-smearing, dissection, and microphotography. Written for the beginner in simplified, clear style.

A163 White, Geoffrey W. Introduction to microscopy. 1966. Butterworth. 255p. illus. $3.60.

Intended for anyone learning to use the microscope whether they are students, technicians, or laymen. Enthusiastically written text on microscopy which deals specifically with the history of the microscope, construction of the instrument and how to set it up, preparation of specimens, measurements, illumination, and photomicrography.

Parasitology

A164 Burt, David R. Platyhelminthes and parasitism. 1970. Amer. Elsevier. 150p. illus. $8.75.

Uses Platyhelminthes and other particular parasitic groups to exemplify the types of associations between parasites and hosts. An especially exciting book for the undergraduate biology student because the author includes many original ideas and details from his own experiences.

A165 Crofton, Harry D. Nematodes. 1966. Hutchinson. 160p. illus. $5; paper, $2.

Acquaints the student and instructor with the general characteristics and activities of nematodes. Concentrates on the basic morphology, structure of the body wall, various organ systems, metabolism, embryology and growth, life history, and the differences between the free-living and the parasitic forms. Gives a good supplementary book list and a nematode classification outline.

A166 Jones, Arthur W. Introduction to parasitology. 1967. Addison-Wesley. 458p. illus. $12.50.

Systematically describes the parasitic organisms beginning with the lowest form, the Protozoa, and ending with the Arthropoda. Additional features are chapters on pathology, host-parasite relationships, immunity, evolution, and health. Provides the science and the non-science student with a sound background in parasitology because it emphasizes the general principles instead of the technical details.

A167 Lee, Donald L. The physiology of nematodes. 1965. Freeman. 154p. illus. $2.50.

Completely surveys nematode physiology and offers a carefully chosen list of further reference materials for in-depth study. Reports all recent knowledge on the subject including facts about nematode metabolism, morphology, oxygen transport, digestion and feeding, excretion, locomotion, behavior, sense organs, hatching and moulting, and nervous systems.

A168 Sprent, J. F. Parasitism; an introduction to parasitology and immunology for students of biology, veterinary science, and medicine. 1963. Williams & Wilkins. 145p. $3.95.

The author first gives some basic definitions and principles of parasitism and then devotes the remainder of the book to immunity. Other factors involved with parasitic activity are hardly mentioned, and thus this book will be of more use for an introduction to immunology.

A169 Thorne, Gerald. Principles of nematology. 1961. McGraw-Hill. 553p. illus. $16.50.

Approaches nematology from the taxonomic, ecological, biological, and the controllability angles. Although about 100 free-living forms are reviewed, the plant nematodes are nevertheless emphasized. Accurately documented and beautifully illustrated with instructive photographs and other items. This book reveals the great fervor which the author has for his field.

Ecology

A170 Amos, William H. The life of the pond. 1967. McGraw-Hill. 232p. illus. $4.95.

A thorough examination of the biology of pond life extending from its origin and organisms to its gradual demise. The pond is discussed from different angles such as the seasonal changes and the animal and plant life in freshwater. Has a glossary of freshwater ecology terms.

A171 Benton, Allen H., and **William E. Werner, Jr.** Field biology and ecology. 2d ed. 1966. McGraw-Hill. 509p. illus. $10.50.

Often recommended for use as a college text for introductory ecology or field biology courses. Its 13 chapters handle such topics as marine ecology, energy transfer, natural history, taxonomy, and the fundamentals behind animal behavior and ecological research. The authors' enthusiasm for the subject pervades their work.

A172 Boughey, Arthur S. Ecology of populations. 1968. Macmillan. 135p. illus. paper, $2.50.

A complete study of the ecology of living organisms. Deals with the processes of evolution, population dynamics, organization within communities, ways of altering relationships between organisms and their environment, and the major problems of population ecology.

A173 Clarke, George L. Elements of ecology. rev. ed. 1965. Wiley. 560p. illus. $9.95.

In this carefully organized, well-documented text, Clarke presents a broad survey of ecology for technicians, engineers, and officials responsible for ecological laws and conservation policies. This revised version of the original published in 1954 considers the effects of atomic energy, pesticide damage, and other modern day factors that change the environment.

A174 Henry, S. Mark, ed. Symbiosis: its physiological and biochemical significance. 1966-67. Academic. 2v. illus. v.1 $16.50; v.2 $17.50.

Volume 1 of this 2-volume set is concerned with the relationships of microorganisms, plant life, and water organisms, and the second volume deals with the associations of invertebrates, birds, ruminants, and other animals and plants within given regions. Valuable as a background for those interested in studying the cohabitation of various living forms but not suitable for a text because the information presented is for the most part not related to unifying themes or principles. The detailed bibliographies are very good.

A175 Macan, Thomas T. Freshwater ecology. 1963. Wiley. 338p. illus. $7.50.

Although intended for the senior in college or the graduate student, this treatise is also quite useful for the student on lower levels because there is so little written on freshwater ecology. Discusses communities, behavior, transport, physical and chemical factors, and other topics important in this field. Filled with factual information and presented in an admirably clear style.

A176* Odum, Eugene P. Ecology. 1963. Holt. 152p. illus. $3.25.

This entire account is based on 2 concepts, the one-way flow of energy and the circulation of materials, which the author believes to be the basic laws of ecology. The 7 chapters are on the scope of ecology, ecosystems, energy flow and nature's metabolism, biogeochemical cycles, limiting factors—Leiberg's Law Extended, ecological regulation, and ecosystems of the world. Serves the high school, college, or lay reader. A well-developed, effectively illustrated work which concentrates more on the function rather than the details of ecology.

A177 _____, and Howard T. Odum. Fundamentals of ecology. 2d ed. 1959. Saunders. 546p. illus. $8.50.

Requires a knowledge of biology and chemistry for complete understanding, but can be valuable for the student and professional alike. Portrays the fundamental concepts and principles of ecology and then describes the specific ecological phases and the actual application of basic ecology in practical situations.

A178 Platt, Robert B., and **John F. Griffiths.** Environmental measurement and interpretation. 1964. Van Nostrand. 235p. illus. $10.50.

Acquaints the reader, whether a researcher, ecologist, engineer, horticulturist, agriculturist, or student, with the techniques needed for successful study of the environment. Covers selection, measurement, and analysis of existing environmental conditions.

A179 Read, Clark P. Parasitism and symbiology: an introductory text. 1970. Ronald. 316p. illus. $10.

Parasitism is treated as a symbiotic form in this undergraduate level work of extremely wide scope. In Read's book, symbiosis is presented as the association between all types of organisms, harmful and nonharmful. The chapters on nutrition and metabolism and adaptations for development in the host are excellent, and in order to span the conventional boundaries between organisms, examples which include viruses, bacteria, fungi, helminths, etc., are given.

A180 Sears, Paul B. Lands beyond the forest. 1969. Prentice-Hall. 206p. illus. $7.95.

A composite view of the ecology, evolution, history, anthropology, and economics of the terrestrial environment. Describes all types of regions ranging from the grasslands to the desert and the factors which affect the land such as continental development and inhabitants. A cultural anthropological approach to ecology.

A181 _____. The living landscape. 1966. Basic. 199p. illus. $5.95.

Sears' underlying theme in this universal introduction to ecology is the fallibility and insignificance of man in relation to nature and the need for him to abide by nature's laws. Pertinent in other countries as well as in the United States, this is a book for the nonbiologist written in conversational style.

A182 Shelford, Victor E. The ecology of North America. 1963. Illinois. 610p. illus. $10.

Comprehensive account of North American ecology in the period immediately preceding European colonization, and the subsequent use and exploitation of the resources available.

A183 Smith, Robert L. Ecology and field biology. 1966. Harper. 702p. illus. $14.95.

Covers population biology, behavior, energy flow in communities, cycles of elements, and other important aspects of ecology. Excellently illustrated and written in fairly elementary style for the undergraduate reader.

A184 Whittaker, Robert H. Communities and ecosystems. 1970. Macmillan. 164p. illus. paper, $3.95 (approx.)

May be used with Boughey's book, entitled *Ecology of populations,* for ideally complete coverage of the community and ecosystem concepts of ecology. Discusses the characteristics and fate of these communities and includes an examination of the effect of their demise on human survival. Presents the most recent ideas and theories.

PLANT SCIENCES

General Works

A185 Alexopoulous, Constantine, and **Harold C. Bold.** Algae and fungi. 1967. Macmillan. 135p. illus. paper, $2.50.

A concise description of morphology and reproduction in algae and fungi. Accompanied by excellent illustrations of common varieties representing each important group or class of these primitive plants. This book, too compact to be considered a textbook, has real value as a reference source or for supplementary reading.

A186 Anderson, Edgar. Plants, man, and life. 1952, reprint 1967. California. 251p. illus. $6.50.

A preface, epilogue, and short glossary have been added to this work originally published in 1952. The author uses a semipopular approach in this unusual book concerning man and his useful or economic plants. He attempts to trace the biological and geographical origin and history of plants man has altered through hybridization or change of natural habitat. Theories or suppositions are clearly identified for readers unfamiliar with economic botany.

A187* Baker, Herbert G. Plants and civilization. 2d ed. 1970. Wadsworth. 194p. illus. $2.95.

A condensed, up-to-date presentation of plants in relation to man's social and economic development. Traces the origin and evolution of the most important food and fiber plants and discusses man's role in their alteration and geographical distribution. An excellent treatment of economic botany.

A188 **Bold, Harold C.** The plant kingdom. 3d ed. 1970. Prentice-Hall. 190p. illus. $6.95.

An inexpensive, clearly written, and well-illustrated survey of the plant kingdom. Unity and diversity of plants, life history patterns, reproductive systems, growth patterns, adaptations, and metabolic systems are some of the topics covered.

A189 **Brook, Alan J.** The living plant: an introduction to botany. 1964. Aldine. 529p. illus. $10.

The American edition of a British textbook on botany, providing the beginning student with an introduction to the field and a survey of the major divisions of the plant kingdom. Extensive consideration is given to subjects such as plant growth and development, genetics, nutrition, metabolism, and reproduction. Only minor attention is paid to plant form and structure and plant classification and identification.

A190 **Chapman, Valentine J.** The algae. 1961. St. Martin's. 472p. illus. $8.

A broad survey of the algae, ranging from microscopic forms to the large seaweeds, and covering the origin, evolution, physiology, reproduction, ecology, distribution, and utilization of these primitive plants. Intended for junior college or university students who have had a basic course in botany.

A191 **Corner, Edred J. H.** The life of plants. 1964. Mentor. 319p. illus. paper, $1.50.

A prominent British naturalist traces the origin and development of plants from their beginnings in the sea to their gradual distribution across the land and discusses the morphological modifications made by plants in adapting to new environments. Useful as supplementary reading material in a beginning botany course.

A192 **Coulter, Merle C.** The story of the plant kingdom. 3d ed. rev. by Howard J. Dittmer. 1964. Chicago. 467p. illus. $5.95.

An elementary botany textbook written in semipopular style with a minimum use of technical terms. A good source of information on the morphology, physiology, and evolution of the plant.

A193* **Daubenmire, Rexford F.** Plant communities: a textbook of plant synecology. 1968. Harper. 300p. illus. $10.95.

A study of environmental conditions, species adaptations, and interrelationships between species in plant communities. A sequel to the author's earlier book, *Plants and environment,* which deals with plant autecology. Suitable as a text for an introductory course in plant ecology and ecological processes at the first year of college level.

A194* ———. Plants and environment: a textbook of plant autecology. 2d ed. 1959. Wiley. 422p. illus. $9.95.

A concise, scholarly study of the behavior of individual plants and their relation to varying factors of the environment. Factors covered include soil, water, temperature, light, etc. Technical terminology is kept to a minimum for the benefit of persons with little or no background in botany.

A195 **Doyle, William T.** The biology of higher cryptogams. 1970. Macmillan. 192p. illus. paper, $4.95.

An account of the life and evolutionary development of the lower land plants that do not form seeds. Clearly written and easily understood by students with only an elementary background in biology. Can be used for supplementary reading in botany courses.

A196 **Edlin, Herbert L.** Plants and man: the story of our basic food. 1967. Nat. Hist. Pr. 253p. illus. $6.95.

A profusely illustrated and simply written survey of the most important economic plants. Coverage is worldwide and includes grains, legumes, root and fiber crops, fruits, vegetables, sugar, and other food crops. An excellent reference book for students from junior high school to college.

A197 **Eyre, S. R.** Vegetation and soils: a world picture. 2d ed. 1968. Aldine. 328p. illus. $8.95.

A plant geography of the world for advanced high school students and college freshmen. Although vegetation is emphasized, adequate treatment is given to the relationship between soils and plants. Written primarily for British schools but one of the few on this subject at this level.

A198 **Faegri, Knut, and Leendert Van Der Pijl.** The principles of pollination ecology. 1966. Pergamon. 258p. illus. $11.

One of the few comprehensive books available on the subject of pollination ecology. A clearly presented and orderly examination of the ways pollen is dispersed, with emphasis on principles rather than practical applications. Illustrative examples given in the appendix would be useful for classroom discussions and exercises.

A199* **Fernald, Merritt L.** Gray's manual of botany; handbook of the flowering plants and ferns of the central and northeastern United States. 8th ed. Largely rewritten and expanded by M. L. Fernald. 1950. Van Nostrand. 1632p. illus. $24.95.

A standard manual in use since 1848. Although the 5th edition, published in 1867, was the last one prepared by Asa Gray, he is still generally considered the author. An essential reference book for any plant science library.

A200 **Free, John B.** Insect pollination of crops. 1970. Academic. 544p. illus. $21.

A thoroughly documented source of basic information on the use of honeybees and other insects as pollinators. The first part of the book discusses insects man can use as pollinators and the most effective methods of using them. The second part treats individual crops. Pollination requirements, the normal insect pollinators, and production increase possibilities with use of honeybees are reviewed for each crop.

A201* **Fuller, Harry J.,** and **Zane B. Carothers.** The plant world; a text in college botany. 4th ed. 1963. Holt. 564p. illus. $10.50.

Written primarily for students who need only a general knowledge of botany, but with enough detailed information to give an adequate foundation to plant science students. Fundamental principles of botany are emphasized although practical applications are not ignored. The author's use of repetition, chapter summaries, and review questions makes this an excellent beginning botany text for technical school students pursuing a plant science program of studies.

A202* **Gleason, Henry A.,** and **Arthur Cronquist.** The natural geography of plants. 1964. Columbia. 420p. illus. $12.50.

An accurate account of the theories and basic principles of plant geography. Written in a readable style using nontechnical language. The illustrations used are primarily photographs of individual plants and vegetation types found in the United States. Valuable as a reference source for the plant scientist, the student, and the layman.

A203 **Good, Ronald.** The geography of the flowering plants. 3d ed. 1964. Wiley. 581p. illus. $13.

A scholarly treatment of the geography of the angiosperms, using examples from all regions of the world and organized into 2 parts. Part 1 consists of a descriptive account of the facts of distribution. Part 2 contains a theoretical consideration of these facts and the author's explanation of them. Intended for readers with a background knowledge in botany.

A204* **Gould, Frank W.** Grass systematics. 1968. McGraw-Hill. 382p. illus. $14.50.

A prominent grass taxonomist presents the fundamental elements of agrostology, characteristics of the genera of U. S. grasses, and much general information on grasses. Intended as a textbook or reference source for an undergraduate course. Valuable features of the book include its easy-to-follow key to grass genera and the line drawings accompanying the descriptions.

A205 **Greulach, Victor A.,** and **J. Edison Adams.** Plants: an introduction to modern botany. 2d ed. 1967. Wiley. 636p. illus. $9.95.

A modern survey of botany, intended as a textbook for an introductory course in high school or beginning college. Stresses plant form and structure, growth and development, ecology, and important developments resulting from recent research.

A206 **Harrison, Sydney G., G. B. Masefield,** and **Michael Wallis.** The Oxford book of food plants. 1969. Oxford. 206p. illus. $10.

Consists of descriptions and colorful illustrations of more than 400 food plants, grouped according to type of food plant. Origin, botany, distribution, nutritional value, and other items of information presented for each plant. Of particular reference value to plant science or food technology students.

A207 **Heiser, Charles B.** Nightshades: the paradoxical plants. 1969. Freeman. 200p. illus. text ed., $5.95.

A highly readable blend of scientific fact and folk stories about the nightshade family ranging from the economically important food plants to the deadly poisonous plants, the garden ornamentals, and others. The potato, the tomato, henbane, belladonna, tobacco, and the petunia are among those discussed. Useful as supplementary reading in general and economic botany.

A208* **Hill, Albert F.** Economic botany; a textbook of useful plants and plant products. 2d ed. 1952. McGraw-Hill. 560p. illus. $12.50.

Worldwide in scope. Discusses plant products of current economic importance and those important in the historical past. Designed to serve college undergraduates who have a basic background in general botany. One of the few American economic botany textbooks in print.

A209 Hill, John B., Henry W. Popp, and **Alvin R. Grove.** Botany. 4th ed. 1967. McGraw-Hill. 634p. illus. $10.50.

A standard introductory textbook. Covers the entire field and emphasizes the economic importance of plants and plant parts. Especially useful for those who need only a broad general knowledge of botany.

A210 Hutchins, Ross E. Plants without leaves. 1966. Dodd. 152p. illus. $3.75.

An entomologist presents a survey of cryptogamic plants, omitting only the club mosses and ferns. This inexpensive little book is written in popular form with most plants referred to by generalized name. An easy-to-read introduction to plants without leaves.

A211* Hylander, Clarence J. The world of plant life. 2d ed. 1956. Macmillan. 653p. illus. $12.95.

A botany book on the high school and junior college level. Systematically describes the plant kingdom from the most primitive forms of life to the highest. Written in popular style and with a minimum of scientific terminology, it tells about the origin, development, structure, and relationship of our common cultivated flowers and food plants. Very good for quick reference and for a general survey of the whole plant kingdom.

A212 Jamieson, B. G. M., and **John F. Reynolds.** Tropical plant types. 1967. Cambridge. 377p. illus. $8.50.

A tropical botany textbook for advanced secondary school and beginning university students. Six chapters cover botanical organisms from bacteria to gymnosperms, with 9 chapters devoted to the angiosperms. The final chapter is concerned with vegetation types and plant communities.

A213 Kingsbury, John M. Deadly harvest. 1965. Holt. 128p. illus. $4.95.

A layman's guide to common poisonous plants of America. Describes the botany and chemistry of each plant and discusses symptoms of poisoning. Includes only those plants definitely known to be poisonous to domestic animals or harmful to man.

A214* ———. Poisonous plants of the United States and Canada. 3d ed. 1964. Prentice-Hall. 626p. illus. $15.

An easy-to-read presentation of technical information, with emphasis on plants poisonous to animals rather than those dangerous to man. Contains botanical data for plant identification and toxicological data for recognition of poisoning symptoms. This approach makes the book especially useful to the veterinarian for diagnosis of poisoning. Treatment, however, is not within the scope of this book.

A215* Küchler, August W. Vegetation mapping. 1967. Ronald. 478p. illus. $15.

A detailed and timely guide to vegetation mapping. Written in an informal style by a leading authority on the subject. Logically organized, it begins with basic ecological considerations followed by technical aspects of vegetation mapping, methods of mapping, and finally, with explanations of its possible applications.

A216 Mayer, A. M., and **A. Poljakoff-Mayber.** Germination of seeds. 1963. Pergamon. 236p. illus. $6.50.

A competent review of the major aspects of seed germination. Of most value to those who want only a broad general knowledge of the subject. Covers structure, factors in germination, dormancy and chemical controls, metabolism and chemical factors affecting it, and ecology of germination. Treatment is limited to the angiosperms.

A217 Meeuse, Bastiaan J. D. The story of pollination. 1961. Ronald. 242p. illus. $7.50.

This story of how plants are pollinated was written for lay naturalists of all ages. Includes many unusual facts about the relationship of insects and other pollinating agents to flowering plants. Beautiful illustrations in color and black and white add to the book's appeal.

A218 Menninger, Edwin A. Fantastic trees. 1967. Viking. 304p. illus. $8.95.

This truly fascinating book is the first to give complete coverage to extraordinary trees. The book is divided into sections: trees with peculiar parts; trees peculiar all over; trees unable to live without animals; trees of peculiar behavior; trees famous because of size, age, or superstition; and the rugged individualists. Anyone involved with trees should read this book.

A219 Milne, Lorus J., and **Margery Milne.** Plant life. 1959. Prentice-Hall. 283p. illus. $8.95.

A well-illustrated basic botany textbook intended to provide a broad introduction to plants and their relationship to man. Includes a minimum of scientific terminology and emphasizes those terms likely to be familiar to the general reader.

A220 Muller, Walter H. Botany: a functional approach. 2d ed. 1969. Macmillan. 400p. illus. $9.25.

A popular introductory textbook intended for a 1-semester course in botany. Gives emphasis to the dependence of all organisms upon green plants for food and to the application of fundamental botanical principles to man's significant problems such as conservation, plant diseases, population growth, and food supply.

A221* Northen, Henry T. Introductory plant science. 3d ed. 1968. Ronald. 586p. illus. $9.50.

A balanced, up-to-date presentation of all aspects of modern botany. Pays particular attention to the plant as a whole and to its relationship to man and the environment. Primarily a beginning botany textbook but also valuable as an introductory level reference source on modern plant science.

A222 _____, and Rebecca T. Northen. Ingenious kingdom: the remarkable world of plants. 1970. Prentice-Hall. 274p. illus. $8.95.

A portrayal of the plant world as a realm of constant action and dynamic change. Examines the reasons behind longevity of individual plants, views primitive plants from the immortal diatoms to the giant kelp, discusses flowering plants with unusual means of reproduction, and reveals recently discovered secrets of how plants know when to bloom, etc. Useful as a reference source or for leisure reading.

A223 Novak, Frantisek A. The pictorial encyclopedia of plants and flowers. 1966. Crown. 589p. illus. $10.

An encyclopedia of pictures covering the botanical world from the most primitive to the highest level of plant life. The illustrations are grouped according to families with short descriptions of structure and geographic distribution included. A useful aid in identifying types of plants in the various families.

A224 Oosting, Henry J. The study of plant communities. 2d ed. 1956. Freeman. 440p. illus. $7.50.

A clearly written and amply illustrated introduction to the ecology of plant communities. Author's purpose is to provide a perspective of plant ecology as seen primarily through plant community relationships.

A225* Raven, Peter, and Helena Curtis. Biology of plants. 1970. Worth. 706p. illus. $11.95.

A clearly written college text for beginning botany courses. Uses a minimum of scientific terminology for the benefit of those with no prior knowledge of the field. Understanding of the text is increased by the author's use of many excellent illustrations. Covers concepts, techniques, and other important aspects of plant biology.

A226 Schery, Robert W. Plants for man. 1952. Prentice-Hall. 564p. illus. $11.50.

The commercial uses of plants is the central theme of this easily read and logically organized economic botany book. Forests and woods, wood products, beverage plants, medicinal plants, and vegetables are some of the subjects treated. The chapters on forests and wood products would be useful in a beginning course in general forestry.

A227 Sinnott, Edmund W., and Katherine S. Wilson. Botany: principles and problems. 6th ed. 1962. McGraw-Hill. 528p. illus. $9.50.

Presents the basic concepts of botany and considers some of the unsolved problems of plant science. This well-known standard textbook for beginning college students has the usual added features of most other textbooks.

A228 Stephenson, William A. Seaweed in agriculture and horticulture. 1968. Faber. 231p. illus. $11.25.

An accountant who became successful at producing liquid seaweed fertilizer gives an exciting account of seaweeds. He describes what they are, where they grow, and their possible uses as fertilizers, foods, medicines, and other products. Should be of particular interest to gardeners, livestock producers, and botanists.

A229 Tiffany, Lewis H. Algae: the grass of many waters. 2d ed. 1968. Thomas. 199p. illus. $7.50.

A nontechnical review of all aspects of algae. Useful to the general botanist or amateur naturalist as a reference or for leisure reading. Describes the various classes of algae, discusses the ecology of algae of various habitats, explains their evolutionary history and their importance to man, and gives directions and techniques for collecting and studying these plants.

Plant Sciences

A230 Tortora, Gerald J., Donald R. Cicero, and Howard I. Parish. Plant form and function: an introduction to plant science. 1970. Macmillan. 608p. illus. $10.95; paper, $3.95.

This general botany text maintains a good balance between the traditional and molecular approaches. Develops the basic theme of form and function in 3 steps. First, considers form and structure, then analyzes the functional activities of plants, and, finally, views the whole plant kingdom in relation to form and function. Designed for use in a beginning college botany course.

A231 Uphof, Johannes C. Dictionary of economic plants. 2d ed. 1968. Stechert-Hafner. 591p. $19.25.

A concise manual of economic plants. For each plant included gives its common, scientific, and family names, its geographical distribution, and its principal economic uses. Arranged alphabetically and thoroughly indexed.

A232 Usher, George. Dictionary of botany. 1966. Van Nostrand. 404p. $11.95.

Defines the phyla, classes, and orders but omits genera and species. In addition to botanical terms, includes many used in biochemistry, soil science, and statistics. A useful dictionary for students on both the secondary and the college level.

A233* Weier, Thomas E., C. Ralph Stocking, and Michael G. Barbour. Botany: an introduction to plant science. 4th ed. 1970. Wiley. 708p. illus. $12.50.

An almost classical text, noted for its outstanding illustrations and its comprehensive coverage of all major aspects of general botany. Includes both the traditional material and the newer concepts resulting from recent research. Designed as a beginning college text and assumes little or no prior knowledge of plant life principles. Senior author for earlier editions of this book was Wilfred W. Robbins. A laboratory manual is available.

A234 Weisz, Paul B., and Melvin S. Fuller. The science of botany. 1962. McGraw-Hill. 562p. illus. $10.95.

An analytical and dynamic approach, with an experimental outlook, is used for this scholarly introductory text oriented toward molecular biology. Gives comprehensive coverage to the historical development of plants, the anatomy of different types of plants, plant control factors, reproduction, and adaptation. Instructor's manual and laboratory manual available.

A235* Wilson, Carl L., Walter E. Loomis, and Taylor A. Steeves. Botany. 5th ed. 1971. Holt. 752p. illus. $12.95.

A well-balanced account of plant life with emphasis on plant diversity and the importance of plants to man. Designed as a 2-semester text which can be adapted to a shorter course. Provides the traditional information on structure, development, function, reproduction, etc., and includes the newer advances in cell biology. New editions of this popular textbook published every 4 or 5 years.

Taxonomy

A236 Abrams, Leroy. An illustrated flora of the Pacific states: Washington, Oregon, and California. v.1 1923, v.2 1944, v.3 1951, v.4 1963. Stanford. 4v. illus. $20 each v.

The standard guide for identifying the flora of the Pacific states, with descriptive keys and a glossary. Contents include: v.1 ferns to birthworts; v.2 buckwheats to kramerias; v.3 geraniums to figworts; and v.4 bignonias to sunflowers. Volume 4 is by Roxana J. S. Ferris.

A237 Bailey, Liberty H. How plants get their names. 1933. Macmillan. 209p. illus. paper, $1.35.

Explains how the rules of botanical nomenclature were developed and how one should apply them. Special features include a list of generic terms common in horticulture and a list of Latin words with their English botanical meaning. Good for background information.

A238* ―――. Manual of cultivated plants most commonly grown in the continental United States and Canada. 1949. Macmillan. 1,116p. illus. $19.95.

This horticultural classic is the standard descriptive key to cultivated plants of the United States. It is a ready means of identification of ornamental shrubs, garden flowers, greenhouse plants, grains, grasses, fruits, vegetables, and other plants.

A239 Berry, James B. Western forest trees. 1924, reprint 1964. Dover. 212p. illus. paper, $2.

Describes 74 species of the Pacific Coast forest and the Rocky Mountain forest complex. Includes keys for the woods, the needleleaf trees, and the broadleaf trees. Arranged by characteristic of leaf and illustrated with life-size drawings.

A240 Bessey, Ernest A. Morphology and taxonomy of fungi. 1950, reprint 1964. Hafner. 791p. illus. $16.95.

Presents an orderly outline of the fundamentals of mycology. Gives the major characteristics of all of the commonly recognized orders of fungi, accompanied by keys to the orders in each class. A special feature is its nearly 100 pages of bibliography. One of the best references for descriptive purposes.

A241 Boom, Boudewijn K., and H. Kleijn. Glory of the tree. 1966. Doubleday. 128p. illus. $8.95.

Fully half of the trees featured are native to America. The authors describe the origin, identifiable characteristics, geographical distribution, practical uses, etc., for each tree included. Special features include a list of trees shown on postage stamps and an explanation of the Latin names of species.

A242 Brightman, Frank H. Oxford book of flowerless plants. Illus. by B. E. Nicholson. 1967. Oxford. 208p. $10.50.

Written for amateur naturalists of all ages with the primary purpose of helping the beginner identify the flowerless plants. Nearly 700 plants are grouped according to their natural habitats, described, and illustrated. These outstanding colored illustrations, drawn from actual specimens, are valuable identification aids. Although the book covers the common flowerless plants of Great Britain, most of the species included are known in America.

A243 Britton, Nathaniel L., and Joseph N. Rose. Cactaceae. 2d ed. 1937. Dover. 4v. in 2. illus. $10 each v.

Clearly and fully describes and illustrates plants of the cactus family. Arranged by orders, families, and tribes with keys to species. Includes 124 genera and 1,235 species found in the Americas, Mexico, or the West Indies.

A244* Brockman, Christian Frank. Trees of North America. 1968. Golden. 280p. illus. $5.95; paper, $3.95.

A field guide consisting of concise descriptions, arranged by families, of the important native and introduced trees of North America. Contains illustrations of the whole tree and close-ups of leaves, buds, fruits, flowers, and bark for identification purposes. Includes a list of commonly used semitechnical terms and an index of both common and scientific names. Recommended for the young amateur as well as the authority.

A245 Clark, Robert B. Flowering trees. 1963. Van Nostrand. 241p. illus. $6.95.

Comprised of detailed illustrations and accurate descriptions of hardy ornamental trees for cold-winter regions of the United States. Includes practical advice on planting, pruning, soils, fertilizers, etc., for the layman.

A246* Collingwood, George H., and Warren D. Brush. Knowing your trees. rev. and ed. by Devereux Butcher. 1964. Amer. Forestry Assoc. 349p. illus. $7.50.

A guide to 170 important American trees, written in popular style for the layman and the scientist. Describes the botanical features of each tree and discusses its economic contributions to man. The illustrations aid in identifying the tree in winter and summer by its leaf, flower, fruit, or bark. Often used as a reference book in beginning forestry courses.

A247 Dallimore, William, and Albert B. Jackson. A handbook of Coniferae and Ginkgoaceae. 4th ed. rev. by Sidney G. Harrison. 1967. St. Martin's. 729p. illus. $35.

This classic handbook containing concise descriptions of the conifers and taxads of the world is kept up-to-date by revisions. Unlike most botanical manuals, arrangement is alphabetical rather than by families. Valuable reference for anyone working with this group of woody plants.

A248 Gleason, Henry A. The New Britton and Brown illustrated flora of the northeastern United States and adjacent Canada. rev. ed. 1968. Hafner. 3v. illus. $40.

This book, sometimes called the "botanist's Bible," describes and illustrates the botanically distinct forms of flora growing wild in the northeastern section of the United States. Covers the visible characteristics of nearly 5,000 species. Special features include finding keys and an extensive index.

A249 Grimm, William C. The book of trees. 1962. Stackpole. 487p. illus. $7.95.

This popular identification guide fills the gap between the pocket book and the scientific treatise. Includes trees native to eastern North America, omitting only those of subtropical Florida. For each tree, it lists distinguishing botanical features accompanied by detailed illustrations of both summer and winter leaf and twig characteristics. Also discusses the economic value of these trees, with special emphasis on their relationship to wildlife.

A250 ———. Recognizing flowering wild plants. 1968. Stackpole. 348p. illus. $7.95.

Like most botanical manuals, this identification guide is arranged by families and genera. Scope is limited to the flowering wild plants native to the eastern half of the United States. Uses a minimum of technical terms. Drawings are used to supplement the descriptions.

A251 ———. Recognizing native shrubs. 1966. Stackpole. 319p. illus. $7.95.

A bridge between the too simple book for the young amateur and the highly technical manual for the scientist. Provides both summer and winter identification of some 461 species and subspecies of shrubs and vines of eastern United States. An unusual feature of this book is its list of derivations for generic and specific plant names.

A252 Harlow, William M. Fruit key and twig key to trees and shrubs. 1946. Dover. 143p. illus. paper, $1.35.

The subtitle, *Fruit key to northeastern trees; twig key to the deciduous woody plants of eastern North America,* explains the scope and purpose of this book. Most native tree species of the area are included. Valuable for identifying detached tree fruits.

A253 ———. Trees of the eastern and central United States and Canada. 1957. Dover. 288p. illus. paper, $1.50.

The standard semipopular guide to trees of the eastern and central United States. Has key plus descriptions and illustrations of the major features of each of the 140 different common trees included. Other matters connected with these trees, such as commercial uses, are also discussed. A valuable source of information for the layman.

A254 Harrar, Ellwood S., and Jacob G. Harrar. Guide to southern trees. 1962. Dover. 709p. illus. paper, $3.

Describes more than 350 trees with facts about their major botanical features, their growth, distribution, economic importance, etc. The standard reference for trees native to the southern states. Illustrated by line drawings.

A255* House, Homer D. Wild flowers. 1934, reprint 1961. Macmillan. 362p. illus. $17.95.

A reprint, with new color plates, of a classic first published in 1934. Begins with a useful introduction defining the terms used and follows with botanical descriptions, illustrations, and the geographic distribution of the major wild flowers of the United States. Outstanding illustrations.

A256 Hutchinson, John. Key to the families of flowering plants of the world. rev. and enl. 1968. Oxford. 117p. illus. paper, $3.75.

Contains key to the families of most of the flowering plants in any part of the world. Should be of value to botanists and to those working in the field and the herbarium as a working key.

A257 Hylander, Clarence J. The Macmillan wild flower book. 1954. Macmillan. 480p. illus. $12.95.

A big book of beautiful, colored drawings of wild flowers growing east of the Rocky Mountains. Includes brief botanical descriptions of the some 500 flowers pictured.

A258 Lawrence, George H. M. An introduction to plant taxonomy. 1955. Macmillan. 179p. illus. $6.95.

The amateur botanist or the college student studying local flora will find this concise book a valuable reference. Presupposes a basic background course in general botany. Plant classification, plant structures, terminology peculiar to the field, historical review of taxonomy in North America, and distinguishing characteristics of some of the more dominant families of vascular plants are among the topics presented.

A259 Martin, Alexander C., and William D. Barkley. Seed identification manual. 1961. California. 221p. illus. $10.

Devoted exclusively to the identification of wild plant seeds found in farmlands, wetlands, and woodlands. The reader must depend on a combination of identification clues and photographs for recognizing seeds since the book contains no keys. Primarily intended for agriculturists, foresters, wildlife biologists, and others concerned with the identification of wild plant seeds.

A260 Peterson, Roger T., and Margaret McKenny. A field guide to wildflowers of northeastern and northcentral North America. 1968. Houghton-Mifflin. 420p. illus. $4.95.

This nontechnical, authoritative guide to the most common flowering plants in the regions covered uses a pictorial approach with arrangement by color and form. For the amateur naturalist with no botanical background or training.

A261 Petrides, George A. A field guide to trees and shrubs. 1958. Houghton-Mifflin. 431p. illus. $5.95.

A popular guide to the wild trees and shrubs of northeastern and northcentral sections of the United States. The author has presented a vast amount of information in simple, descriptive terms. Attractive tree silhouettes serve as identification aids.

A262 Pohl, Richard W. How to know the grasses. 2d ed. 1968. Brown. illus. 244p. $4.

A key to the common American grasses most often encountered by the beginner and those of major importance in weed control, range management, farming, and gardening. Distribution maps and line drawings help to clarify the descriptions.

A263 Porter, Cedric L. Taxonomy of flowering plants. 2d ed. 1967. Freeman. 472p. illus. $8.25.

Covers the history, theory, terminology, and methodology of taxonomy of the flowering plants. A useful textbook or reference for the amateur, the student in taxonomy, the student in the agricultural sciences, or the specialist.

A264* Preston, Richard J. North American trees (exclusive of Mexico and tropical United States). 2d ed. 1961. Iowa State. 395p. illus. $4.50.

One of the few native-tree manuals covering most of the United States instead of just 1 region. Uses technical terminology only when necessary for scientific accuracy. Consists of descriptions of species with line drawings and distribution maps on the facing page. Designed for use by students as well as the specialist.

A265* Rehder, Alfred. Manual of cultivated trees and shrubs hardy in North America, exclusive of the subtropical and warmer temperate regions. 2d ed. 1940. Macmillan. 996p. $14.95.

The standard reference guide for identification of the cultivated trees and shrubs of the United States. Contains descriptive keys to the families, genera, and species. Includes a glossary. This book is the tree counterpart of Bailey's *Manual of cultivated plants.*

A266 Rydberg, Per A. Flora of the prairies and plains of central North America. 1932, reprint 1965. Hafner. 969p. illus. $14.95.

Dr. Rydberg originally planned this flora as a complete manual of the cryptogams and seed plants of Kansas, Nebraska, Iowa, Minnesota, the Dakotas, and parts of Canada. However, it also includes most of the species in the prairies and the plains regions of the United States. Still considered one of the best books ever published in its field.

A267 ———. Flora of the Rocky Mountains and adjacent plains. 2d ed. 1922. Hafner. 1143p. $17.50.

The standard reference for identification of the flora of the Rocky Mountains and the adjacent plains. Describes 1,055 genera and 6,029 species. Descriptions include Latin and common names, botanical features, natural habitat, and distribution. An extensive index facilitates use of the book.

A268* Sargent, Charles S. Manual of the trees of North America (exclusive of Mexico). 2d ed. 1922. Smith. 2v. illus. $15. Dover. paper, $6.

This comprehensive reference was written by the country's foremost dendrologist. Contains detailed descriptions of botanical features and notes on distribution for 717 tree species. Line drawings are used as identification aids.

A269* Smith, Alexander H. The mushroom hunter's field guide. rev. ed. 1963. Michigan. 264p. illus. $6.95.

An authoritative yet practical guide to the collection and identification of mushrooms. In mostly nontechnical language, the author tells when, where, and how to find edible mushrooms and how to avoid the poisonous ones. Contains outstanding illustrations of most of the edible mushrooms, even the uncommon ones.

A270 Sudworth, George B. Forest trees of the Pacific slope. 1908, reprint 1967. Dover. 445p. illus. paper, $4.

Reprint of a classic, written by a dendrologist and first issued as a government publication. The first fully illustrated tree manual published in the United States. Contains nontechnical botanical descriptions, detailed data on range, and highly accurate line drawings.

A271 Symonds, George W., and Stephen M. Chelminski. The tree identification book. 1958. Barrows. 208p. illus. $15.

The subtitle, *A new method for the practical identification and recognition of trees,* indicates the purpose of this guide. Uses a maximum of photographic illustrations and a minimum of text to identify 119 common native species and 17 introduced species of the northeastern states. Will aid beginners and experts whatever the season.

A272 ———, and **A. W. Merwin.** The shrub identification book. 1963. Barrows. 379p. illlus. $15.

Employs the visual method for practical year round identification of shrubs, woody vines, and ground covers. This unusual book contains separate sections showing flowers, fruit, twigs, leaves, thorns, and bark of each species. Another section brings all diagnostic features of each genus together. A practical supplement to traditional methods for the enthusiast, the forester, and the dendrology student.

A273 **Willis, John C.** A dictionary of the flowering plants and ferns. 7th ed. rev. by Herbert K. A. Shaw. 1966. Cambridge. 1214p. $18.50.

Endeavors to bring together most of the information needed by nonspecialists on the structure and form, history and development, classification, and economic uses of the flowering plants and ferns. Omits common names and horticultural notes. Of value to students who have a basic knowledge of botany.

Anatomy, Physiology, and Morphology

A274 **Asimov, Isaac.** Photosynthesis. 1969. Basic. 193p. illus. $5.95.

Technical but readily understandable account of biochemical, photochemical, and biophysical aspects of this vital process. Author uses simplified presentation for the benefit of persons with no biochemical background. Gives the history of our discovery and early knowledge of photosynthesis, discusses solar energy, etc.

A275* **Audus, Leslie J.** Plant growth substances. 2d ed. rev. and enl. 1959. Wiley. 553p. $14.50. (3d ed. forthcoming)

A presentation of the nature of plant growth, the effects of natural and artificial plant growth substances, and the use of growth regulating substances in agriculture and horticulture. Well illustrated. Language kept as nontechnical as possible. An important feature is the appendix showing the sensitivities of a great number of weed and crop plants to hormone herbicides.

A276 **Bell, Peter R.,** and **Christopher L. F. Woodcock.** The diversity of green plants. 1968. Arnold. 374p. illus. $12.50; paper, $6.50.

An up-to-date narrative of the structure, development, reproduction, and theories of evolution of each group of the green plants from the algae to the angiosperms. Written in easy-to-read style and illustrated with diagrams. Items of information not found in similar textbooks are occasionally inserted.

A277* **Black, Michael,** and **Jack Edelman.** Plant growth. 1970. Harvard. 193p. illus. $4.25.

A basic introduction to plant growth and development for secondary and beginning college students. Deals with the nature of growth, internal growth regulators, the effect of the environment on growth and development, and the use of artificial controls. Illustrated with line drawings. One of the few texts dealing with experimental work in this subject at this level.

A278* **Bold, Harold G.** Morphology of plants. 2d ed. 1967. Harper. 541p. illus. $14.95.

A discussion of morphology and reproduction of the more important plant types of the entire plant kingdom, ranging from the most primitive to the highest evolved groups. Of value not only as an introductory college textbook but to anyone interested in the structure of plants. The common descriptive style of writing is used. Laboratory manual available.

A279* **Bonner, James F.,** and **Arthur W. Galston.** Principles of plant physiology. 1952. Freeman. 499p. illus. $7.

A 1-semester elementary plant physiology textbook assuming a basic science background possessed by the student. Good for introducing principles and concepts of plant growth and development, nutrition, metabolism, etc., to students in the applied fields of plant science. Each chapter is accompanied by supplemental reading references.

A280 **Briggs, George E.** Movement of water in plants. 1967. Blackwell Scientific. 153p. $6.50.

The basic physical principles determining water movement in plants is the concern of this book. Traces water movement from its entrance into the root to its escape from the leaf. A discussion of the rate of water movement and a consideration of the plant as a whole are included.

A281* **Buvat, Roger.** Plant cells; an introduction to plant protoplasm. 1969. McGraw-Hill. 256p. illus. paper, $4.95.

A great amount of information is presented in this concise, readable book on cell biology for senior high school or first year college students. Contains a short historical introduction to cell and tissue concepts, an outline of cell physical and chemical properties, and a detailed treatment of each organelle and its function within the cell.

A282 Clowes, Frederick A. L., and **Barrie E. Juniper.** Plant cells. 1969. Blackwell Scientific. 560p. illus. $14.40.

A comprehensive plant cell biology book based on a wide survey of the latest literature. Emphasis is on the cells of higher plants and their detailed structure at the light and electron microscope level. An advanced text useful for supplementary reading for students who want a deeper understanding of the subject.

A283 Cutter, Elizabeth G. Plant anatomy: experiment and interpretation. pt.1 Cells and tissues. 1969. Addison-Wesley. 168p. illus. $8.50; paper, $3.95.

A concise survey of plant anatomy, using a developmental approach and stressing both classical and modern experimental evidence. Presents carefully selected characteristics of each of the plant cell types, together with factors determining their development. A useful text for an interdisciplinary course in cell biology.

A284* Devlin, Robert M. Plant physiology. 2d ed. 1969. Van Nostrand. 564p. illus. $11.95.

A well-documented, scholarly plant physiology sourcebook. Deals with water relations, photosynthesis, metabolism, nutrition, growth hormones, and growth and development of the plant. A valuable reference manual for students preparing for greenhouse operations, nursery work, farm crop work, or horticultural occupations at the technician level.

A285 Eames, Arthur, and **Laurence H. MacDaniels.** Introduction to plant anatomy. 2d ed. 1947. McGraw-Hill. 427p. illus. $11.

A standard textbook covering the structure and development of the plant, with each theory supported by results of research. Although one of the older books, it is still of value for the basic information it contains.

A286* Esau, Katherine. Plant anatomy. 2d ed. 1965. Wiley. 767p. illus. $15.95.

An up-to-date general treatment of the structure and development of seed plants, particularly the angiosperms. The subject is approached from a standpoint of developmental anatomy, considering first the cell and tissue types and then the arrangement of the structural elements within the plant organ. Basic terms and concepts are explained. Considered one of the best plant anatomy books available. The comprehension level of this book ranges from high school to professional.

A287 Fahn, Abraham. Plant anatomy. 1967. Pergamon. 534p. illus. $15.

A comprehensive introduction to general plant anatomy designed as a college undergraduate text. The book begins with a consideration of the cell followed by chapters on meristems, tissues, the primary and secondary plant body, and concludes with a study of reproductive organs. A good source book for students of plant anatomy and for botanists.

A288 Fogg, Gordon E. Photosynthesis. 1968. Amer. Elsevier. 116p. illus. $3.95.

Reviews early historical findings and introduces recent discoveries in the field of photosynthesis. Requires a background in the basic physical sciences for full understanding of this text. May be best used as a supplementary source of information.

A289* Galston, Arthur W. The life of the green plant. 2d ed. 1964. Prentice-Hall. 116p. illus. $4.95.

This expert's survey and evaluation of the current status of plant physiology can be used as a text for an introductory course or for a review course in the subject. The material covered is relevant to most all green plants although the angiosperms are stressed. An interesting feature of this book is the author's scattered comments on such subjects as the population crisis and the case against organic gardening.

A290* Hayward, Herman E. Structure of economic plants. 1938, reprint 1967. Stechert-Hafner. 674p. illus. $22.

Using a developmental approach, the author has presented a comprehensive anatomical study of corn, wheat, onion, hemp, and other important economic plants. Each plant is discussed in detail, with emphasis on the special features which make that crop of economic significance. A useful book for collateral reading in garden and field crop courses.

A291 Heath, Oscar V. The physiological aspects of photosynthesis. 1969. Stanford. 310p. illus. $8.50.

Stresses areas of botanical plant physiology rather than the biochemistry and biophysics or studies of photosynthesis by communities of plants. Topics are covered from a historical viewpoint and discussed in relation to higher plants. Gives equal treatment to both sides of controversial questions.

A292 Jensen, William A. The plant cell. 2d ed. 1970. Wadsworth. 136p. illus. $2.50.

A short historical review of research on cell biology is followed by a more detailed presentation of cell walls and membranes, chloroplasts, ribosomes, the nucleus, etc. Clearly presents the biochemical processes involved in respiration, photosynthesis, protein synthesis, and the problems of cell development and differentiation.

A293 Juniper, Barrie E., and others. Techniques for plant electron microscopy. 1970. Blackwell Scientific. 108p. illus. $2.40.

A simple account of plant electron microscopy techniques with special attention given to problems encountered in preparing plant material specimens. The type of book which should be kept in the laboratory, not the library. Written by practicing electron microscopists.

A294 Kozlowski, Theodore T., ed. Tree growth. 1962. Ronald. 442p. illus. $12.

Includes papers presented at a conference on tree growth and brief discussions by 31 well-known contributors. Treats all aspects of tree growth. A valuable reference source for forestry students and a basis for class discussions in genetics, physiology, mensuration, and silviculture.

A295 ———. Water metabolism in plants. 1964. Harper. 227p. illus. paper, $5.95.

A useful book for horticulturists, foresters, botanists, and anyone responsible for efficient use of water. Written by a knowledgable forester who includes evidence from his own researches. Covers plant-water balance, water relations of cells and tissues, theories of water transport, effects of water deficits on plants, etc.

A296* Kramer, Paul J. Plant and soil water relationships. 1969. McGraw-Hill. 482p. illus. $16.

A comprehensive study of the basic precepts of soil and plant-water relationships. Concerned with soil-moisture relations, plant-water relations, external factors affecting water absorption, nutrient absorption, and effects of absorption deficits on plants, with principal attention paid to internal plant processes. A worthwhile source of information for technicians, students, and others concerned with water absorption by plants.

A297 ———, and **Theodore T. Kozlowski.** Physiology of trees. 1960. McGraw-Hill. 642p. illus. $15.50.

An ecological rather than a biochemical approach is used in this presentation of the important physiological processes of trees. Intended for the reader with an elementary knowledge of plant physiology and therefore, does not explain all of the fundamental principles concerning the facts presented.

A298 Leopold, Aldo C. Plant growth and development. 1964. McGraw-Hill. 466p. illus. $13.

In-depth treatment of the growth of the whole plant rather than details of the processes themselves. Topics presented include assimilation, growth and development, effects of environmental conditions on physiological processes, and chemical modification of plant growth and development. Suitable text for more advanced horticulture and field crop students and a good reference source for the less advanced.

A299 Levitt, Jacob. Introduction to plant physiology. 1969. Mosby. 304p. illus. $10.50.

An introductory book on the beginning college level. Covers the whole subject of plant physiology including biochemistry, biophysics, and growth and development. A special feature is the excellent selection of references given at the end of each chapter.

A300* Meyer, Bernard S., Donald B. Anderson, and **Richard H. Bohning.** Introduction to plant physiology. 1960. Van Nostrand. 541p. illus. $9.75.

A concise, yet comprehensive treatment of plant-water relationships, plant biochemistry, and plant growth and reproduction. Subject matter presented at an elementary level but does require some prior knowledge of general botany and basic chemistry. Has become a standard textbook in many colleges.

A301 O'Brien, Terence P., and **Margaret E. McCully.** Plant structure and development. 1969. Macmillan. 114p. illus. $9.95; paper, $5.50.

Uses a pictorial and physiological approach to present all levels of plant structure from subcellular organelles to the whole plant and growth at all stages from the seedling to the mature plant. Useful for a wide variety of biology or botany courses at the college level or even in some advanced courses at the high school level.

A302 Rabinowitch, Eugene, and **Govindjee.** Photosynthesis. 1969. Wiley. 274p. illus. $8.95; paper, $5.95.

The authors use a conversational style of writing to present this survey of the field. Suitable as an introduction to photosynthesis for students with varying backgrounds in the sciences. Organized for step-by-step guidance of the reader rather than by logic of subject matter.

A303* Rosenberg, Jerome L. Photosynthesis. 1965. Holt. 127p. illus. $2.50.

Up-to-date description of the basic food making processes in green plants. Tells what photosynthesis is, how it works, and its importance to plant life. The research and discoveries of early investigators are skillfully woven into the discussions. Technical terminology is kept to a minimum for the benefit of students with only a minor academic background in the basic sciences.

A304 Sass, John E. Botanical microtechnique. 3d ed. 1958. Iowa State. 228p. illus. $5.95.

A standard book on plant microtechnique. Divided into 2 sections: (1) general methods; and (2) special methods for the various phyla of the plant kingdom. A good handbook for students studying to become technicians in a plant science laboratory.

A305 Sire, Marcel. Secrets of plant life. 1970. Viking. 239p. illus. $19.50.

A painless introduction to plant classification but not a textbook. Consists of true, brilliant color photographs of plant life, ranging from tree portraits to minute details of anatomy and development in a wide variety of plants. Many of the illustrations are accompanied by explanatory descriptions.

A306 Sporne, K. R. The morphology of gymnosperms. 1965. Hillary. 216p. illus. $4; paper, $2.

A comprehensive treatment of vascular seed plant form and structure suitable for reading at all student levels. Chapters treat individually the 9 orders of gymnosperms in a readable style. Author attempts to give impartial treatment to controversial issues.

A307 Steward, Frederick C. Plants at work. 1964. Addison-Wesley. 184p. illus. paper, $3.25.

This beginning textbook on plant physiology will be of most value as supplementary reading in elementary agriculture courses. The author has used the experimental viewpoint in presenting the subject matter but has tied in the relevant aspects of plant history and plant structure and form.

A308* Stewart, William D. P. Nitrogen fixation in plants. 1966. Oxford. 168p. illus. $4.50.

A comprehensive discussion in clear, concise language of the physiology, biochemistry, and cytology of all groups of nitrogen fixers. Concludes with a chapter on the importance of nitrogen fixation and its significance in soil fertility. Should fill the needs of high school and college level students in the agricultural sciences programs.

A309 Stiles, Walter. Trace elements in plants. 3d ed. 1961. Cambridge. 249p. illus. $9.

A practical, descriptive account of trace elements in plants and some of the effects of their deficiency or excess on grazing animals. Manganese, zinc, boron, copper, molybdenum, and chlorine are the trace elements covered. Only minor attention paid to the influence of soils on availability of trace elements.

A310 Street, Herbert E., and **Helgi Öpik.** The physiology of flowering plants. 1970. Arnold. 263p. illus. $8.40; paper, $4.20.

An intermediate text for those who already have some knowledge of plant biochemistry, biophysics, and anatomy. Discusses the usual topics in an easy-to-read manner with sufficient detail to meet the needs of a first or second year college student.

A311 Torrey, John G. Development in flowering plants. 1967. Macmillan. 184p. illus. paper, $3.95.

The author has collected the more important facts on the development of flowering plants into a useful introductory book for the nonspecialist. Includes all factors of development and differentiation from seed to fruit wtih stress on the causal aspects.

A312 Wareing, P. F., and **Irving D. Phillips.** The control of growth and differentiation in plants. 1970. Pergamon. 303p. illus. $7.

A pleasant introductory text on plant growth and cell differentiation, intended for students at the college level. Emphasizes control and gives special attention to the part played by natural regulators in the growth and differentiation processes.

A313 Wilson, Brayton F. The growing tree. 1970. Massachusetts. 152p. $6.50.

The complicated subject of tree growth presented in a manner calculated to appeal to all tree lovers. Focuses on the morphological characteristics peculiar to the tree, the basic growth processes, growth regulators, etc. The author shows how the processes interact with each other and with the environment, and points up interesting variations.

A314 Wilson, Charles M. Roots: miracles below. 1968. Doubleday. 234p. illus. $5.95.

An ecologist gives an interesting botanical account of plant roots and a discussion of man's diverse uses of roots. Topics discussed include the relation of roots to soil types, soil temperatures, water, fungi, nematodes, viruses, and bacteria.

Plant Breeding

A315 Allard, Robert W. Principles of plant breeding. 1960. Wiley. 485p. illus. $12.95.

A textbook for agriculture students who are already familiar with genetics, biometry, and experimental design at the elementary level. Although the book is mainly concerned with fundamental principles, breeding methods for a great number of cultivated plants are included.

A316* Briggs, Fred N., and Paulden F. Knowles. Introduction to plant breeding. 1967. Reinhold. 426p. illus. $12.95.

The authors present a study of the genetic basis of plant breeding and proceed to explore principles and practices. Intended as an elementary textbook for students with little training in basic biology and who do not intend to specialize in plant breeding.

A317 Hayes, Herbert K., Forrest R. Immer, and David C. Smith. Methods of plant breeding. 2d ed. 1955. McGraw-Hill. 551p. illus. $13.50.

Presents basic information on cytology and genetics, relates plant breeding to other fields of plant biology, outlines fundamental plant breeding methods, and classifies crops according to origin and mode of reproduction.

A318* Poehlman, John M. Breeding field crops. 1959. Holt. 427p. illus. $12.50.

An interesting and accurate introduction to the breeding of field crops intended as a text in a beginning plant breeding course. For each crop plant treated, the author gives some general facts about the crop and discusses its history and origin, its types and varieties, its botany and genetics, breeding methods and objectives in breeding it, a distribution map, and a bibliography. Last chapter devoted to seed production.

A319 Williams, Watkin. Genetical principles and plant breeding. 1964. Blackwell Scientific. 504p. illus. $12.75.

A lucid and attractively written book directed principally toward general agriculture students.

Author's purpose is to provide a general understanding of genetical principles controlling plant life and of the methods leading to plant improvement.

Plant Diseases and Control

A320 Agrios, George N. Plant pathology. 1969. Academic. 629p. illus. $14.

An account of general principles of plant pathology followed by descriptions of specific plant diseases and their causal organisms. Contains references at end of each chapter and a glossary of terms. This stimulating and up-to-date introduction to plant pathology emphasizes those diseases occurring mainly in North America.

A321* Anderson, Harry W. Diseases of fruit crops. 1956. McGraw-Hill. 501p. illus. $14.50.

Detailed information on cultivated fruit crop diseases occurring in the temperate zones of the world are provided in this textbook. Diseases of each fruit crop are presented in order of their importance and then diseases are examined according to geographical distribution, symptoms, causal agents, etc. Intended for research workers and practicing horticulturists as well as students.

A322 Barnes, Ervin H. Atlas and manual of plant pathology. 1968. Appleton. 325p. illus. paper, $9.75.

A manual with a 2-fold purpose. Uses photographs and line drawings to give the student visual experience in recognizing diseases, specimens, and structures of causal organisms. Provides experiments to give the student actual experience in studying diseases of living plants.

A323 Bawden, Frederick C. Plant viruses and virus diseases. 4th ed. 1964. Ronald. 361p. illus. $8.75.

This book has long been regarded as the basic work in plant virology and the standard English-language textbook on the subject. The author has presented a clear, logical discussion of material which will give the student an understanding of nearly all areas of plant virology.

A324 Baxter, Dow V. Disease in forest plantations: thief of time. 1967. Cranbrook. 251p. illus. $8.50.

The author begins by tracing the historical development of U.S. forest tree plantations and ends with an appraisal of the past, present, and future of forest plantings. In the main body of the text he presents pertinent information concerning sites, relation of site to disease, effects of disease, and reforestation.

A325* **Boyce, John S.** Forest pathology. 3d ed. 1961. McGraw-Hill. 572p. illus. $15.50.

This well-known book, belonging to the American Forestry series, consists of a basic discussion of disease with emphasis given to fungi. This is followed by a thorough treatment of seedling diseases, root, foliage and stem diseases, timber decays, and diseases causing distortion. Principles and disease control recommendations are also presented. The authoritative illustrations add much to the book's value as a textbook or a general reference source.

A326 **Carefoot, Garnet L.,** and **Edgar R. Sprott.** Famine on the wind: man's battle against plant disease. 1967. Rand McNally. 693p. illus. $5.95.

Presents, in popular scientific style, an account of pertinent facts concerning 10 important plant diseases and man's effort to control them. Also, dramatizes the immediate and long-range effects of plant diseases on a rapidly increasing world population. Good supplemental reading for students enrolled in food technology, agricultural economics, and plant science programs.

A327 **Carter, Walter.** Insects in relation to plant disease. 1963. Wiley. 705p. illus. $28.50.

Plant diseases and insects as carriers of plant diseases make up the central theme of this book written by a well-known authority on the subject. Treatment is limited almost entirely to those diseases of economic importance in world food production. Of most value to students concerned with plant science or economic entomology.

A328* **Chupp, Charles,** and **Arden F. Sherf.** Vegetable diseases and their control. 1960. Ronald. 693p. illus. $14.

A textbook on diseases of more than 40 major vegetable crops (excludes potatoes) and on some tropical and minor vegetables. For each disease, authors give distribution and frequency of occurrence, describe symptoms, and introduce recommended control measures. The index lists diseases separately and by crops, gives common names of diseases, and scientific names of organisms. Written simply with a minimum of technical terms to increase the book's value to commercial growers, farm managers, home gardeners, and students.

A328a **Couch, Houston B.** Diseases of turfgrass. 1962. Reinhold. 289p. illus. $10.

A comprehensive, practical presentation of information on the identification, nature, and control of various fungus- and nematode-incited diseases affecting turfgrasses. Special features include an extensive bibliography and an appendix containing various lists of grass species and diseases. Useful as a text for specialized courses in turfgrass management and disease control and as a manual for those concerned with turfgrass management.

A329 **Darlington, Arnold.** The pocket encyclopedia of plant galls in colour. 1968. Philosophical Lib. 191p. illus. $7.50.

Only book in color on this subject and the most up-to-date. Uses a combination of illustrations and descriptions to identify galls, gall-causing agents, and the relationship of these agents with other organisms. Suggests investigations into the biology of galls which can be done in the field or the laboratory.

A330* **Dickson, James G.** Diseases of field crops. 2d ed. 1956. McGraw-Hill. 529p. illus. $11.50.

Discusses diseases of field crops on a worldwide basis. Under each crop plant diseases are arranged according to primary causes. Covers: (1) nonparasitic or environmental diseases; (2) viruses or infectious diseases transmitted by insects or other means; (3) diseases caused by bacteria; and, (4) diseases caused by fungi. A textbook for horticulture and plant pathology classes on the college level and a reference source for less advanced classes.

A331 **Edgerton, Claude W.** Sugarcane and its diseases. 2d ed. rev. 1959. Louisiana State. 301p. illus. $7.50.

An excellently illustrated treatise in simple, clear language. Easily understood by the nonspecialist. The author discusses classification, breeding, varieties, and other aspects of sugarcane, with most attention given to its different diseases. An added value is the many references cited from world literature.

A332 **Evans, Elfed.** Plant diseases and their chemical control. 1968. Blackwell Scientific. 288p. illus. $12.25.

Covers the biological, chemical, and physical aspects of chemical control of plant disease and reviews the history of plant disease and the progress made in disease control. Primarily aimed at students who have a basic background in biology and chemistry and who are pursuing special courses in crop protection.

A333 Felt, Ephraim P. Plant galls and gall makers. 1940, reprint 1965. Hafner. 364p. illus. $11.95.

A 1965 reprint of the standard U.S. guide to plant galls and their makers. The first part of the book consists of a general discussion of plant galls and the insects and mites responsible for making them. The second part contains the keys to plant galls arranged according to plant families. Illustrated with black and white drawings and photographs.

A334* Jenkins, William R., and Donald P. Taylor. Plant nematology. 1967. Van Nostrand. 270p. illus. $12.95.

A beginning text for a general course on nematodes and nematode diseases of plants. More than half of the book is an encyclopedia of the important parasitic genera. Not a book of techniques although recommendations for control are not completely omitted.

A334a Large, Ernest C. The advance of the fungi. 1940, reprint 1964. Peter Smith. 488p. $5.50; Dover. paper, $3.50.

A history of man's attempts to control plant diseases. Similar to A326, but more comprehensive in coverage and more technical in treatment. Good background reading for students and others interested in plant pathology.

A335 Lucas, George B. Diseases of tobacco. 2d ed. 1965. Scarecrow. 778p. illus. $18.

An advanced textbook-manual on the diseases of an important commercial crop. In tobacco-growing areas, it is a useful reference for teachers, growers, manufacturers, and students. Contains numerous black and white photographs illustrating the distinctive features of a disease.

A336 Matthews, Richard E. F. Plant virology. 1970. Academic. 778p. illus. $29.50.

An up-to-date account of plant viruses and all important aspects of virus diseases of plants. Includes a comprehensive treatment of general principles. Useful reference for advanced students, research workers, and teachers of less advanced courses in plant virology.

A337 Plakidas, Antonios G. Strawberry diseases. 1964. Louisiana State. 195p. illus. $5.

Not an amateur gardening book but a comprehensive and authoritative report of research findings for the research worker, the student, and the library. Covers diseases caused by bacteria, nematodes and viruses, fruit rots, and fungus diseases of foliage, roots, and crowns.

A338 Sharvelle, Eric G. Chemical control of plant diseases. 1969. Univ. Pub. 340p. illus. $7.95.

A handbook compiled to aid everyone from the research worker to the professional farmer and amateur gardener. Covers principles and economics of plant disease control, chemical control of all types of economic plant diseases, a dictionary of fungicides, and of plant disease control terminology.

A339* Shurtleff, Malcolm C. How to control plant diseases in home and garden. 2d ed. 1966. Iowa State. 649p. illus. $10.50.

This textbook, written in lay language, is designed to serve the student, the home gardener, and the dealer as an encyclopedia of information on plant diseases and how to control them. Arrangement is alphabetical by common name of host with diseases listed and described under each host. In addition to various types of disease problems of plants, describes the different control measures, various types of pesticides and equipment, and how to handle pesticides safely.

A340* Smith, Kenneth M. Plant viruses. 4th ed. 1968. Methuen. 166p. illus. $5.

An up-to-date text in plant virology written in an introductory manner and reflecting recent advances in the field. Suitable as a textbook for students who need only a general knowledge of the field. Discusses the usual topics such as disease symptoms, serology testing for viruses, transmission by vectors, etc.

A341 Smith, William H. Tree pathology: a short introduction. 1970. Academic. 309p. illus. $12.36.

A highly readable introduction to tree pathology for those unfamiliar with the subject. Of more value as collateral reading than as a textbook since it is mainly concerned with important pathological stress factors. A comprehensive account of forest tree diseases was not intended by the author. One of the few new books on forest pathology.

A342* Sprague, Howard B., ed. Hunger signs in crops. 3d ed. 1964. McKay. 390p. illus. $12.50.

A concise, easily understood account of the visible symptoms of malnutrition in economic crops. Each chapter is devoted to a single crop such as potatoes or group of crops such as vegetables or forage grasses. Presents crop requirements and deficiency symptoms for calcium, magnesium, sulfur, the trace elements, and the major elements. A valuable feature of this book

A343 **Stakman, Elvin C.,** and **Jacob G. Harrar.** Principles of plant pathology. 1957. Ronald. 581p. illus. $10.

A comprehensive study of plant diseases, their causes, effects, and control. Main attention is focused on diseases of food crops and on those of wide distribution. More advanced in content than Walker's book.

A344 **Strobel, Gary A.,** and **Don E. Mathre.** Outlines of plant pathology. 1970. Van Nostrand. 465p. illus. $12.95.

A modern approach is used in this introduction to plant pathology covering important principles and concepts. Discusses the 6 major groups of plant pathogens, shows the relationship of plant diseases to man, describes the host-parasite relationship, and presents control measures.

A345* **Tuite, John F.** Plant pathological methods: fungi and bacteria. 1969. Burgess. 239p. illus. $8.50.

This manual on methods of plant pathology is not written for the research worker, but primarily for the student and for those without access to adequate library resources. Most of the methods included have been successfully used by the author or endorsed by other investigators in the field.

A346* **Walker, John C.** Plant pathology. 3d ed. 1969. McGraw-Hill. 819p. illus. $15.

A standard textbook using the traditional approach to present fundamental concepts and principles of plant pathology for the beginning college student. The book is mainly devoted to chapters on diseases caused by specific groups of pathogens, such as bacteria. Other chapters deal with history, environmental aspects, host-parasite relations, and methods of disease control.

A347 **Wallace, H. R.** The biology of plant parasitic nematodes. 1963. St. Martin's. 279p. illus. $9.50.

A comprehensive and detailed biological description of plant parasitic nematodes and the world in which they live. Discusses many basic principles which apply to forest soils as well as to cultivated soils. A useful general reference and a review of plant nematode behavior for the student or nematologist.

A348 **Wallace, Thomas.** The diagnosis of mineral deficiencies in plants by visual symptoms: a colour atlas and guide. 2d ed. enl. 1961. Chemical Pub. 125p. illus. $14.50.

A useful book for reference because of its excellent color photographs and its descriptions of the deficiency symptoms shown by plants. Covers 42 species of plants including vegetables, tree and small fruits, small grain, field crops, etc.

A349* **Westcott, Cynthia.** Plant disease handbook. 3d ed. 1970. Van Nostrand. 843p. illus. $20.

A valuable reference manual for both amateur and technically trained gardeners. Arrangement is alphabetical by common name of plant diseases. Includes a list of host plants and the diseases found on each, a glossary of terms, and an extensive bibliography. A companion volume to this text is the author's *The gardener's bug book*.

A350 **Wheeler, Bryan E. J.** An introduction to plant diseases. 1969. Wiley. 374p. illus. $12.75.

A pleasant and uncomplicated introduction to plant pathology including diseases found in temperate and tropical regions. Describes disease symptoms, the biology of causal organisms, and methods of control for each plant disease included.

Weeds and Weed Control

A351 **Crafts, Alden S.** The chemistry and mode of action of herbicides. 1961. Wiley. 269p. illus. $13.50.

A semitechnical book on herbicides which would require some basic chemistry for complete understanding. Gives a simple definition and applications as well as chemical composition and reactions for the different types of herbicides discussed.

A352* _____, and **Wilfred W. Robbins.** Weed control. 3d ed. 1962. McGraw-Hill. 660p. illus. $16.50.

A college textbook on weed control and a reference for those who are practicing weed control in crop production. Consists of a general treatment of weeds, biological and ecological considerations, control of weeds using herbicides, tillage methods and biological control methods, and special weed problems.

A353 **Fogg, John M.** Weeds of lawn and garden. 1945. Pennsylvania. 215p. illus. $9.

A handbook for identification of some common lawn and garden weeds of eastern temperate North America. Arranged in the manner of most other botanical manuals. Common and scientific names, origin and geographic distribution, structural characteristics, and methods of control are given for each species.

A354 Gilkey, Helen M. Weeds of the Pacific Northwest. 1957. Oregon State. 441p. illus. $6.

A weed identification manual intended for the layman. Although scientific names are used, technical terminology is kept to a minimum. Very good illustrations of plant parts and simple keys are used for the 46 families covered. Will serve a wider area than the title implies.

A355 Isely, Duane. Weed identification and control in the north central states. 2d ed. 1960. Iowa State. 400p. illus. $5.95.

A reference manual and textbook covering principles and methods of both weed identification and weed control. Designed especially for students with little or no background in botany.

A356 King, Lawrence J. Weeds of the world: biology and control. 1966. Wiley. 526p. illus. $18.

As indicated by the title, subject coverage is limited to the biology and control of weeds. The material is presented in terms of general principles and theories with little attention paid to individual species. Intended for a wide audience ranging from the farmer to the scientist.

A357* Klingman, Glenn C. Weed control. 1961. Wiley. 421p. illus. $9.95.

Information necessary for successful weed control is presented clearly and concisely in this classroom textbook. Weed damage, general weed control methods, application equipment, herbicides, and practical methods for weed control in specific types of crops are some of the topics discussed.

A358 Muenscher, Walter C. Weeds. 2d ed. 1955. Macmillan. 560p. illus. $10.

A standard guide and reference book on weed identification and control, written by a well-known weed specialist. Contains clear, accurate descriptions of the major botanical features, the growth, distribution, and control of several hundred weeds. Descriptions accompanied by illustrations of various plant parts.

A359 Muzik, Thomas J. Weed biology and control. 1970. McGraw-Hill. 273p. illus. $12.50.

A book for the reader who wants to know more about the current methods of weed control. Written in semitechnical language. Emphasis placed on general principles of chemical and physical methods of control with less attention given to specific techniques.

Field and Forage Crop Production

A360* Ahlgren, Gilbert H. Forage crops. 2d ed. 1956. McGraw-Hill. 536p. illus. $10.50.

A systematic presentation of the basic fundamentals of forage production. Covers the major legumes, principal grasses including dryland grasses, seeding requirements, fertilizers, mineral nutrition, production and pasture management, weed control, etc. Useful as a reference for vocational agriculture teachers, students, soil conservationists, and others concerned with forage plants.

A361 Akehurst, B. C. Tobacco. 1968. Humanities. 551p. illus. $15.

A well-written, timely book covering the cultivation and processing of tobaccos grown in various parts of the world. Deals with the botany of tobacco, the different types of tobacco, its diseases and pests, nutrition, growth, and processing, and finally the tobacco consumer.

A362* Aldrich, Samuel R., and Earl R. Leng. Modern corn production. 1965. Farm Qtly. 308p. illus. $9.75.

An authoritative account of every aspect of corn production from planting to marketing plus a look into the future of this major farm crop. Written in lay language for the farmer and the student. Authors have stressed those principles which will aid the corn grower in solving his particular problems.

A363 American potato yearbook. 1948–   . Annual. MacFarland. illus. $2.50.

New edition published each year. Consists of the latest information on the potato industry including production, marketing, and utilization. Also contains a list of available publications on potatoes and a list of periodicals important to the potato industry. An important publication for everyone connected with the potato industry.

A364 Archer, Sellers G., and Clarence E. Bunch. The American grass book; a manual of pasture and range practices. 1953. Oklahoma. 330p. illus. $5.95.

A how-to book that gives complete and authoritative coverage of grassland management. The last section of the book gives descriptions, illustrations, and distribution maps of grasses and legumes. Practices described are principally those of the southern prairies and plains.

A365* Barnard, Colin, ed. Grasses and grasslands. 1964. St. Martin's. 269p. illus. $10.

Consists of facts and theories on grass biology, grassland ecology, grass use and management, plant breeding, etc. Uses Australian examples for illustration but seldom uses local terminology. Should be in agriculture and plant science libraries as a reference for anyone studying grasses or grasslands.

A366 Barnes, Arthur C. The sugar cane. 1965. Wiley. 456p. illus. $17.

Presents in readable form a general account of sugarcane as an economically important world crop. Treats production, harvesting, processing, and utilization of sugarcane. Should appeal to a wide audience including students of food technology, sugarcane growers, and researchers.

A367* Brickbauer, Elwood A., and William P. Mortenson. Approved practices in crop production. 1967. Interstate. 397p. illus. $7.50; text ed., $5.75.

A survey of the latest agricultural practices in crop production. Written especially for the vocational/technical student. Organized into 3 parts: (1) feed, forage crops, and grain; (2) grain, fiber and root crops, and tobacco; and, (3) weeds, insects, and diseases. Includes a glossary of commonly used crop production terms.

A368 Burgess, Abraham H. Hops: botany, cultivation, and utilization. 1964. Wiley. 300p. illus. $15.95.

The scope of subject material covered in this semitechnical book, combining theory and practice, is indicated by the title. An important work for the researcher, the brewer, and especially for the hop producer.

A369 Coffman, Franklin A., ed. Oats and oat improvement. 1961. Amer. Soc. of Agron. 650p. illus. $11.50.

The author has brought together the most relevant facts available on oats. Covers botanical features, oat breeding techniques, environmental influences, and production of oats. A valuable contribution in oat literature for the specialist, the student, and the layman.

A370 Cook, Arthur H., ed. Barley and malt; biology, biochemistry, technology. 1962. Academic. 740p. illus. $23.

An exhaustive treatment of current knowledge on barley and malt. Covers the botanical and agricultural aspects of barley, malting technology, and a summary of analytical concepts, structural chemistry, and enzymic makeup of barley and malt. Immensely valuable to anyone interested in barley or malt.

A371 Coursey, Donald G. Yams. 1967. Humanities. 230p. illus. $9.50.

A timely and practical manual on the botany, production, utilization, nutritive value, economic importance, etc., of yams. This comprehensive account will be of interest to agriculturists, food technologists, and economic botanists.

A372* Delorit, Richard J., and Henry L. Ahlgren. Crop production. 3d ed. 1967. Prentice-Hall. 672p. illus. $9.75.

A high school or junior college textbook of the encyclopedic type presenting the basic principles of crop production in lay terms. Specific, practical information concerning nutritional requirements, diseases and pests, culture, uses, economic importance, etc., is presented for each of 22 major crops.

A373 Eddowes, Maurice. Crop technology. 1969. Hutchinson. 219p. illus. $6.

A definitive introduction to crop technology combining modern concepts with a practical approach. Attempts to give the reader the understanding necessary to attain maximum crop yields at minimum cost.

A374 Grist, Donald H. Rice. 4th ed. 1965. Humanities. 548p. $13.

A balanced treatment of all aspects of rice as an important world crop. Presentation is in 3 parts: (1) rice as a plant; (2) the production of rice; and, (3) rice as a product. The bibliography serves as an excellent guide to world literature on the subject.

A375* Hughes, Harold D., and others, eds. Forages: the science of grassland agriculture. 2d ed. rev. 1966. Iowa State. 707p. illus. $8.95.

Authoritative, current information on U.S. forage crops for instruction and reference at all educational levels. Chapters written by authors with extensive knowledge and experience. Covers forage plant botany, breeding, production, utilization, and management. Each of the major grasses and legumes are covered in individual chapters.

A376* ———, and Edwin R. Henson. Crop production: principles and practices.

rev. ed. 1957. Macmillan. 627p. illus. $9.95.

Presents the basic principles, examines practical methods, and reports on recent research in crop production in this definitive text for instruction on the vocational/technical level. Specific topics covered include such factors as plant morphology, crop management, soil conservation, irrigation, and drainage, plus chapters on production and improvement of grain, corn, and other important field crops.

A377 **Inglett, George E.** Corn: culture, processing, products. 1970. AVI. 369p. $17.50.

Uses a multidisciplinary approach to present the production, processing, and utilization of corn. Intended to provide students, food technologists, economists, industrialists, and others with a broad understanding of the important agricultural, economical, and scientific aspects of corn and its products.

A378 **Janick, Jules,** ed. Plant agriculture; readings from *Scientific American.* 1970. Freeman. 246p. illus. $10; paper, $4.95.

A collection of 25 articles previously appearing in *Scientific American*. Covers the historical, physiological, ecological, production, and economic aspects of crop plants. Could be effectively used as collateral reading for students in agricultural science programs at the junior college and technical school level.

A379 _____, and others. Plant science: an introduction to world crops. 1969. Freeman. 629p. illus. $12.

A well-planned survey of pertinent technological, economical, and scientific aspects of world crops and their production. Excellent supplementary reading for a general introductory course in crop science or forestry.

A380 **Johnson, William H.,** and **Benson J. Lamp.** Principles, equipment, and systems for corn harvesting. 1966. Agr'l Consulting Assoc. 370p. illus. $9.95.

A comprehensive treatment of corn harvesting. Filled with information covering every aspect of the subject from basic principles to financial considerations. Textual material supplemented by illustrations consisting of pictures, charts, tables, and graphs.

A381 **Jones, Henry A.,** and **Louis K. Mann.** Onions and their allies. 1963. Wiley. 286p. illus. $11.25.

The authors deal with 7 economic species but focus their primary attention on the common garden onion. They describe the botany of the onion plant, discuss breeding methods for developing new and better varieties, and examine crop production methods.

A382 **King, George H.** Pastures for the South. 4th ed. 1963. Interstate. 310p. illus. $6.50; text ed., $5.

A textbook for instruction for agriculture students on the vocational/technical level. Gives complete coverage to pasture management and improvement in the southern states. Includes pasture establishment, adaptation of perennial grasses and legumes, fencing, establishment of a water supply, etc.

A383* **Kipps, Michael S.** Production of field crops. 6th ed. 1970. McGraw-Hill. 790p. illus. $15.50.

This well-known book has been in use since 1924. A lengthy discussion of the latest available information on the botany of common crop plant species, breeding methods for plant improvement, principles of crop production, and practical production methods for specific forage, grain, and cash crops. Information on other crops presented in less detail. Thomas K. Wolfe coauthored earlier editions of this book.

A384 **Klingman, Glenn C.** Crop production in the South. 1957. Wiley. 416p. illus. $8.50.

In a simple, readable style, the author presents a general treatment of botany, plant physiology, chemistry, and other basic sciences as they apply to the major aspects of crop production in the southern states. Amply illustrated with pictures, maps, diagrams, and tables. Review questions at the end of each chapter.

A385* **Leonard, Warren H.,** and **John H. Martin.** Cereal crops. 1963. Macmillan. 824p. illus. $10.95.

A conscientious presentation of general principles and factual information necessary for scientific production of cereal crops. Describes the distribution, structure, growth and development, genetics and breeding, and production methods for each of the cereals. Important aspects of the subject are illustrated with pictures, graphs, and tables.

A386* **McVickar, Malcolm H.,** and **John S. McVickar.** Approved practices in pasture management. 2d ed. 1963. Interstate. 332p. illus. $6.25; text ed., $4.75.

An elementary textbook on the principles and practices of pasture management for all types of livestock. Includes chapters on fertilizers, irrigation practices, weed control, diseases and pests, etc. Full use is made of research data from all parts of the United States.

A387* **Martin, John H.,** and **Warren H. Leonard.** Principles of field crop production. 2d ed. 1967. Macmillan. 1044p. illus. $11.95.

This well-known classroom text gives exhaustive treatment to the basic principles and pertinent information necessary for full comprehension of modern field crop production. Covers new machinery developed, improved crop varieties, new agricultural chemicals, scientific methods of crop production, and crop species and classes. Significant aspects are accompanied by illustrations.

A388 **Mitchell, Roger L.** Crop growth and culture. 1970. Iowa State. 349p. illus. $8.50.

A valuable contribution to crop science literature. The author presents the basic principles governing practices in crop plant culture and applies these principles to production. Plant growth and development, physical and chemical growth regulators, diseases and pests, harvesting, and storing are examples of material covered.

A389 **Norman, Arthur G.,** ed. The soybean. 1963. Academic. 239p. illus. paper, $6.30.

The subtitle, *Genetics, breeding, physiology, nutrition, management,* identifies the scope of this concise report on one of man's most valuable crops. An excellent source of information on the soybean crop in North America.

A390 **Painter, Reginald H.** Insect resistance in crop plants. 1968. Kansas. 520p. illus. paper, $4.75.

Paperbound copy, with a new preface, of a classic. Author discusses insect response to color, odor, and leaf surfaces. Devotes chapters to insect resistance in specific crops. Should be on library shelves. Not practical for textbook use because some chapters are out-of-date.

A391* **Pearson, Lorentz C.** Principles of agronomy. 1967. Van Nostrand. 434p. illus. $9.50.

Emphasis is on principles and precepts rather than specific practices in the production of food, feed, and fiber. Author attempts to provide the student with the background knowledge necessary to adapt to changing practices and situations. Author's approach and style of presentation makes the book an excellent choice for students studying agriculture at the technical level.

A392* **Peterson, Rudolph F.** Wheat: botany, cultivation, and utilization. 1965. Wiley. 422p. illus. $16.95.

Not a textbook but a general survey of the botany, production, and utilization of wheat as a world crop. Written in nontechnical language to appeal to a wide range of readers from high school to professional levels. One of the best references available on wheat.

A393 **Prince, Ford S.** Grassland farming in the humid Northeast. 1956. Van Nostrand. 441p. illus. $9.50.

Concerned with forage production, varieties of grasses and legumes, production problems and management, utilization of forage crops, and other aspects of grassland farming and economics. Suitable for supplementary use in high school agricultural courses and for study by college students.

A394 **Quisenberry, Karl S.,** ed. Wheat and wheat improvement. 1967. Amer. Soc. of Agron. 560p. illus. $10.

A detailed examination of the most important information concerning the botanical aspects, the production, marketing, and utilization of wheat. Each chapter is carefully documented. Contains a great deal of valuable information for farmers, processors, food technologists, and others concerned with wheat and its products.

A395 **Rose, Graham J.** Crop protection. 2d ed. 1963. Chemical Pub. 490p. illus. $11.

Covers general considerations, biological and chemical protection methods, the breeding of resistant varieties of crops, application machinery, livestock protection, and protection of stored products. A practical manual for use on the job as well as for student reference.

A396 **Semple, Arthur T.** Grassland improvement. 1970. Chemical Rubber. 400p. illus. $33.

An up-to-date account of the background and current progress in grassland improvement. Purpose of book is to show how to get maximum use of grasslands. Clearly written and well illustrated. Places more emphasis on improving tropical pasture than those of temperate regions.

A397 **Smith, Ora.** Potatoes: production, storing, processing. 1968. AVI. 642p. illus. $25.

Detailed treatment of every phase of potato producing, storing, processing, and utilization. Of more value to the grower than to food technologists and others since greater emphasis is placed on agricultural production than on other aspects of the potato industry. A worthwhile contribution to potato literature.

A398 **Suter, Robert C.** The courage to change. 1965. Interstate. 294p. illus. $5.75.

An analysis of technical advances in storing forages and medium-moisture feeds and the effects of these changes on harvesting of forage crops and on livestock feeding. Includes an account of the Harvestore system of farming. A well-illustrated, valuable reference for advanced students studying agriculture on the technical level.

A402* **Wheeler, William A.** Forage and pastance: the story of corn in America. 1966. Harper. 199p. illus. $6.95.

Introductory chapters trace the history of corn. Remainder of book devoted to economic and industrial importance of this crop. Not concerned with production of corn but gives in detail its uses following production. Written in a popular style which will appeal to the consumer as well as those involved in the corn industry.

A400 **Wall, Joseph S.**, and **William K. Ross**, eds. Sorghum production and utilization. 1970. AVI. 702p. illus. $30.

The first major work on sorghum. Brings together in 1 volume the contributions of a great many specialists. Covers the history, physiology, production, economics, and future of grain sorghum. A well-organized and timely book with something of value for everyone from the grower to the scientist.

A401 **Walton, Earnest V.**, and **Oris M. Holt.** Profitable southern crops. 1959. Prentice-Hall. 504p. illus. $9.08.

Written for the young farmer and vocational/technical school groups. Supplies valuable information on all aspects of important field crops widely grown in the South and Southwest. Excludes commercial crops of localized importance or of highly specialized nature. A separate chapter is devoted to each crop.

A402* **Wheeler, William A.** Forage and pasture crops, a handbook of information about the grasses and legumes grown for forage in the United States. 1950. Van Nostrand. 752p. illus. $9.75.

Stresses forage crop management and the problems and uses of legumes and grasses. Selected reference sources are given in detail. Chapters are reviewed by experts in the field. One of the older publications still considered of great value.

A403 _____, and **Donald D. Hill.** Grassland seeds. 1957. Van Nostrand. 734p. illus. $15.

Contains a vast amount of useful information and more than 200 photographs and diagrams concerning grass and legume seeds. The book first treats general processes such as testing and harvesting; then proceeds to specific crops and their seed problems. Valuable to everyone involved in seed industry and trade.

A404 **Whyte, Robert O.** Crop production and environment. 2d ed. 1960. Faber. 392p. illus. $6 (approx.).

A wide survey of the interrelations of plants and their environment. Covers first the general considerations of common importance and then deals with types of plants in groups. Reviews important research between the years of 1946 and 1960. A clear, concise presentation for the agricultural botanist, crop ecologist, plant breeder, and agronomist.

A405 **Wilsie, Carroll P.** Crop adaptation and distribution. 1962. Freeman. 448p. illus. $9.50.

Designed for a course in crop ecology for advanced students in agricultural curriculums. Organized into 3 parts: (1) general principles and concepts; (2) environmental factors; and (3) crop distribution on a climatic basis. Carefully written but does not always define technical terms used.

A406 **Wilson, Harold K.** Grain crops. 2d ed. 1955. McGraw-Hill. 396p. illus. $11.50.

A systematic presentation of the botany, production, harvesting, and marketing of each of the principal grain crops. Considers the more general aspects as related to all crops followed by detailed discussions of individual crops. Planned primarily as a college text.

A407* _____, and **A. Chester Richer.** Producing farm crops. 1960. Interstate. 336p. illus. $6.50; text ed., $5.

Deals with the entire field of crop production and its relationship to other branches of agriculture such as botany, entomology, and agricultural economics. Gives considerable attention to reasons behind farm crop production practices. Written by men with extensive experience in agricultural education. Highly recommended as a textbook for vocational/technical students studying this branch of agriculture.

A408 **Woodroof, Jasper G.** Peanuts: production, processing, products. 1966. AVI. 291p. illus. $14.50.

A concise but complete description of all phases of the peanut industry. Deals with the production, harvesting, storing, processing, and utilization of peanuts. Contains good illustrations and extensive references.

General Horticulture

A409 Abraham, George. The green thumb book of fruit and vegetable gardening. 1970. Prentice-Hall. 344p. illus. $7.95.

A discussion in conversational style of common fruits and vegetables. Covers soils and fertilizers, plant propagation, fruit and vegetable production, greenhouse gardening, tropical fruits for indoor cultivation, evergreens, and many other topics. Written by a greenhouse operator.

A410 Adriance, Guy W., and Fred R. Brison. Propagation of horticultural plants. 2d ed. 1955. McGraw-Hill. 298p. illus. $10.50.

A senior high school or beginning college text for basic courses in horticulture and related fields as well as a practical guide for commercial growers and amateur gardeners. Mainly concerned with propagation methods but briefly discusses the history of horticulture and the structure of plants.

A411 American tomato yearbook. 1949– . Annual. Macfarland. illus. $2.50.

New edition published each year. Contains a vast amount of information of value to those concerned with the tomato industry. Includes a list of helpful books and pamphlets, a buyers' guide, a list of periodicals of interest to the tomato industry, a section on tomato products, directions for obtaining market news information on tomato production, etc.

A412 Atkins, Fred C. Mushroom growing today. 5th ed. 1966. Macmillan. 188p. illus. $5.95.

A practical comprehensive treatment of mushroom growing by an internationally known authority. He presents the procedures and problems involved in all phases of commercial production of mushrooms. Added features include a chapter for the amateur grower and a chapter on food value of mushrooms.

A413 Bailey, Liberty H., ed. The standard cyclopedia of horticulture. 1935. Macmillan. 3v. illus. $65.

This famous encyclopedia covers more than 40,000 fruit, vegetable, and ornamental plants grown in America. Describes the plants, gives their scientific and common names, and briefly discusses their culture and propagation.

A414 _____, and Ethel Z. Bailey. Cultivated conifers in North America; comprising the pine family and the taxads. 1933. Macmillan. 404p. illus. $13.95.

This estimable work has systematic keys and botanical descriptions of conifers plus a comprehensive section on their production and handling.

A415* _____, and _____. Hortus second: a concise dictionary of gardening, general horticulture, and cultivated plants in North America. 1941. Macmillan. 778p. illus. $14.95.

The authoritative standard 1-volume encyclopedic dictionary of horticulture. Descriptions are accurate and nontechnical with both botanical and common names. Coverage is more than dictionary length for many subjects. Still useful although old.

A416* Baumgardt, John P. How to prune almost everything. 1968. Morrow. 192p. illus. $6.50.

An up-to-date popular book on pruning. Contains an alphabetical list of trees, shrubs, and other plants with specific directions on how to prune each. A large number of illustrations are used to clarify the subject matter for students and home owners. Not intended for the professional horticulturist or landscape architect.

A417 Bush-Brown, Louise, and James Bush-Brown. America's garden book. new rev. ed. 1965. Scribner. 752p. illus. $8.95.

One of the standard, general purpose gardening encyclopedias. Covers: construction of special garden features; care of trees and shrubs; culture of flowers, fruits, and vegetables; soil problems; pest, disease, and weed control; greenhouse maintenance; and many other aspects of gardening.

A418 Carleton, R. Milton. Vegetables for today's gardens. 1967. Van Nostrand. 180p. illus. $5.95.

A practical guide to vegetable growing for the home gardener or student. An unusual book, in that the author emphasizes selecting varieties for good flavor. A weak point in the book is the lack of good illustrations for all of the plants described.

A419* Chandler, William H. Deciduous orchards. 3d ed. 1957. Lea & Febiger. 492p. illus. $7.50.

A well-planned book for the student concerned with deciduous orchards. Covers the general nature and botanical features of deciduous trees and their fruit, their physiology, environmental factors and conditions affecting the orchard, propagation problems, planting and pruning procedures, the processes and responses of deciduous orchards, and a discussion of deciduous tree species.

A420* ———. Evergreen orchards. 2d ed. rev. 1958. Lea & Febiger. 535p. illus. $8.50.

The standard book in its field. Brings together a vast amount of information concerning evergreen orchard crops and their potentialities. Oranges, tea, coffee, and cacao are examples of the fruits covered. Climate, nutrition, propagation, and fruiting are examples of important aspects discussed. Although characteristics of many tropical and subtropical fruits are described, the book is not intended primarily for students concerned with these areas.

A421 Child, Reginald. Coconuts. 1964. Humanities. 216p. illus. $7.50.

Pertinent research and technical information on the coconut is presented in condensed form for the grower, the researcher, and the merely curious. Informative and readable chapters cover the botanical features and general characteristics of the plant, gives a complete account of production from seed selection to harvesting, and discusses various coconut products.

A422* Childers, Norman F. Modern fruit science: orchard and small fruit culture. 4th ed. 1969. Rutgers Hort. Pubns. 912p. illus. $15.80.

A college textbook of immense reference value to the grower, the high school student, and others. Presently used in many state universities. Covers all phases of production for the apple, pear, quince, peach, nectarine, apricot, almond, cherry, plum, edible nuts, grape, strawberry, and bush fruits. An added feature is the discussion of judging contests and exhibiting.

A423 ———, ed. Nutrition of fruit crops: tropical, subtropical, temperate, tree, and small fruits. 2d ed. 1966. Rutgers Hort. Pubns. 888p. illus. $17.95.

A well-produced reference source on the nutrition of individual fruit crops. Primarily consists of reviews by individuals who are authorities on their respective crops. Book's subtitle indicates its scope of subject coverage. Also includes chapters discussing the use of leaf analysis as a nutrition guide for deciduous and citrus fruits.

A424* Christopher, Everett P. Introductory horticulture. 1958. McGraw-Hill. 482p. illus. $10.50.

A fairly easy horticulture text for beginning college students and a sourcebook for high school students and for vegetable, fruit, and flower producers. Explains practical horticultural procedures and the basic principles underlying these procedures. Provides excellent basic background knowledge in such areas as soils, fertilizers, propagation, and plant growth.

A425* ———. Pruning manual. 1954. Macmillan. 320p. illus. $6.95.

This comprehensive book, based on L. H. Bailey's *The pruning manual,* is a useful handbook for those with no background in botany or horticulture. Deals with pruning fruit trees, shade trees, and ornamental shrubs. Contains sections devoted to grafting, wound treatment, root-pruning, and basic pruning equipment.

A426 Darrow, George McM. The strawberry: history, breeding, and physiology. 1966. Holt. 447p. illus. $15.

An American authority on fruit presents a scholarly, semitechnical study on strawberries for both amateur and professional horticulturists. Although Darrow wrote most of the book, he has included several chapters written by other specialists. The bibliography will serve as an excellent reference source for those who wish to do further study.

A427 Denisen, Ervin L. Principles of horticulture. 1958. Macmillan. 509p. illus. $9.95.

This elementary textbook is suitable for use in secondary, vocational, and post-high school technical school courses. Describes basic horticultural principles and applies these principles to practical techniques. Discusses all general aspects of home gardening and includes vegetable gardens, lawns, fruit orchards, and ornamental plants.

A428 Dictionary of gardening. 2d ed. ed. by Patrick M. Synge. 1956. Oxford. 4v. illus. $52.

———. Supplement. ed. by Patrick M. Synge. 1969. Oxford. 554p. $24.

An exhaustive, encyclopedic treatment of gardening, not restricted to any geographic area. Illustrations are designed to be helpful to the gardener rather than merely ornamental. Deals with fundamentals and with widely grown plant species and hybrids. The supplement treats new plants and hybrids and is revised periodically.

A429 Dowdell, Dorothy, and **Joseph Dowdell.** Careers in horticultural sciences. 1969. Messner. 222p. illus. $3.95.

A vocational guidance book written on the high school level and covering every aspect of horticulture as a profession. Concerned with the technician levels as well as the scientific levels of horticultural production, processing, distribution, retailing, and research. Outlines the possible problems to be encountered as well as the benefits.

A430 Duckworth, Ronald B. Fruit and vegetables. 1966. Pergamon. 306p. illus. $9; paper, $6.50.

A general textbook covering fruit and vegetables grown for processing purposes. Presents a broad treatment of the chemistry and biology of fruits and vegetables, their production, marketing, processing, and utilization. Suitable for students of food science in technical schools and colleges. Not intended for use by food operating personnel or plant managers.

A431 Eck, Paul, and **Norman F. Childers,** eds. Blueberry culture. 1967. Rutgers. 378p. illus. $15.

A definitive compilation of the accumulated scientific and cultural knowledge on all phases of the blueberry industry from growing to processing and marketing. Contains a great amount of relevant information for the grower, the student, and the amateur gardener.

A432 Edmond, Joseph B., T. L. Senn, and **F. S. Andrews.** Fundamentals of horticulture. 3d ed. 1964. McGraw-Hill. 476p. illus. $10.50.

A thorough study of the background, development, and application of horticultural procedures. Includes discussions of important commercial fruit, flower, and vegetable crops. Written for use as a basic horticulture text on the vocational/technical school level.

A433 Encyclopedia of organic gardening. Jerome I. Rodale and staff members of *Organic gardening* magazine. 1969. World. 1152p. illus. $9.95.

An encyclopedia covering the whole field of horticulture from the organic point of view. Even people unacquainted with the organic idea or disagreeing with it will find this book a useful reference work.

A434 Free, Montague. Plant pruning in pictures. 1961. Doubleday. 283p. illus. $5.95.

The subtitle, *How, when, and where to prune, and with what tools,* explains the scope of this authoritative handbook. Covers the pruning of various groups of plants ranging from house plants to trees. The techniques of pruning are explained largely by using photographs and line drawings accompanied by short descriptions.

A435 Gardener's almanac of timely hints for home gardeners. 13th ed. 1967. Mass. Hort. Soc. 176p. illus. $3.95.

A when-to-do-what book listing gardening tasks for each month of the year. Emphasis is mainly on gardening in the northeastern states with small sections covering other regions of the United States.

A436 Gardner, Victor R. Principles of horticultural production. 1966. Michigan State. 583p. $15.

Presents the fundamental principles of fruit production for the student and the producer. Does not attempt to deal with specific practices. Mostly concerned with edible fruit crops but frequently refers to crops grown for a specific plant part such as seeds, roots, or flowers.

A437* Garner, Robert J. The grafter's handbook. 3d ed. 1968. Oxford. 260p. illus. $5.75.

Generally considered the most comprehensive book available on grafting. Gives clear and detailed descriptions of the various grafting techniques. A large number of photographs and line drawings are used to illustrate the methods described. Primarily concerned with horticultural and orchard trees but many of the techniques can be used with forest species. Of immense value to both the student and the practical horticulturist.

A438 Greenfield, Ian. Turf culture. 1962. Hill. 364p. illus. $10.25.

A comprehensive treatment of the subject on a worldwide basis. Discusses turf composition, weed and pest control, soils, relevant scientific principles, and the practical application of these principles to turf production. Contains a great deal of valuable information for the student or technician interested in turf culture and management.

A439 Greensill, T. M. Tropical gardening. 1966. Praeger. 272p. illus. $8.95.

A well-illustrated, practical book for gardeners and students in tropical and subtropical climates. The majority of the book deals with house plants, flowers, vegetables, and fruit trees. Other sections cover soils, plant nutrition, plant propagation, and pest and disease control.

A440 Hanson, Angus A., and **Felix V. Juska,** eds. Turfgrass science. 1969. Amer. Soc. of Agron. 715p. illus. $12.50.

This book is number 14 of the society's series of agronomy monographs. Filled with scientific information on all aspects of turfgrass science. Covers grass genetics, grass ecology, history of turf usage, physiology of growth, production, turf tools and equipment, and the management of turfgrass in both private and public grounds.

A441* Hartmann, Hudson T., and **Dale E. Kester.** Plant propagation: principles and practices. 2d ed. 1968. Prentice-Hall. 702p. illus. $14.50.

An outstanding basic text and reference work suitable for both beginning and advanced students. Mainly concerned with plants of economic importance. Covers the fundamental plant propagation principles and general and special propagation methods, including techniques and equipment.

A442 Hulme, A. C., ed. Biochemistry of fruits and their products. 1970– . Academic. 2v. v.2 forthcoming. illus. v.1 $30; v.2 price not set.

Volume 1 discusses the principal constituents of fruits and examines hormonal control of growth and ripening and the metabolism of harvested fruits. The second volume will deal with the biochemistry of individual members of commercially important fruits and important aspects of commercial processing. A good source of information for growers, processors, food technologists, and students.

A443 Hume, H. Harold. Citrus fruits. rev. ed. 1957, reprint 1967. Macmillan. 444p. illus. $11.95.

Provides a wide variety of information covering all important aspects of citrus fruit culture, marketing, and processing. Most emphasis is placed on citrus culture in Florida but other citrus regions of the United States are discussed. Can serve as a guide for citrus producers, a textbook for students, and a reference for botanists.

A444 ———. Gardening in the lower South. rev. ed. 1954. Macmillan. 377p. illus. $7.95.

A standard gardening reference in use since 1929. Written by a garden authority with years of research and experience in the area. Covers every aspect of flower gardening and contains a considerable amount of information on growing fruits and vegetables.

A445 Hyams, Edward S., and **Alan A. Jackson,** eds. The orchard and fruit garden; a new pomona of hardy and subtropical fruits. 1961. Longmans. 207p. illus. $25.20.

A classroom textbook with each chapter written by an authority in the field. Covers the common fruit crops plus tomatoes, cucumbers, nuts, and melons. Outstanding color photographs accompany each fruit discussed. Can also be used as a reference source by orchardists and gardeners.

A446* Janick, Jules. Horticultural science. 1963. Freeman. 472p. illus. $9.50.

Specifically written for use as a textbook in a beginning horticulture course at post-high school levels. Text is divided into 3 major areas: the biology, the technology, and the industry of horticulture. Emphasizes basic principles necessary for successful horticultural practices and uses a large number of illustrations to help clarify the subject matter. Provides a broad background for the student who wishes to study more detailed aspects of the field.

A447 Jaynes, Richard A., ed. Handbook of North American nut trees. 1969. Northern Nut Growers' Assn. 421p. illus. $7.50.

One of the best books available on the subject of nut trees and their culture. Each of the book's contributors has had considerable experience in his field. Should serve as a working manual for those involved in nut tree culture as well as a source of information for students.

A448 Johns, Glenn F., ed. The organic way to plant protection. 1966. Rodale. 399p. $4.95.

Especially written for gardeners who want to grow fruits, vegetables, flowers, and lawns without using chemical pesticides and weed killers. Covers general information on insects and follows this with specific fruits, vegetables, and flowering plants, giving details on insect and disease control for each. Worthwhile reading even for people who don't agree with the organic method.

A449* Knott, James E. Handbook for vegetable growers. rev. ed. 1962. Wiley. 245p. illus. $6.95.

Virtually an encyclopedia, bringing together in a concise form much of the widely scattered information on vegetable production. Concerned with general aspects of the subject rather than specific information or recommendations. An excellent handbook for the home gardener and the commercial grower and a ready reference for the student in horticulture.

A450 Kraft, Ken, and **Pat Kraft.** Fruits for the home garden. 1968. Morrow. 287p. illus. $6.95.

A simple, easy-to-understand handbook designed primarily for the home gardener but useful also to the beginning technical student. Covers all phases of fruit culture with special emphasis on dwarf and semidwarf varieties. A chapter is devoted to each of the important fruits, berries, and nuts.

A451 McDonald, Elvin. Garden ideas A to Z. 1970. Doubleday. 196p. illus. $7.95.

A book of practical hints and unusual ideas for flower gardeners. Entries are arranged alphabetically from annuals to winter tasks. Uses a large number of photographs to help explain the text. By using the index students can find much useful information on flower gardens.

A452 MacGillivray, John H. Vegetable production. 1961. McGraw-Hill. 397p. illus. $11.50.

An introductory college textbook containing principles and procedures of vegetable production and their application to major crops or crop groups. Since it contains special reference to western crops, it should also be useful as a practical manual for the western vegetable grower.

A453* Mahlstede, John P., and **Ernest S. Haber.** Plant propagation. 1957. Wiley. 413p. illus. $9.50.

A comprehensive textbook concerned with the fundamental principles and practical techniques necessary for successful plant propagation. Book's weakest point is its bibliography. Written for the beginner in the field as well as the experienced nurseryman or seedsman.

A454 Murphy, Richard C., and **William E. Meyer.** The care and feeding of trees. 1969. Crown. 164p. illus. $5.95.

A useful reference book for laymen and students concerned with tree care. Contains practical information on stock selection, planting, pruning, fertilizing, control of insects and disease, cavity repairs, trunk cabling, etc.

A455 Musser, Howard B. Turf management. 2d ed. rev. 1962. McGraw-Hill. 356p. illus. $10.95.

A comprehensive guide on turf management for parks, golf courses, home grounds, and other large turf areas. Covers turfgrass propagation, maintenance, weed and insect control, fertilizer, drainage, and other maintenance and management problems. Includes a section on golf course design and operation.

A456 New illustrated encyclopedia of gardening. Thomas H. Everett, ed. 1964. Greystone. 14v. illus. $56.95.

An up-to-date reference work on all aspects of gardening. Style of writing is popular without losing any of its authority. Has no index to the set although thorough use is made of self-indexing cross references. Other commendable features include the step-by-step method of instruction used throughout and the coverage devoted to pests and diseases.

A457* Pirone, Pascal P. Tree maintenance. 3d ed. 1959. Oxford. 483p. illus. $15.

A college textbook and manual written by a well-known plant pathologist. Covers all aspects of tree maintenance including transplanting, feeding, pruning, insect and disease control, and detection of tree troubles. Written in a semitechnical style so that it can be used by the beginning worker in tree care as well as the student and the specialist.

A458 Reed, Clarence A., and **John Davidson.** The improved nut trees of North America and how to grow them. 1954. Devin-Adair. 404p. illus. $10.

An older book written primarily for the layman and still useful as a reference guide for the nut grower. Deals with nut trees and all phases of their culture. Covers selected native nut trees as well as standard named varieties.

A459 Roper, Lanning. On gardens and gardening. 1969. Harper. 238p. illus. $15.

A well-known horticulturist presents a practical approach to gardening. Covers a wide range of topics from house plants to shrubs and vegetables, including many smaller aspects often overlooked in other books on the subject. Should appeal to all gardeners. Outstanding illustrations.

A460* Scheer, Arnold H., and **Elwood M. Juergenson.** Approved practices in fruit production. 1964. Interstate. 504p. illus. $7.50; text ed., $5.75.

A textbook covering all the more important practices in fruit production. First part of book concerned with general information followed by chapters on specific fruits. Gives coverage to nut trees but not to citrus or bush fruits. Appears on many reference lists for students at the high school or technical school level.

A461 Schery, Robert W. The lawn book. 1961. Macmillan. 207p. illus. $5.95.

An authoritative, easily understood manual covering every phase of lawn care from planning to maintenance. Includes helpful information on the major lawn grasses and their utilization. Significant aspects are illustrated by excellent line drawings.

A462 Schneider, George W., and **Clarence C. Scarborough.** Fruit growing. 1960. Prentice-Hall. 307p. illus. $8.64.

An elementary presentation of the fundamental principles of plant growth and their application to practical fruit production. Includes all major fruit crops except nuts and citrus fruits. Includes questions at the end of each chapter and a glossary of terms in the appendix.

A463 Schuler, Stanley. Gardening from the ground up. 1968. Macmillan. 308p. illus. $5.95.

A sound treatment of gardening written in nontechnical language for the beginning gardener. Discusses the usual topics including planting, pest control, soil improvement, vegetable growing, pruning trees and shrubs, etc.

A464 _____. Gardening in the East. 1969. Macmillan. 407p. illus. $7.95.

A practical manual covering all aspects of gardening in the eastern region. Of immense value to the beginning gardener. Possibly its only weak point is the frequent use of DDT in control recommendations.

A465 Shoemaker, James S. Small fruit culture. 3d ed. 1955. McGraw-Hill. 447p. illus. $11.50.

A practical guide to the selection, production, harvesting, and handling of small fruit crops in the various geographical regions. Covers grapes, strawberries, bramble-fruits, gooseberries, cranberries, blueberries, and currants.

A466* _____. and **Benjamin J. E. Teskey.** Practical horticulture. 1955. Wiley. 374p. illus. $5.75.

A comprehensive general horticulture textbook written primarily for students interested in applied agriculture and business management on the vocational/technical level. Includes ornamental trees, flowers, lawns, fruits, and vegetables. Written in a practical easy-to-read style with suggestions for further reading listed at the end of each chapter.

A467 Simmonds, Norman W. Bananas. 2d ed. 1966. Humanities. 512p. illus. $10.50.

An up-to-date sourcebook dealing with banana production, marketing, and utilization. One of the best books available on the subject. Intended to meet the needs of everyone interested in banana culture from the planter to the researcher.

A468 Singer, Rolf. Mushrooms and truffles: botany, cultivation, and utilization. 1961. Wiley. 272p. illus. $10.95.

A comprehensive, practical guide to the botany, production, and uses of mushrooms and truffles. Much of the information is of a basic nature and should appeal to students of botany as well as the commercial grower.

A469 Sprague, Howard B. Turf management handbook: good turf for lawns, playing fields, and parks. 1970. Interstate. 253p. illus. $6.95.

An authoritative guide covering all phases of turf management. Written in a simple style and systematically organized so that the reader can go directly to any particular topic. Includes illustrations of types of grasses, equipment, and planting techniques. A valuable source of information not only for the beginner but also for the working turf manager.

A470 Steffek, Edwin F. Pruning manual. rev. ed. 1969. Van Nostrand. 137p. illus. $4.95.

A popular guide to the pruning of trees, shrubs, and hedges. The author explains when and how to prune these plants and gives suggestions on selecting and using various pruning tools. He uses drawings to help clarify the pruning methods described.

A471 Sunset western garden book. Editors of *Sunset Magazine.* new 3d ed. 1967. Lane. 448p. illus. $5.95.

An elementary gardening encyclopedia for western gardeners. Covers the selection of ornamental plants for this climatic region and gives detailed information on their culture.

A472 Talbert, Thomas J. Growing fruit and vegetable crops. 1953. Lea & Febiger. 350p. illus. $4.50.

A handbook for the gardener, truck farmer, and fruit grower and a textbook for a beginning course in horticulture. Written in everyday language with a minimum of theory and scientific discussions. Considers fundamentals and basic problems of all of the common fruit and vegetable crops.

A473* **Taylor, Norman,** ed. Taylor's encyclopedia of gardening, horticulture, and landscape design. rev. ed. 1961. Houghton Mifflin. 1329p. illus. $12.95.

An authoritative, nontechnical, 1-volume encyclopedia edited by an eminent botanist. Arrangement is alphabetical with ample use of cross references and significant illustrations. Contains a vast amount of useful information for the layman, the student, and the specialist.

A474* **Thompson, Homer C.,** and **William C. Kelly.** Vegetable crops. 5th ed. 1957. McGraw-Hill. 611p. illus. $12.50.

Considers those principles necessary for successful vegetable production and handling. Can serve as a textbook in vegetable gardening courses and a reference guide for all concerned with the vegetable industry. Gives comprehensive treatment to all the economically important vegetables and many of the minor ones grown in the temperate regions. Devotes 1 chapter to tropical and subtropical vegetable crops.

A475 **Ticquet, C. E.** Successful gardening without soil. 1956. Chemical Pub. 176p. illus. $5.

A practical manual on gardening in gravel, sand, or water for both the beginner and the professional grower. Gives a clear, concise account of the principles and techniques of soilless culture. Points out possible problems and difficulties and gives recommendations for their solution.

A476 **Tukey, Harold B.** Dwarfed fruit trees. 1964. Macmillan. 562p. illus. $15.

The author's purpose is seen in the book's subtitle, *For orchard, garden, and home, with special reference to the control of tree size and fruiting in commercial fruit production*. Generally considered the most authoritative book available on dwarfed fruit trees. Covers all aspects from the history of dwarfing rootstocks to orchard management and plants grown under glass.

A477* **Vengris, Jonas.** Lawns: basic factors, construction, and maintenance of fine turf areas. 1969. Thomson. 229p. illus. $7.50.

Textbook specifically designed for students studying lawn construction and management. Covers species of turfgrasses, grass seeds, soils and fertilizers, propagation, mowing, watering, lawn problems, weeds, disease, and insect control, etc.

A478 **Voykin, Paul N.** A perfect lawn the easy way. 1969. Rand McNally. 124p. $3.95; paper, $1.95.

A professional golf course superintendent provides a simple, month-by-month guide to all aspects of planting and maintaining lawns in the different climatic areas. Written from a very personal viewpoint and gives excellent hard-to-find tips about lawns.

A479 **Wagner, Philip M.** A wine-grower's guide. 2d ed. 1965. Knopf. 224p. illus. $6.95.

The standard manual of wine-grape cultivation in America. Uses a minimum of technical language to present authoritative information on the cultivation and utilization of wine grapes. Intended for the amateur as well as the commercial grape grower.

A480* **Ware, George W.,** and **John P. McCollum.** Producing vegetable crops. rev. ed. 1968. Interstate. 558p. illus. $10.75; text ed., $7.95.

A combination handbook and encyclopedia written by 2 outstanding men in the field of vegetable production. Covers the basic principles of vegetable crop production and applies these principles to cultural methods and practices. Written primarily as a textbook on the vocational/technical school level but will serve equally well as a practical guide for the vegetable gardener.

A481 **Webster, Helen N.** Herbs; how to grow them and how to use them. 5th ed. rev. 1959. Branford. 224p. illus. $4.50.

A standard work on herbs, their folklore, culture, and uses in cooking and medicine. Was first published as a pamphlet in 1933 and has gone through numerous revisions and printings.

A482 **Wells, James S.** Plant propagation practices. 1955. Macmillan. 344p. illus. $7.95.

A practical manual for the beginning nurseryman. Most of the book deals with practices and the underlying principles governing propagation. Covers mainly woody plants, both deciduous and evergreen.

A483 **Williamson, Joseph.** Lawns and ground covers. 3d ed. 1964. Lane. 112p. illus. $1.95.

A thorough treatment of the culture and maintenance of lawns and groundcovers suitable for use on the West Coast and in other milder areas. Amply illustrated with photographs.

A484 Winkler, Albert J. General viticulture. 1962. California. 633p. illus. $10.

A broad treatment of the subject beginning with man's earliest uses of grapes and continuing up to modern viticulture. A vast amount of information on cultural practices and techniques is presented in a clear, concise manner that is easily understood by the beginner.

A485 Wittwer, Sylvan H., and S. Honma. Greenhouse tomatoes: guidelines for successful production. 1969. Michigan State. 180p. illus. $5.75.

A how-to book on growing tomatoes in a greenhouse. Covers planting and cropping schedules, seeds and seed treatment, growing, and harvesting tomatoes. The authors, long experienced in tomato cultivation, have produced a guide of value to any tomato grower.

A486 Woodroof, Jasper G. Tree nuts: production, processing, products. 1967. AVI. 2v. illus. $37.

A consideration of the general economic importance of American tree nuts followed by special chapters on the major kinds. In the second volume special consideration is given to the many uses of English walnuts and pecans. Useful to anyone actively involved in any phase of the nut industry plus students, teachers, and food scientists.

A487 ———, ed. Coconuts: production, processing, products. 1970. AVI. 241p. illus. $18.50.

Specialists from various parts of the world collaborated in preparing this book. Molds into 1 volume current information on coconuts, their production, processing practices, and their utilization.

A488 Work, Paul, and John Carew. Vegetable production and marketing. 2d ed. 1955. Wiley. 537p. illus. $7.95.

A comprehensive book designed as an introduction to vegetable production and marketing. A concise but good treatment of individual vegetables although does not attempt to provide specific recommendations for each situation. Although some techniques are dated, the book is still useful as a supplementary reference.

Ornamental Horticulture

A489 Abraham, George. The green thumb book of indoor gardening. 1967. Prentice-Hall. 304p. illus. $6.95.

Handbook of practical information written by a greenhouse operator with many years of experience. Includes such information as plant selection, temperature, soils, fertilizers, watering procedures, and transplanting. Contains a great deal of information valuable to the layman, the student, and the professional greenhouse operator.

A490 Aul, Henry B. How to plan modern home grounds. 1959. Sheridan. 312p. illus. $7.

Gives a modern approach to designing home grounds and gardens. Covers such home development problems as parking, driveways, terraces, etc. Uses illustrations to show how to build various home ground features.

A491* Ball, Vic. The Ball red book. 11th ed. 1965. Ball. 367p. illus. $3.50.

A practical handbook on growing and selling flowers and potted plants. Gives comprehensive coverage for each type of flower and plant. Written for the retail greenhouse operator by the staff of a wholesale commercial greenhouse. Excellent source of information for students interested in ornamental horticulture.

A492 Bartrum, Douglas. Water in the garden. 1968. Branford. 174p. illus. $5.50.

A British how-to book which shows how water can be used to best advantage in the garden. It tells how to build garden pools and ponds, describes the use of fountains, and explains which plants are best suited to water gardens. One of the few books published on this subject.

A493* Behme, Robert L. Bonsai, saikei, and bonkei; Japanese dwarf trees and tray landscapes. 1969. Morrow. 225p. illus. $12.95.

Gives complete information on creating miniature trees by dwarfing (bonsai) and on designing tray landscapes using either tiny living plants (saikei) or artificial materials (bonkei). Includes details on buying, training, and caring for miniature trees.

A494 Benz, Morris. Flowers: free form-interpretive design. 1960. San Jacinto. 247p. illus. $15.

A discussion of the modernistic trend in Oriental and Occidental floral art. Explains the principles of design, the use of modern containers, the use of color, the mechanics of flower arranging, and the care of cut flowers.

A495 ———. Flowers: geometric form. 3d ed. 1966. San Jacinto. 320p. illus. $20.

Comprehensive handbook on design, principles, methods, mechanics, color harmony, and foliage types. Gives step-by-step directions for arrangements for all occasions, all types of corsages, wedding bouquets, funeral wreaths, etc.

A496   Black, Arthur. Landscape sketching. 1952. McGraw-Hill. 109p. illus. $6.50.

A step-by-step presentation of the basics of drawing as applied to preparation of landscape plans. Author's aim is to provide the reader with a background knowledge of perspective values, planes, shadows, and composition.

A497   Blake, Claire L. Greenhouse gardening for fun. 1967. Barrows. 256p. illus. $7.95.

A complete guide to practical, simple, and inexpensive methods of greenhouse gardening. Covers types of greenhouses, climate control, equipment, types of plants, soils, and plant foods. Includes many shortcuts to make greenhouse gardening more fun and less work.

A498   Brilmayer, Bernice. All about begonias. 1960. Doubleday. 223p. illus. $5.95.

Describes how to select, grow, and propagate several hundred types of begonias. Lists of species and varieties are given for each major group described. These lists are divided into those for the beginner, the experienced grower, and the collector.

A499   Canham, Allan E. Electricity in horticulture. 1964. MacDonald. 199p. illus. $3.50.

A British manual intended for commercial greenhouse operators. Filled with authoritative practical information on using electricity to save time and labor while producing larger and better yields. Useful in helping students appreciate principles and practical applications of electricity in horticulture.

A500   Carleton, R. Milton. Your lawn: how to make it and keep it. 1959. Van Nostrand. 165p. illus. $4.95.

An authoritative practical account of dos and don'ts for amateur lawn makers. Covers all phases of constructing and maintaining lawns and gives methods for renovating and correcting previous errors.

A501*   Carnations: elegance in floral arrangements. 1968. Colo. Flower Growers Assn. 183p. illus. $7.53.

Consists mainly of full color illustrations of carnation arrangements with part color diagrams showing step-by-step methods of arrangement. Excellent for use in teaching flower arrangement in floriculture.

A502   Cherry, Elaine C. Fluorescent light gardening. 1965. Van Nostrand. 256p. illus. $6.95.

An authoritative treatment of all facets of gardening under lights. Presented in a precise but popular manner. Designed for the beginner or the moderately advanced gardener with problems.

A503   Chidamian, Claude. The book of cacti and other succulents. 1958. Doubleday. 243p. illus. $5.50.

A detailed account of succulents written by a man with considerable experience in the field. Covers their description, history, culture, care, and decorative uses.

A504   Conover, Herbert S. Grounds maintenance handbook. 2d ed. 1958. McGraw-Hill. 501p. illus. $17.50.

Provides a guide to all important aspects of grounds maintenance. Written primarily for use by those responsible for public, industrial, and commercial grounds. The author includes material gleaned from a wide range of sources as well as his own experience.

A505   Cumming, Roderick W. The chrysanthemum book. 1964. Van Nostrand. 301p. illus. $7.95.

A comprehensive, well-written manual on chrysanthemums. In simple language it covers the history, culture, breeding, development of new varieties, pest and disease control, and exhibiting of this flower.

A506   Cyphers, Emma H. Design and depth in flower arrangement. 1958. Hearthside. 118p. illus. $3.95.

Explains the why and how of designing a flower arrangement and of creating depth or third dimension. Discusses the importance of supporting space, selecting a skeleton contour, potentialities in plant materials, selection of containers and accessories, etc.

A507   Eckbo, Garrett. The art of home landscaping. 1956. McGraw-Hill. 248p. illus. $7.95.

A how-to book on planning, building, planting, and problems one might encounter in home landscaping. Although written primarily for the home owner, the book is a useful reference for the student studying landscape architecture.

A508   _____. Urban landscape design. 1964. McGraw-Hill. 248p. illus. $17.

Urban landscape improvement is the central theme of this book. Covers buildings in groups, streets and squares, parks and playgrounds, maintenance and design, etc. Uses actual case studies to show how various problems have been solved.

A509 Feldmaier, Carl. Lilies. 1970. Arco. 228p. illus. $9.95.

A comprehensive manual on lilies covering every aspect of their breeding, propagation, cultivation, and use in the garden or home. Useful for both amateur and professional gardeners.

A510 Foster, Frank G. The gardener's fern book. 1964. Van Nostrand. 226p. illus. $7.95.

The subtitle, *A guide for the gardener, a reference for the nature lover,* explains the author's purpose in writing this book. Covers the botany and identification of ferns but focuses primary attention on their cultivation and propagation.

A511 Foster, H. Lincoln. Rock gardening. 1968. Houghton Mifflin. 438p. illus. $7.

Bulk of the book is a descriptive catalog of nearly 2,000 species of plants suitable for rock gardens. Includes planning and constructing rock gardens and growing alpines and other wildflowers. Designed for American gardeners and especially those in the Northeast.

A512 Frazier, John B., and **Richard J. Julin.** Your future in landscape architecture. 1967. Rosen. 124p. $3.78.

A definitive study of the potentials and challenges of landscape architecture as a profession. Designed primarily for high school students but equally of value to technical school students.

A513 Free, Montague. All about house plants. 1946. Doubleday. 328p. illus. $6.50.

Not one of the juvenile all-about books. This standard encyclopedia on house plants covers their selection, propagation, culture, and decorative uses. More than 1,000 plants are discussed.

A514 Fukuda, Kazuhiko. Japanese stone gardens. 1970. Tuttle. 312p. illus. $22.50.

Uses famous gardens of Japan as examples in presenting the aesthetics of stone gardening. Also presents guidelines for construction. Added features include a glossary of terms.

A515 Gardiner, George F. Greenhouse gardening. 1968. Chemical Pub. 224p. illus. $7.50.

A manual on selecting and growing plants to provide flowers in the greenhouse or for room decoration during all seasons of the year. Excludes plants of little decorative value.

A516 Genders, Roy. The rose; a complete handbook. 1966. Bobbs-Merrill. 623p. illus. $12.50.

One of the most comprehensive of all rose books. Primary attention is focused on rose varieties and their culture. Includes a chapter on history and one on miscellaneous information.

A517 Graf, Alfred B. Exotic plants illustrated. 1966. Roehrs. 840p. illus. $27.50.

Condensed version of the famous cyclopedia, *Exotica,* containing more than 4,000 illustrations of exotic plants. Arranged by important families and in practical groupings. Gives basic information concerning their botanical features, their care, and use. Useful to horticultural students for quick visual identification of indoor ornamental plants.

A518* _____. Exotica 3: pictorial cyclopedia of exotic plants. 1970. Roehrs. 1834p. illus. 1v. ed., $60; 2v. lib. ed., $75.

Probably the most completely illustrated book of house plants available. Illustrates and describes more than 12,000 indoor ornamental plants. The text covers the botany of these plants, their propagation, their culture, and related aspects. Worthy of special note is the horticultural color chart and the common name index.

A519 Graff, M. M. Flowers in the winter garden. 1966. Doubleday. 203p. illus. $4.95.

A description, written in popular style, of plants for the winter garden. Covers both summer and winter care of these plants. Includes an index, a glossary, a bibliography, and a source list of nurseries.

A520 Hawkes, Alex D. Orchids, their botany and culture. 1960. Harper. 602p. illus. $7.95.

A definitive treatment of orchid growing, including a description of the most common varieties, cultural instruction, colored photographs, and line drawings. Of value to readers at all levels of expertise.

A521* Hay, Roy, and **Patrick M. Synge.** The color dictionary of flowers and plants for home and garden. 1969. Crown. 373p. illus. $15.

This book, written by 2 British gardening experts, contains more than 2,000 photographs of flowers and plants, followed by an A-Z dictionary. Covers alpine and rock garden plants, annual and biennial plants, indoor ornamentals, perennials, trees, and shrubs. Includes only a few species not well known in the United States.

A522 Hemp, Paul E. Fifty laboratory exercises for vocational ornamental horticulture students. 1968. Interstate. 174p. illus. $3.50.

The laboratory exercises are grouped into 6 instructional areas: flowers and house plants; turf maintenance and management; landscape management; plant propagation; plant growth and development; and horticulture mechanics. Designed to fit the needs of beginning horticulture students.

A523 ———. Ornamental horticulture source units for vocational teachers. 1968. Interstate. 134p. illus. paper, $4.

Provides course outlines, study questions, and bibliographies for the instructor's use. Covers turf management, landscaping, greenhouse management, nursery management, plant propagation, trees, flowers, and floral arrangements.

A524 Hoover, Norman K., and Elwood M. Juergenson. Approved practices in beautifying the home grounds. 1966. Interstate. 285p. illus. $6.25; text ed., $4.75.

A how-to manual and high school textbook on landscaping. Contains concise information on sound practices and procedures in all phases of landscaping. Materials and examples were used from all major U.S. regions.

A525 Hull, George F. Bonsai for Americans. 1964. Doubleday. 265p. illus. $6.95.

A practical guide to using domestic plants and familiar gardening techniques to create miniature potted trees. Written by an expert in gardening and landscape design.

A526* Ishimoto, Tatsuo, and Kiyoko Ishimoto. The art of shaping shrubs, trees, and other plants. 1966. Crown. 125p. illus. $3.95.

This topiary book is devoted to the pruning and shaping of trees and shrubs. Includes step-by-step explanations of methods ranging from simple pruning to elaborate trimming to form odd or ornamental shapes. Authors have selected the best methods used in various parts of the world. Essentially a picture book of ideas.

A527 Kenfield, Warren G. The wild gardener in the wild landscape. 1966. Hafner. 232p. illus. $7.50.

This book is concerned with the art of naturalistic landscaping of home grounds which include forested areas. Covers methods of eliminating unwanted plants, adding desirable plants, and maintaining forest openings. An easily read book written by an experienced caretaker of New England estates.

A528* Kilvert, B. Cory. Informal gardening: the new homeowner's guide to planting his property. 1969. Macmillan. 286p. illus. $5.95.

A basic guide for the beginner which contains information not usually found in gardening books. Author begins with conserving the land's natural resources and systematically proceeds through the various stages of construction to the finished product. Valuable as a reference source for students studying landscaping on the vocational/technical level.

A529 Kolaga, Walter A. All about rock gardens and plants. 1966. Doubleday. 385p. illus. $1.98.

Shows how beautiful rock gardens can be built in almost any part of the United States. Discusses culture of basic rock garden plants, types of gardens, sites, construction, planting, soils, care and maintenance.

A530* Laurie, Alexander, Donald C. Kiplinger, and Kennard S. Nelson. Commercial flower forcing. 7th ed. 1968. McGraw-Hill. 514p. illus. $14.75.

Presents the fundamentals of flower forcing and the practical application of these fundamentals to the culture of greenhouse crops. An authoritative but simply written book primarily designed as a college text and a reference for the commercial greenhouse operator.

A531 ———, and **Victor H. Ries.** Floriculture; fundamentals and practices. 2d ed. 1950. McGraw-Hill. 525p. illus. $12.50.

A standard text dealing with all major aspects of the culture and use of ornamental plants. Emphasis is on the herbaceous plants. Useful as an authoritative reference source for floriculturists as well as a text for floriculture students.

A532 Lawson, Alexander H. Bamboos: a gardener's guide to their cultivation in temperate climates. 1968. Taplinger. 192p. illus. $12.

Emphasizes practical gardening and gives detailed instructions for selection, planting, propagation, and care of bamboos. Also includes descriptions of the various hardy species and discusses their suitability for the garden.

A533 Lee, Frederick P. The azalea book. 2d ed. 1965. Van Nostrand. 448p. illus. $12.50.

A definitive handbook incorporating azalea history, botany, selecting, planting, fertilizing, pruning, disease control, and other important aspects. Contains information for the scientific expert as well as the amateur.

A534 Luxton, George E. Flower growing in the North; a month-by-month guide. 1956. Minnesota. 313p. $4.95.

A gardening book for those who live in regions where very cold winters and short growing seasons present special problems. Covers the entire year with a chapter for each month.

A535 McDonald, Elvin. Complete book of gardening under lights. 1965. Doubleday. 215p. illus. $5.50.

An authoritative guide to successful gardening under artificial lights. Covers various types of lights, seed and plant propagation, soil, potting, pests and diseases, etc. Although written for garden enthusiasts, it can easily be adapted to classroom use.

A536 McFarland, John H., and **Robert Pyle.** How to grow roses. rev. ed. 1968. Macmillan. 176p. illus. $4.95.

A revised edition of the classic guide to rose gardening. Essentially a how-to book covering all aspects of rose growing in clear, easily read language.

A537 Marsh, Warner L. Landscape vocabulary. 1964. Miramar. 315p. illus. $8.50 (approx.)

Defines the terms currently used by landscape architects. Covers the whole field from the individual garden to regional design. Will be of practical value to all professions concerned with environment.

A538 Mastalerz, John W., ed. Bedding plants. 1966. Pennsylvania Flower Growers (order from the editor, 101 Tyson Bldg., Pennsylvania State University, University Park, Pa. 16802). 121p. illus. $2.

A manual on the economics, breeding, culture, insects, diseases, and uses of bedding plants. Written by members of the agricultural extension and research staff of Pennsylvania State University.

A539 Menninger, Edwin A. Flowering vines of the world. 1970. Hearthside. 450p. illus. $25.

A fully illustrated encyclopedia of climbing plants by a prolific horticultural writer. Describes types of vines, their suitability for gardens, their culture, and care.

A540* Moore, Stanley B. Ornamental horticulture as a vocation. 1969. Mor-Mac. 364p. illus. $6.50.

The author, a vocational horticulture teacher, prepared this book as a guide in teaching an introductory or survey course in ornamental horticulture. Describes all plant materials from annuals to trees and discusses their culture, propagation, and special requirements. Written for use in all parts of the United States but of most value in areas with climates similar to that of Kentucky and Ohio.

A541 Morse, Harriet K. Gardening in the shade. 1962. Scribner. 242p. illus. $7.95.

A guide to special problems of gardening in the shade. Discusses full light and half shade in all types of gardens from the formalized bed to the woodland opening.

A542 Nehrling, Arno, and **Irene Nehrling.** Easy gardening with drought-resistant plants. 1968. Hearthside. 320p. illus. $6.95.

Contains everything for the dry land gardener from designing to planting and maintaining a garden. Lists and discusses drought-resistant trees, shrubs, vines, ground covers, and other plants.

A543 _____, and _____. Picture book of annuals. 1966. Hearthside. 288p. illus. $6.95.

Illustrates all of the common garden annuals and gives a brief description plus cultural information for each. Special aspects of annual culture such as propagation, potting, landscaping, etc., are treated in individual chapters. A most useful book for both amateur and commercial growers.

A543a _____, and _____. The picture book of perennials. 1964. Hearthside. 286p. illus. $6.95.

Similar to A543, but about perennials. Special features include a calendar of things to do each month, lists of perennials for special purposes, blooming time of perennials in northeastern United States, and a glossary.

A544* Nelson, Kennard S. Flower and plant production in the greenhouse. 1967. Interstate. 335p. illus. $7.25; text ed., $5.75.

A basic text for training greenhouse workers. Designed for use on the high school or junior college level. Gives a broad view of the whole industry followed by detailed instructions for each of the major functions of greenhouse plant and flower production.

A545 **Northen, Henry T.,** and **Rebecca T. Northen.** Complete book of greenhouse gardening. 1956. Ronald. 366p. illus. $8.50.

An authoritative book written from a do-it-yourself point of view. Covers almost every aspect of the subject including selection, building, management, and operation of the greenhouse plus specific information on growing the wide variety of plants suitable for greenhouse production.

A546 **Northen, Rebecca T.** Home orchid growing. 3d ed. 1970. Van Nostrand. 374p. illus. $15.

A new edition of a classic text covering every major species and tribe of orchids. Emphasis is on greenhouse production but other methods are included. Detailed treatment of all phases of growing from pollination to care of cut flowers.

A547 **Ortloff, Henry S.,** and **Henry B. Raymore.** The book of landscape design. 1959. Barrows. 316p. illus. $4.95.

A detailed discussion of the history and basic principles of landscape design. Drawings, photographs, and plans are used to illustrate significant aspects. Intended for the landscaping student, the designer, and the general public.

A548* **Pfahl, Peter B.** The retail florist business. 1968. Interstate. 435p. illus. $9.25; text ed., $6.95.

A basic text for students and a reference for those already in the retail florist business. Gives definitive treatment to all phases of establishing, operating, and managing the business from financing to marketing. One of the few books available on this subject.

A549* **Pinney, John J.** Operating a garden center. rev. ed. 1967. Amer. Nurseryman. 152p. illus. $3.95.

A discussion of the many problems facing the operator of a garden center. Increased income, discount competition, and changes in the industry are examples of problems treated.

A550* ―――. Your future in the nursery business. 1969. Rosen. 160p. illus. $3.78.

A practical, authoritative account of the nursery business as a career. Discusses its requirements, opportunities, disadvantages, and rewards and explains the differences between the nursery business and the florist business.

A551 **Price, Molly.** The iris book. 1966. Van Nostrand. 204p. illus. $7.95.

This book, one of the few available on the subject, is filled with practical information for the beginning or experienced grower. Gives detailed information on iris species and varieties, botany, culture, and exhibition.

A552 **Reusch, Glad,** and **Mary Noble.** Corsage craft. 2d ed. 1960. Van Nostrand. 189p. illus. $6.95.

A discussion of the principles underlying corsage making followed by sketches showing different methods and procedures. Contains numerous ideas on using flowers as jewelry and other ornaments.

A553 **Rubenstein, Harvey M.** Guide to site and environmental planning. 1969. Wiley. 190p. illus. $14.95.

A concise treatment of the technical information necessary for environmental designing. Intended for landscaping students and others involved in city or regional planning and environmental design.

A554 **Shields, Phyllis G.** Guide to flower arranging in 10 easy lessons. 1967. Branford. 117p. illus. $5.

Written primarily for the teacher although easily understood by the novice. The author, a well-known lecturer and judge in arranging circles, guides the reader from the first basic steps to sophisticated designs.

A555 **Simonds, John O.** Landscape architecture. 1961. McGraw-Hill. 244p. illus. $16.

A noted landscape architect outlines and analyzes man's shaping of his natural environment. Begins with site selection and systematically proceeds to the completed project.

A556 **Steffek, Edwin F.** Wild flowers and how to grow them. 1954. Crown. 192p. illus. $4.50.

A well-organized book on wild flowers. Contains complete information on finding, identifying, transplanting, and cultivating more than 300 species. Illustrations are carefully keyed to descriptions.

A557 **Stowell, Jerald P.** Bonsai: indoors and out. 1966. Van Nostrand. 134p. illus. $4.95.

The author explains the principles of bonsai art in America and tells how to begin, maintain, and display these plants. A special feature is the list of places from which plants and supplies may be obtained.

A558 **Sutter, Anne B.** New approach to design principles: a comprehensive analysis of design elements and principles in floral design. 1967. (order from the author, 12311 Conway Rd., Creve Coeur, Mo. 63141). 196p. illus. $7.95.

Explains basic design principles in a clear, concise style. Covers components, design elements, inherent principles of design, coordinating principles of designs, etc. Includes a color chart. A standard text in flower-judging schools.

A559 **Synge, Patrick M.** The complete guide to bulbs. 1962. Dutton. 319p. illus. $7.95.

A scholarly, authoritative book for the more serious amateur gardener and the specialist. Has an introductory chapter on the culture of bulbs in general, followed by an alphabetical encyclopedia of bulbs with descriptions and planting directions for the different species.

A560 **Taylor, Norman.** The guide to garden shrubs and trees. 1965. Houghton Mifflin. 450p. illus. $8.95.

An identification guide to the more popular cultivated trees and shrubs. Includes shrubs introduced from abroad and usually not mentioned in the fundamental American guides. Intended primarily for the nonprofessional gardener.

A561* ———, ed. Guide to garden flowers. 1958. Houghton Mifflin. 315p. illus. $5.95.

A compact guide to the identification and culture of garden flowers. Over 400 species are discussed in simple language with a minimum of technical terms. Includes not only all the popular garden species but also some of the less familiar ones. Both common and scientific names are given. Includes a good index and a short bibliography.

A562 **Techniques of landscape architecture.** ed. for the Institute of Landscape Architects by A. E. Weddle. 1967. Amer. Elsevier. 226p. illus. $16.75.

A handbook dealing with practical aspects of landscape architecture. Attempts to cover the techniques needed to handle most types of landscape problems. In this book landscape is defined as the total outdoor environment in both urban and rural areas.

A563 **Threlkeld, John L.** The camellia book. 1962. Van Nostrand. 204p. illus. $7.95.

Practical and authoritative advice on all the important particulars of camellia culture is presented in a clear, easy-to-read manner. Contains information of value to the novice, the advanced amateur, and the specialist.

A564 **Way, Marjorie S.** Essential forms for flower arrangement. 1969. Greene. 184p. illus. $6.95; paper, $4.95.

An excellent how-to book, with each photograph illustrating a special basic abstract quality and form or line used in flower arranging. Each point illustrated is accompanied by a clear explanation.

A565 **Wilson, Helen V. P.,** and **Leonie Bell.** The fragrant year. 1967. Barrows. 306p. illus. $10.

The book's scope is indicated by its subtitle, *Scented plants for your garden and your house.* Classifies and defines 9 basic fragrances: balsamic, spicy, heavy, sweet, honeyed, fruited, violet, rose, and unique. Examples are given for each fragrance.

A566* **Wyman, Donald.** Ground cover plants. 1970. Macmillan. 175p. illus. $5.95.

A comprehensive yet concise reference book for gardeners and students dealing with the description and culture of plants used for ground covers. A valuable feature is the chapter containing lists of plants used for special purposes such as sunny or shady areas.

A567* ———. Shrubs and vines for American gardens. 2d ed. 1969. Macmillan. 613p. illus. $14.95.

A standard reference text on the selection and planting of ornamental shrubs and vines. Contains complete descriptions and detailed information on the landscape usefulness of each shrub or vine discussed. Written for nurserymen, park superintendents, landscape architects, and home gardeners.

A568* ———. Trees for American gardens. rev. and enl. ed. 1965. Macmillan. 502p. illus. $10.95.

Presents more than 1200 species and varieties of trees recommended for American gardens. Gives detailed botanical and cultural information for each of the trees listed. An unusual feature is the list of trees to be considered as second or third choices. At the beginning of the list an explanation is given as to why these trees are secondary instead of recommended.

A569 **Zion, Robert L.** Trees for architecture and the landscape. 1968. Van Nostrand. 284p. illus. $25.

Portraits and descriptions of trees, along with how to use them in landscape designing, and how to care for, protect, and transplant them. A special feature of the book is its list of trees recommended for every state except Hawaii and Alaska. Expensive, but practical and useful.

A570 Zucker, Isabel. Flowering shrubs. 1966. Van Nostrand. 380p. illus. $17.95.

A reference book on flowering shrubs. About half of the book covers using, growing, and acquiring shrubs. The other half is an encyclopedia of recommended shrubs and small trees. Has photographs for each species discussed, showing flower and/or fruit. Includes shrubs in the U.S. plant hardiness zones 6 to 1.

Mycology

A571* Alexopoulous, Constantine. Introductory mycology. 2d ed. 1962. Wiley. 613p. illus. $13.95.

A standard introductory textbook on mycology designed to follow a basic botany course. Author has focused primary attention on morphology and taxonomy with elementary level physiology and genetics interwoven wherever possible. Laboratory manual available.

A572 Barnett, Horace L. Illustrated genera of imperfect fungi. 1960. Burgess. 225p. illus. $6.

Includes brief, concise descriptions and diagrammatic drawings of more than 450 genera of imperfect fungi. Designed primarily as an aid to identification of the common imperfect fungi and intended for students studying in agricultural or medical fields. (The third edition, published in 1972, appeared after the cutoff date set for this list.)

A573 Burnett, John H. Fundamentals of mycology. 1968. St. Martin's. 546p. illus. $13.95.

An encyclopedic work covering a vast amount of facts on mycology. Covers structure and growth, function, recombination, speciation, and evolution. A detailed account interesting to both the beginner and the specialist.

A574 Christensen, Clyde M. The molds and man. 3d ed. 1965. Minnesota. 284p. illus. $6.

An introduction to the biology of the fungi written mainly for the beginner and the layman. Covers fungi as plant and animal parasites and as destroyers of stored foods, textiles, and building materials. Devotes 1 chapter to industrial uses of fungi.

A575 Gray, William D. The relation of fungi to human affairs. 1959. Holt. 510p. illus. $12.50.

A discussion of both the beneficial and harmful roles of fungi in human affairs. Presents a vast amount of information in a clear, concise manner. A large number of illustrations are used to clarify the subject matter and make it more interesting. Intended as a textbook of industrial mycology.

A576 Hazen, Elizabeth L., Morris A. Gordon, and **Frank C. Reed.** Laboratory identification of pathogenic fungi simplified. 3d ed. 1970. Thomas. 253p. illus. $18.25.

A practical, profusely illustrated guide to the identification of fungi that cause disease in man and animals. This standard in its field is especially intended for those persons engaged in examination of laboratory specimens.

A577 Kavaler, Lucy. Mushrooms, molds, and miracles. 1965. Day. 256p. $7.50; New Amer. Lib. paper, 75¢.

An informal presentation of both beneficial and harmful fungi and their importance to man. Contains an extensive bibliography. Useful as collateral reading for high school and junior college students.

A578 Lange, Morten, and **Frederick B. Hora.** Guide to mushrooms and toadstools. 1963. Dutton. 257p. illus. $6.95.

An illustrated field guide to approximately 600 species of fungi. Contains brief descriptions for each fungus included but depends mainly on illustrations for identification. Written primarily for Denmark and England but most of the species included are also found in America.

A579 Phaff, Herman J., M. W. Miller, and **E. M. Mrak.** The life of yeasts: their nature, activity, ecology, and relation to mankind. 1966. Harvard. 186p. illus. $5.50.

A simple but wide-ranging, authoritative account of yeasts written particularly for the nonspecialist. Does not pretend to be comprehensive but contains a wealth of up-to-date information not readily found elsewhere.

A580 Webster, John. Introduction to fungi. 1970. Cambridge. 424p. illus. $10.50.

A clearly written, modern textbook of mycology. Primarily concerned with fungi as organisms but draws attention to their economic and ecological importance. Contains an extensive bibliography and a large number of illustrations useful for identification purposes.

A581 Wolf, Frederick A., and **Frederick T. Wolf.** The fungi. 1948, reprint 1969. Hafner. 2v. illus. $25.

The first modern textbook of mycology to cover the chemistry, biology, and activities of fungi as well as their taxonomy and morphology. Still useful as a reference work.

Forestry and Forest Products

A582 **Ainsworth, John H.** Paper, the fifth wonder. 3d rev. ed. 1967. Thomas Print. 370p. illus. lib. ed. $10.

Fascinating story of papermaking including how-to information as well as theoretical. Includes lists of paper markets, printing terms, graphic art terms, and paper terms which add to its usefulness as a reference source.

A583* **Allen, Shirley W.,** and **Grant W. Sharpe.** An introduction to American forestry. 3d ed. 1961. McGraw-Hill. 466p. illus. $9.75.

A well-known and widely used text designed for beginning students in forestry. It clearly explains basic principles and practices of silviculture, forest management, protection of forests, economics of forestry, and utilization of forests. The book will appeal to readers beyond the classroom because it is about forestry as well as of it.

A584* **Avery, Thomas E.** Forest measurements. 1967. McGraw-Hill. 290p. illus. $10.50.

A sound, practical treatment of the field of applied forest mensuration with emphasis on forest inventory. Designed for a beginning mensuration course with a how-to method of presentation. Assumes the reader will have a good mathematical background. Useful to the teacher as a framework around which he can build his own course.

A585 **Baker, Frederick S.** Principles of silviculture. 1950. McGraw-Hill. 414p. illus. $10.50.

Essentially a book on silvics rather than silviculture. Discusses the principles which make up the biological basis for silviculture practice. Includes the nature of forest trees, their growth and reproduction, and their ecological relationship to the environment. An older but still useful book.

A586 **Bakuzis, Egolfs V.,** and **Henry L. Hansen.** Balsam fir: a monographic review. 1965. Minnesota. 445p. illus. $9.50.

A readable account of balsam fir covering its botany, ecology, microbiology, silviculture, and utilization. Special features include extensive references grouped with the chapters concerned and lists of associated fungi and insects. Of value to forest biologists as well as those involved in the pulp and paper industry.

A587 **Barrett, John W.,** ed. Regional silviculture of the United States. 1962. Ronald. 610p. illus. $13.50.

Not a complete reference on the silviculture of each forest region but rather a guide to the most important biological, physical, and economic aspects as well as the silvicultural practices. Each regional chapter is written by an area specialist. A valuable book for both forestry students and practicing foresters.

A588 **Britt, Kenneth W.,** ed. Handbook of pulp and paper technology. 1964. Van Nostrand. 357p. illus. $25.

An authoritative handbook presenting the most important facts concerning pulp and paper technology. Written primarily for pulp and paper manufacturers but easily understood by anyone with an elementary background in chemistry.

A589* **Bromley, Willard S.** Pulpwood production. 2d ed. 1969. Interstate. 231p. illus. $3.95.

Specifically designed as a textbook in pulpwood harvesting for vocationally-oriented students at the high school and subprofessional levels. Covers the principles underlying pulpwood harvesting practices, business and economic aspects of pulpwood production, equipment and its operation, and production management.

A590 **Brown, Nelson C.** Logging. 1949. Wiley. 418p. illus. $7.95.

An older book concerned with timber harvesting principles and methods and the interrelations of forestry and logging. Covers felling and bucking of timber, minor log transport, log loading for transport, major land and water transport, and financial aspects of logging engineering.

A591* **Browning, Bertie L.** Methods of wood chemistry. 1967. Wiley. 2v. illus. v.1 $18; v.2 $19.75.

Author uses a simple, concise style to describe commonly used procedures and techniques in wood chemistry plus several less common methods not usually found in books. A useful feature is the list of organizations which issue standards relating to wood chemistry. Of value to anyone interested in wood technology.

A592 ──────, ed. Chemistry of wood. 1963. Wiley. 689p. illus. $29.95.

Has 2 unusual chapters, 1 dealing with the chemistry of developing wood and the other concerned with the chemistry of bark. The remaining chapters discuss the usual topics including structure, supply, chemical utilization of wood, etc. Primarily useful to the scientist and technologist.

A593 Casey, James P. Pulp and paper. 2d ed. 1960-61. Wiley. 3v. illus. v.1 $24.50; v.2 $29.50; v.3 $33.50.

An excellently organized reference tool with a detailed index and an extensive list of references. Covers pulping and bleaching, fundamental operations of the industry, properties of paper, etc. Should be in the technical library collection of any institution concerned with paper technology.

A594* Clar, C. Raymond, and **Leonard R. Chatten.** Principles of forest fire management. rev. ed. 1966. Calif. State Bd. of Forestry. 274p. illus. $1.

Book is elementary in style and content but exceedingly valuable to individuals responsible for preventing and controlling forest fires. Deals with causes of forest fires, their behavior, fire danger signals, fire fighting techniques, tools for the fire manager, aircraft uses, smokejumpers, remote sensing, photography, and fire management organization. Much of this work is the result of experience of the California fire control organization.

A595 Clepper, Henry, and **Arthur B. Meyer,** eds. American forestry—six decades of growth. 1960. Soc. of Amer. Foresters. 319p. $5.

An overall view of the birth and growth of forestry in the United States and a report on progress in the various areas of forestry. Includes forestry education, forestry research, timber production and utilization, forest land management, silviculture, wildlife management, watershed management, etc.

A596 Collier, John W. Wood finishing. 1967. Pergamon. 306p. illus. $5.75.

A practical book on finishing timber for interior use. Fills the gap between the book for the individual craftsman and that for finisher in a modern production plant. A valuable reference book for everyone from the apprentice to the foreman and manager.

A597 Coombs, Charles I. High timber: the story of American forestry. 1960. World. 223p. illus. $4.95.

Although written on the young adult level, this book includes material of value to any layman interested in the forestry and natural resources field. Treats history of forestry but puts major emphasis on the practice of forestry today and its importance to the nation's welfare.

A598 Davis, Kenneth P. Forest fire: control and use. 1959. McGraw-Hill. 584p. illus. $4.50.

The first hard-cover textbook devoted solely to forest fire control and use. Treats social, organizational, and technical aspects in a realistic and practical manner. A handy and useful reference for the field worker as well as a textbook in forest fire control courses of instruction.

A599 ──────. Forest management: regulation and valuation. 2d ed. 1966. McGraw-Hill. 519p. illus. $13.

A technical timber management book paying special attention to forest management, forest organization, and forest valuation. A standard text and reference, carefully organized, well written, and fully documented.

A600* Dowdell, Dorothy, and **Joseph Dowdell.** Tree farms: harvest for the future. 1965. Bobbs-Merrill. 164p. illus. $4.

Concerned with tree farm management from seedlings to the finished product with primary attention focused on corporate tree farms and western tree farms. Reading level ranges from grade school to college, and content from mediocre to excellent depending on the reviewer. Although opinions of reviewers vary widely, most consider it a very useful reference for forestry students, tree farmers, vocational counselors, and general readers.

A601 Duerr, William A. Fundamentals of forestry economics. 1960. McGraw-Hill. 579p. illus. $12.50.

Author intersperses economic theory and principles with practical applications to present a multitude of topics in forestry economics. More comprehensive than Worrell's *Economics of American forestry* but not as easy for beginning students.

A602 Feininger, Andreas. Tree. 1968. Viking. 116p. illus. $22.50.

A scientifically accurate yet easily read book dealing with trees as an essential part of our natural environment. Intended for all readers interested in the appreciation, management, and use of our environment. Contains a considerable number of botanical facts and outstanding color illustrations.

A603 Findlay, Walter P. K. Timber pests and diseases. 1967. Pergamon. 280p. illus. $6.

A concise, semitechnical book giving major emphasis to the joint impact of insects and fungi on wood. Written primarily for British use but most of the information applicable anywhere timber deterioration is a problem.

A604* Forbes, Reginald D., and Arthur B. Meyer, eds. Forestry handbook. 1955. Ronald. 1201p. illus. $15.

A widely known authoritative guide to working methods, techniques, and important facts and data in all phases of applied forestry and related fields. Intended for both field and office use by professional foresters, forestry technicians, and resource managers.

A605 Freeman, Orville L., and Michael Frome. The national forests of America. 1968. Putnam. 194p. illus. $12.95.

A well-illustrated history of the national forest movement and a guide to the U.S. national forests. The appendixes, listing the national forests, areas under review for the National Wilderness Preservation Systems, etc., are of more reference value than the text.

A606 Goor, Amihud Y., and Charles W. Barney. Forest tree planting in arid zones. 1968. Ronald. 409p. illus. $15.

Two eminent authorities present basic principles and practical procedures for establishing productive forests in arid and semiarid zones. Includes a description of arid zones, techniques of collecting and handling tree seeds, nursery growing of trees, description of more than 100 tree species suitable for areas of low rainfall, etc.

A607 Guthrie, John A., and George R. Armstrong. Western forest industry: an economic outlook. 1961. Johns Hopkins. 324p. illus. $9.

A fully documented study of the economic outlook for western forest product industries. Mainly concerned with the lumber industry, the pulp and paper industry, and the plywood industry. A useful reference for anyone concerned with forestry and forest products.

A608* Hackett, Donald F., and Patrick E. Spielman. Modern wood technology. 1968. Macmillan. 757p. illus. $10.72.

An introductory textbook on wood technology useful on the post-high school academic level. Covers the history of woodworking processes, practical information on woodworking tools, discussions of lumbering, and wood-using industries, etc.

A609 Hardman, H., and E. J. Cole. Papermaking practice. 1960. Toronto. 334p. illus. $7.50.

A practical treatment of papermaking, beginning with pulp manufacture and proceeding to bleaching, screening, beating, sizing, etc. Includes the type of material not readily found in other papermaking texts. Especially designed for the paper technology student and the working papermaker.

A610 Harlow, William M. Inside wood: masterpiece of nature. 1970. Amer. Forestry Assn. 120p. illus. $6.50.

A well-known authority on wood technology presents a highly readable account of wood anatomy. Beginning with visible features such as tree rings he proceeds to features visible only with the aid of a microscope. Includes discussions of wood structure, development of wood and bark, fungi damages, wood as a material, etc.

A611* _____, and Ellwood S. Harrar. Textbook of dendrology. 5th ed. 1968. McGraw-Hill. 512p. illus. $13.50.

The standard dendrology textbook for beginning forestry students and a useful reference source for others. Covers silvicultural characteristics, taxonomic descriptions, photographs of the diagnostic features, and distribution maps for the most important forest trees of North America.

A612 Hillis, W. E., ed. Wood extractives and their significance to the pulp and paper industries. 1962. Academic. 513p. illus. $17.50.

A study of the biochemistry of wood extractives and their possible effects on pulp and paper manufacture. A useful reference for pulp and paper technologists, foresters, and others concerned with wood utilization.

A613 Hilterbrand, Luther R. An introduction to forestry. 1967. Balt. 235p. illus. $4.

A very elementary forestry text concerned primarily with methods of forestry practices such as tree planting, silvicultural practices, and harvesting methods. Although written on high school level, it contains information of value to those interested in tree farms.

A614 Hunt, George M., and George A. Garratt. Wood preservation. 3d ed. 1967. McGraw-Hill. 433p. illus. $13.50.

The latest edition of a standard wood preservation textbook first published in 1938. Covers preservatives, equipment, methods and techniques, painting and gluing characteristics of treated woods, and other information valuable to both the student and the practitioner.

A615 Husch, Bertram. Forest mensuration and statistics. 1963. Ronald. 474p. illus. $11.50.

An introductory text providing a detailed presentation of the basic principles of forestry measurement and the statistical methods for applying them. Coverage includes photogrammetric procedures and data processing. Presupposes basic courses in college algebra and physics.

A616 Jane, Frank W. The structure of wood. 2d ed. rev. 1970. Fernhill. 478p. illus. $18.

A well-written, authoritative source of information for every wood anatomy student and research worker. Primary attention is focused on the structure of wood as a plant tissue although frequent references are made to economic aspects.

A617 Jepsen, Stanley M. Trees and forests. 1969. Barnes. 155p. illus. $6.95.

A concise account of forestry emphasizing the economic aspects of the coniferous forest. Arranged in logical sequence, beginning with seeding and planting and ending with wood-using industries and wood products.

A618 Ketchum, Richard M. The secret life of the forest. 1970. Amer. Heritage. 108p. illus. $7.95.

A detailed account of the complex life within the woodland community. Describes the growth and life cycle of the tree and its importance to the natural forest environment. A pleasure-reading type of book with no table of contents and no index. Produced in cooperation with the St. Regis Paper Company.

A619 Koch, Peter. Wood machining processes. 1964. Ronald. 530p. illus. $15.

A handbook and reference on wood-machining processes. Begins with background information and proceeds to an analysis of cutting processes, a description of machines, a discussion of current wood-machining practices, and a section on veneer-cutting equipment and practices.

A620 Lutz, Harold J., and **Robert F. Chandler.** Forest soils. 1946. Wiley. 514p. illus. $10.95.

An older book still used as a reference source in soils courses taught to forestry students. Covers the basics of soil science and their practical application to forestry. Deals primarily with the temperate zone forests of the United States and Europe.

A621 McCulloch, Walter F. Woods words: a comprehensive dictionary of loggers terms. 1958. Ore. Hist. Soc. 219p. paper, $2.95.

Defines several thousand authentic terms, slang words, and phrases used by loggers to communicate with each other.

A622 Meyer, Hans A., and others. Forest management. 2d ed. 1961. Ronald. 282p. illus. $8.50.

One of the better books on forest management. Written by men experienced in forestry teaching and practice. A concise account of the history of forest management and a guide to the efficient organization and operation of forests. Gives very good coverage to the uneven-aged forest.

A623 Mirov, Nicholas T. The genus Pinus. 1967. Ronald. 602p. illus. $15.

An account of the current knowledge of pines. Covers such topics as origin and development, taxonomy, distribution, ecology, chemistry, and economic value. An indexed review of relevant literature on the pine rather than a textbook.

A624* Nash, Andrew J. Statistical techniques in forestry. 1965. Lucas. 146p. illus. text ed., $5.50.

A revised and extended version of *Elementary statistics for foresters*, published in 1960. Designed as a text in beginning statistics for forestry students. Clearly discusses and illustrates standard statistical methods. Each chapter includes practice problems and exercises.

A625 Osmaston, Fitzwalter C. Management of forests. 1968. Hafner. 384p. illus. $10.50.

A concise, logically organized text written specifically for forestry students and covering the standard topics of traditional forest management. Primarily concerned with timber and makes only brief references to other forest resources.

A626* Panshin, Alex J., and others. Forest products. 2d ed. 1962. McGraw-Hill. 583p. illus. $14.50.

A well-documented text and reference giving an accurate picture of forest products technology. Covers the economics of forest utilization, wood products, products chemically derived from wood, and miscellaneous forest products.

Suitable for the novice as well as the more experienced person wishing a quick review of the field.

A627* _____, **Carl De Zeeuw,** and **Harry P. Brown.** Textbook of wood technology. 3d ed. 1970– . McGraw-Hill. 2v. v.2 forthcoming. v.1 $18; v.2 price not set.

A useful classroom text for wood technology students and a good reference for workers in the field and interested laymen. Volume 1 covers the commercial woods of the United States, their structure, identification, properties, and uses. Volume 2 of the second edition deals with the physical, chemical, and mechanical properties of U.S. commercial woods.

A628 **Platt, Rutherford H.** The great American forest. 1965. Prentice-Hall. 271p. illus. $7.95.

A highly interesting narrative of the history and growth of American trees and forests from prehistorical to modern times. Covers such subjects as paleobotany, physiology, and ecology. Will appeal to the general public as well as to naturalists and foresters.

A629 _____. 1001 questions answered about trees. 1959. Dodd. 381p. illus. $6.50.

A handy reference tool and a book for leisure reading. Information is given in question and answer form under 6 main topics: history, forestry, home trees, tree products, pests and diseases, and physiology of trees.

A630 **Prodan, Michail.** Forest biometrics. 1968. Pergamon. 447p. illus. $20.

The English translation of a German text intended for advanced students, practical workers, and researchers. Covers a wide field of statistics and includes numerous examples taken from actual forestry problems.

A631* **Rich, Stuart U.** Marketing of forest products: text and cases. 1970. McGraw-Hill. 712p. illus. $18.50.

Especially written for students studying marketing or general management in the forest industries. Defines marketing and its role in the lumber, pulp and paper, and other wood products segments of forest industry. Points brought out are backed up by authentic case studies.

A632 **Rydholm, Sven A.** Pulping processes. 1965. Interscience. 1269p. illus. $40.

A sound, well-organized reference book of considerable value to those interested or involved in wood pulping processes. Concerned with the chemical and physical characteristics of wood, the preparation of bleached and unbleached pulp, and pulp properties and uses.

A633 **Sartorius, Peter,** and **Hans Henle.** Forestry and economic development. 1968. Praeger. 340p. $17.50.

A worldwide regional presentation of the history of the forest as a factor in economic development and its future use as a source of wood, its role in conservation of natural resources, and its value for outdoor recreation. Both authors have international forestry and forest industry experience.

A634 **Shirley, Hardy L.** Forestry and its career opportunities. 2d ed. 1964. McGraw-Hill. 454p. illus. $11.50.

Depicts forestry in relation to the national and world economy and presents the individual employment opportunities within the forest industry. Written principally as a career guide for students.

A635 _____, and **Paul F. Graves.** Forest ownership for pleasure and profit. 1967. Syracuse. 214p. illus. $5.50.

A guide to prospective buyers and owners of forest property, particularly in the eastern region of the United States. A readable, concise discussion of buying forest land, managing forest property, and marketing of timber. Special features include an appendix describing forestry tools and instruments, a list of forestry organizations and associations, and an annotated bibliography.

A636* **Smith, David M.** The practice of silviculture. 7th ed. 1962. Wiley. 578p. illus. $12.95.

A standard text in applied silviculture for nearly 50 years. Written in a clear and easily understood style. Gives comprehensive treatment to practical techniques and where and when to apply them. One of the best available books on the culture and care of forest trees in North America.

A637* **Spurr, Stephen H.** Forest ecology. 1964. Ronald. 352p. illus. $8.50.

A logically organized textbook on basic silviculture. Concerned with the forest environment and the effect of site factors, the forest community, and phytogeography. Stresses the ecosystem concept but generally avoids classical, confusing plant ecology terminology. Spurr's style of writing is clear and easy to understand.

A638 _____. Forest inventory. 1952. Ronald. 476p. illus. $11.50.

An older book concerned with forest mensurational techniques. Stresses simple low-cost methods of measuring trees and stands, estimating their volume, and predicting stand growth. Includes sampling problems involved in forest inventory design, aerial photography, and photographic interpretation techniques.

A639 ———. Photogrammetry and photo-interpretation. 1960. Ronald. 472p. illus. $12.

First edition of this book was published as *Aerial photographs in forestry*. This edition broadened to serve all nonphotogrammetrists who use aerial photographs as a professional tool. Covers what aerial photography is, its uses, its processes, and the basic characteristics of the aerial photograph. Has special section on forestry applications. Written for advanced readers but useful as reference for the nonprofessional.

A640 Stamm, Alfred J. Wood and cellulose science. 1964. Ronald. 549p. illus. $15.

A valuable source of well-documented scientific and technical information about wood, with emphasis on its fundamental physical properties. Also, a well-designed textbook for students studying the use of wood in chemical processing.

A641 ———, and **Elwin E. Harris.** Chemical processing of wood. 1953. Chemical Pub. 595p. illus. $10.

A good starting point for a more detailed study of any phase of chemical utilization of wood. Not primarily intended as a classroom text but can easily be used for that purpose. Technical material is presented as simply as possible, and each chapter is accompanied by extensive bibliographies.

A642* Stoddard, Charles H. Essentials of forestry practice. 2d ed. 1968. Ronald. 325p. illus. $7.

An ideal text for an introductory course in forestry on the vocational/technical level of training. In a clearly written, straightforward style the author presents a basic guide to forest planning and field practices in timber growing, logging, protection, harvesting, and processing. A valuable feature of the book is its appendixes covering forest terminology, timber characteristics, management plans for small private forests, timber sales, and forestry organizations.

A643 ———. The small private forest in the United States. 1961. Johns Hopkins. 171p. illus. paper, $2.

Contains a considerable amount of information on the economics of small private forest operation. Examines problems of the small forest owner and suggests possible programs designed to improve timber production. Many of the statistics quoted are now outdated but the book is still a good reference on small forest ownership and its problems.

A644 Streit, Fred. Paper quality control. 1968. Lockwood. 129p. illus. $10.

Concerned only with quality control in the paper industry. Deals with all phases of the quality function and basic applications of the statistical method. Will appeal to managers, supervisors, technologists, trainees, students, and others concerned with paper quality control.

A645 Strelis, I., and **Robert W. Kennedy.** Identification of North American pulpwoods and pulp trees. 1967. Toronto. 117p. illus. $7.50; paper, $2.50.

Deals with wood anatomy for both softwoods and hardwoods, fiber identification aids, and nonwoody and man-made fibers. Of primary value to workers in the field as a key to fiber identification of North American materials. Published in association with the Pulp and Paper Research Institute of Canada.

A646 Troup, Robert S. Silvicultural systems. [ed. by E. W. Jones]. 2d ed. 1952. Oxford. 216p. illus. $9.60.

A useful handbook first published in 1928. Updated and slightly altered by E. W. Jones. A well-organized and easily read presentation emphasizing systems more than the conditions which led to the systems. International in scope.

A647 Tsoumis, George. Wood as raw material. 1968. Pergamon. 276p. illus. $10.

A brief, general survey of the identification, biological characteristics, and physical properties of wood. Useful as a reference for vocational/technical schools of secondary school level and above. Could be used by an experienced teacher as an introductory course textbook.

A648* Vardaman, James M. Tree farm business management. 1965. Ronald. 207p. illus. $7.50.

A practical guide to the business and management aspects of tree farming. Points up problems and offers solutions for almost every phase from selecting a forester to when and how to sell timber for highest profits. Logically organized so that the reader can easily refer to a specific chapter to find solutions to a problem. Written by the operator of a firm of consulting foresters.

A649* Wackerman, Albert E., William D. Hagenstein, and Arthur S. Michell. Harvesting timber crops. 2d ed. 1966. McGraw-Hill. 437p. illus. $12.50.

An elementary text designed to introduce forestry students to the basic logging principles in every field of timber harvesting. Generally considered one of the standard texts in the study of American logging practices.

A650 Weaver, Howard E., and David A. Anderson. Manual of southern forestry. 1954. Interstate. 368p. illus. $6.75; text ed., $5.

A good book but somewhat outdated. Intended as an aid in teaching vocational or technical forestry in the southern states. Covers the protection, management, and proper utilization of farm timberland.

A651 Wenzl, Hermann F. J. The chemical technology of wood. 1970. Academic. 692p. illus. $35.

A definitive survey of the methods of chemical technology of wood and a discussion of some of the ecological problems resulting from use of these methods. Intended for foresters and forest managers as well as chemists and technicians who work with wood.

A652 Widner, Ralph R., ed. Forests and forestry in the American states. 1968. Assn. of State Foresters. 594p. illus. $5.10 (approx.)

A reference anthology of the important historical events in the development of forests and state forestry in the United States. Deals with exploitation of forest resources, traces the origin of the forestry profession, and discusses the partnership that has developed between the national and state forestry agencies.

A653* Wilde, Sergius A. Forest soils. 1958. Ronald, 537p. illus. $12.

One of the best books available on the properties of forest soils and their relation to silviculture. The first part is concerned with soil minerals, soil organisms, forest humus, and other facets of soil as a medium for tree growth. Part 2 deals with applying soils knowledge in practical forestry. Has a particularly good section on nursery soils.

A654 Worrell, Albert C. Economics of American forestry. 1959. Wiley. 441p. illus. $12.50.

A lucid discussion of economic principles and their application to the forest industry. Uses simple terms easily understood by the less advanced forestry student. Subjects given comprehensive coverage include consumer supply and demand for forest products, forest ownership, labor, capital, and production economics.

A655 Youngberg, Chester T., ed. Forest-soil relationships in North America. 1965. Oregon State. 532p. illus. $8.

Consists of chapters based on papers presented at a forest soils conference. Mainly concerned with the application of forest soils knowledge to forest management. Contributors include foresters and soil scientists from the United States and Canada. A valuable source of information for ecologists, forest land managers, forest research workers, students, and teachers.

A656 Zaremba, Joseph. Economics of the American lumber industry. 1963. Speller. 232p. illus. $7.95.

A painstakingly researched treatment of the importance of the lumber industry in economic development. Deals with the economics of lumber manufacture, distribution, and consumption rather than timber growing. Intended for a mixed group of readers including managers in wood-using industries, foresters, economists, students, and teachers.

ANIMAL SCIENCES

General Zoology

A657 Alexander, R. McNeill. Animal mechanics. 1968. Washington. 348p. illus. $9.50.

Describes clearly and in great detail the application of some of the principles of mechanics to a broad range of zoological investigations. Written for undergraduate students in zoology, ecology, etc., who have very little knowledge of physics and none of engineering.

A658 Anderson, Sydney, and J. Knox Jones, Jr. Recent mammals of the world. 1967. Ronald. 453p. illus. $12.50.

A well-documented outline of each order and family of living mammals giving the kind of information usually found in a book of this type. Intended to provide a ready source of information for mammalogists and others interested in mammals of the world.

A659 Andrewartha, Herbert G. Introduction to the study of animal populations. 1961. Chicago. 281p. illus. paper, $2.45.

A clear discussion of the principles and methods of determining size of animal populations. Environmental conditions determining animal population are considered to be weather, food, other organisms, and a place to live. An excellent introductory book for students of wildlife.

A660 Asdell, Sydney A. Patterns of mammalian reproduction. 2d ed. 1965. Comstock. 670p. $17.50.

An assembly of authoritative information concerning the reproduction patterns of mammals. Author has concentrated on the major studies of each species, especially those consisting of precise and quantitative facts. An effective guide for students of mammalian reproductive physiology.

A661 Bellairs, Angus d'A. Reptiles. 2d ed. 1968. Hillary. 200p. illus. paper, $2.

A modern general account of reptiles. Author's style of writing should increase the book's appeal to the general reader. Useful as a starting point for students interested in the subject.

A662 ———, and Richard Carrington. World of reptiles. 1966. Amer. Elsevier. 153p. illus. $4.75.

An up-to-date account of living reptiles in general, with a closer look at each group. Includes turtles, lizards, snakes, crocodiles, etc. Covers only the common and the most unusual reptiles. A valuable reference source for students learning to operate or work in zoos, reptile gardens, etc.

A663 Blair, W. Frank, and others. Vertebrates of the United States. 2d ed. 1957. McGraw-Hill. 819p. illus. $14.50.

A comprehensive guide to the identification of vertebrates found in the United States. A considerable amount of natural history and ecologic information is included. Covers fishes, amphibians, reptiles, birds, and mammals but omits marine fishes and turtles.

A664 Bodemer, Charles W. Modern embryology. 1968. Holt. 475p. illus. $10.50.

Briefly introduces general basic principles of embryology and follows with a survey of developmental anatomy and physiology. Based on the second edition of Lester G. Barth's well-known book, *Embryology*.

A665 Buchsbaum, Ralph. Animals without backbones. rev. ed. 1948. Chicago. 405p. illus. $9; text ed., $7.

An invaluable introduction to invertebrates written for young adults and laymen. A widely popular book because of its numerous photographs and diagrams.

A666 Burkhardt, Dietrich. Signals in the animal world. 1968. McGraw-Hill. 150p. illus. $10.

A modern approach to a complex subject. Deals with the function of sense organs, ways animals orient themselves to their environment, and how they communicate. A scholarly presentation in terms easily understood by the layman.

A667 Cloudsley-Thompson, J. L. Animal conflict and adaptation. 1965. Dufour. 172p. illus. $8.95.

A stimulating book concerned with animal adaptation to its environment, conflict between species, conflict within species, and problems of human conflict and adaptation. Emphasizes the importance of maintaining balance and variety among living things in the environment.

A668 Elton, Charles S. Animal ecology. 1966. Washington. 207p. illus. $9.50.

Mainly concerned with the ecology of communities although population of individual species is not disregarded. Originally published in 1927 and practically unchanged except for the preface referring to recent trends in animal ecology and the inclusion of more references. Provides basic background reading for students of the subject.

A669 Esmay, Merle L. Principles of animal environment. 1969. AVI. 325p. illus. $15.

Concerned with animal response to environmental factors, dispersal of heat from animal systems, and environmental control in animal housing systems. Of considerable interest to agricultural engineers, animal producers, and poultry producers.

A670 Ewer, R. F. Ethology of mammals. 1968. Plenum. 418p. illus. $26.

A description of the behavior of a variety of mammals with respect to all facets of their environment. Author gives a detailed explanation of traditional ethological theory and relates research findings to nature observations. Adequately documented but entertainingly written. Should appeal to both novice and specialist.

A671 Fisher, James, and **Roger T. Peterson.** The world of birds. 1968. Doubleday. 288p. $12.95.

This guide to ornithology deals with all aspects of bird biology and makes ample use of illustrations in clarifying the subject matter. Contains valuable sections dealing with extinctions and threatened extinctions and useful chapters on technical aids and investigations of such subjects as migration.

A672 **Gilchrist, Francis G.** Survey of embryology. 1968. McGraw-Hill. 426p. illus. $9.95.

A survey of the descriptive, experimental, and biochemical phases of embryology. Intended for college students who need only an introduction to the subject. References at end of each chapter.

A673 **Goin, Coleman J.,** and **Olive B. Goin.** Comparative vertebrate anatomy. 1965. Barnes & Noble. 242p. illus. paper, $1.75.

A concise introduction to the basic principles and concepts of comparative vertebrate anatomy. Based on a survey of several standard textbooks. Useful to students on both high school and college levels.

A674* **Goodnight, Clarence J., Marie L. Goodnight,** and **Peter Gray.** General zoology. 1964. Van Nostrand. 564p. illus. $10.75.

A definitive account of general zoology. Begins with basic discussions of life and matter and proceeds to a review of the animal kingdom with a specific examination of the vertebrate organ systems. Ends with a treatment of genetics, evolution, and ecology. Excellent reference for both high school and college levels.

A675 **Gordon, Malcolm S.,** and others. Animal function. 1968. Macmillan. 640p. illus. $12.95.

An ecological approach to physiology with emphasis on the adaptation of animals to their natural environment. Deals mainly with vertebrates and to particular phases of their physiology.

A676 **Griffin, Donald R.** Bird migration. 1964. Doubleday. 180p. illus. $4.50.

A general treatment of the subject presented by a well-known professor of zoology. Includes history, facts, methods, and experimental results of bird migration. An interesting book for readers at all levels of expertise.

A677 **Hainsworth, Marguerite D.** Experiments in animal behaviour. 1967. Houghton Mifflin. 206p. illus. $4.20.

Draws together many useful ideas and a number of possible experiments on animal behavior. Useful as a teacher's aid as well as a textbook for students studying animal behavior.

A678 **Hamilton, William J., Jr.** American mammals. 1939. McGraw-Hill. 434p. illus. $9.95.

An account of the lives, habits, and economic relations of American mammals is presented from the field naturalist's point of view. Material is well selected and presented in an interesting manner.

A679 **Humason, Gretchen L.** Animal tissue techniques. 2d ed. 1967. Freeman. 569p. illus. $9.50.

An explanation and the basis and application of histological techniques rather than just a list of the various methods. Contains a wealth of detail and information accompanied by useful illustrations. Intended for beginning students and working technicians rather than the experimental specialist.

A680* **Hutt, Frederick B.** Animal genetics. 1964. Ronald. 546p. illus. $12.50.

A simple introduction to elementary animal genetics with primary attention focused on domestic animals and on those concepts of most value to the practical man. Contains numerous good illustrations and a list of references at the end of each chapter. Intended for students in agricultural and veterinary colleges.

A681 **Ingles, Lloyd G.** Mammals of the Pacific states: California, Oregon, and Washington. 1965. Stanford. 506p. illus. $7.50.

Gives descriptive accounts of more than 200 species of mammals found in western America. Intended as a general reference for students, laymen, or field workers concerned with mammals in the Pacific states. Also useful as a reference in an introductory study of mammals.

A682 **Jaeger, Edmund C.** Desert wildlife. 1961. Stanford. 308p. illus. $5; paper, $2.95.

A highly readable series of short descriptive essays on desert wildlife. Deals primarily with selected desert dwelling vertebrates and contains much intriguing information on the natural history of these creatures.

A683 **Kendeigh, Samuel C.** Animal ecology. 1961. Prentice-Hall. 468p. illus. $12.95.

A fairly complete description of the interrelationships between animals and their animate and inanimate environment in the major ecological communities of North America. A highly accurate account written on the high school to college level although somewhat loaded with technical facts and terms.

A684 **Klopfer, Peter H.,** and **Jack P. Hailman.** Introduction to animal behavior. 1967. Prentice-Hall. 297p. illus. $8.50.

A brief introduction to animal behavior using a historical approach. A worthwhile book for wildlife managers and field workers who are interested in behavioral aspects of wildlife populations.

A685* **Larousse encyclopedia of animal life.** ed. by Maurice Burton, and others. 1967. McGraw-Hill. 640p. illus. $25.

The best 1-volume encyclopedia of zoology available. Although authors are British the book contains many American examples and common names. A well-organized reference with entries arranged in taxonomic sequence. Valuable for novice, student, or specialist who needs information about the varied relationships in the animal kingdom.

A686 **MacArthur, Robert H.,** and **Joseph H. Connell.** The biology of populations. 1966. Wiley. 216p. illus. $7.95.

An introductory book of concepts rather than a list of facts. Can be used with companion books on the study of the cell and the organism in a general biology course. Reviewers do not agree on this book's merits. Opinions range from "neither profound nor complete" to "one of the best available."

A687 **Manning, Aubrey.** An introduction to animal behavior. 1967. Addison-Wesley. 208p. illus. $9.50; paper, $3.95.

An introductory text dealing with animal behavior from a biological point of view. Covers reflexes and complex behavior, hormones and behavior, and evolution and learning. A well-illustrated book with a practical approach for the student reader.

A688 **Marchant, Ronald A.** Where animals live. 1970. Macmillan. 170p. illus. $4.95.

Contains a great number of essential facts on the habits of mammals, birds, and reptiles. Explains why some species adapt more readily than others, what happens when populations grow too large, etc.

A689 **Matthews, Leonard H.,** and **Maxwell Knight.** The senses of animals. 1963. Philosophical Lib. 240p. illus. $7.50.

A digest, in nontechnical language, of the causes of animal behavior. Describes what the field naturalist is able to observe; how animal senses work; and, the structural and functional bases for the sensory life of animals. Of most value to the amateur and the general student.

A690 **Mayr, Ernst.** Animal species and evolution. 1966. Harvard. 797p. illus. $15.

A definitive treatment and evaluation of man's current knowledge of population genetics and microevolutionary theory. Contains a useful bibliography for further study. Generally considered one of the better books on species and evolution.

A691 **Meglitsch, Paul A.** Invertebrate zoology. 1967. Oxford. 961p. illus. $11.75.

Presents an urbanely written and well-illustrated, comprehensive treatment of all invertebrates. Considered one of the best books available on this subject. Chosen by a number of instructors as a textbook for their course on invertebrate zoology.

A692 **Milne, Lorus J.,** and **Margery Milne.** The nature of animals. 1969. Lippincott. 225p. illus. $5.95.

A brief discussion of the important aspects of animal life including their habits, heredity, reproduction, similarities, differences, and ecological importance to man. Of interest to zoology students on both the high school and college levels.

A693 **National Geographic Society.** Wild animals of North America. 1960. Nat. Geographic. 400p. illus. $7.95.

A comprehensive discussion of mammals in general and a close examination of the common species of wild animals found in North America. Molds into 1 volume contributions by several prominent naturalists. Contains outstanding illustrations in color and black and white.

A694* **Pennak, Robert W.** Collegiate dictionary of zoology. 1964. Ronald. 583p. $9.50.

A 1-volume dictionary of zoological terms, scientific names, and popular names. A special feature is the taxonomic outline of the animal kingdom.

A695 ———. Fresh-water invertebrates of the United States. 1953. Ronald. 769p. illus. $15.

Essentially a handbook emphasizing the taxonomy, natural history, and ecology of freshwater invertebrates found in the United States. Not intended to be a complete guide to all freshwater invertebrates since it omits cestodes, parasitic nematodes, etc. Includes keys to the genera and to the most common species.

A696 **Pettit, Lincoln C.** Introductory zoology. 1962. Mosby. 619p. illus. $9.50.

A classical zoology textbook with a modern approach. Consists of a description of zoology as a science, a survey of the animal kingdom, and a discussion of ecology and evolution.

A697 Rand, Austin L. Ornithology. 1967. Norton. 311p. illus. $8.50.

A concise introduction to birds, their relationship to their environment, to their ancestry, and to each other. Written for the nonspecialist but not too elementary for the ornithologist to enjoy.

A698* Robbins, Chandler S., Bertel Bruun, and Herbert S. Zim. Birds of North America; a guide to field identification. 1966. Golden. 340p. illus. $5.95; paper, $3.95.

An authoritative, concise description of several hundred birds with illustrations facing the corresponding text page. Includes identification, geographical distribution, bird songs, and other diagnostic features. Index contains both common and scientific names. The only standard field guide that includes both eastern and western birds in 1 volume.

A699 Romer, Alfred S. The vertebrate story. 4th ed. 1959. Chicago. 437p. illus. $7.50.

A standard reference book especially for high school and college students who need only a general introduction to vertebrates. Topics covered include comparative anatomy, embryology, evolution, etc.

A700 Rue, Leonard L. Pictorial guide to the mammals of North America. 1967. Crowell. 299p. illus. $7.95.

A concise pictorial guide to 65 species representative of North American mammals. Provides the usual kind of information found in a guide book. Includes an appendix containing a list of state and national parks and refuges.

A701* Sadleir, R. M. The ecology of reproduction in wild and domestic animals. 1969. Methuen. 324p. illus. $12.

A review of major studies on the effects of the environment and other ecological factors on the various aspects of mammalian reproduction. The detailed information on domestic animals gleaned from authoritative agricultural journals will be especially useful. Represents all major orders of animals.

A702 Saunders, John W. Animal morphogenesis. 1968. Macmillan. 117p. illus. paper, $2.95.

A well-organized, introductory text concentrating on animal development. Easily understood by students with a good basic biology background at the high school level. Contains selected references for further reading.

A703* _____. Patterns and principles of animal development. 1970. Macmillan. 353p. illus. $8.95.

A logically organized presentation of the principles and concepts of animal development. Designed for use as a text in an introductory course in animal embryology or developmental biology. Also suitable for use as supplementary reading material.

A704 Scheer, Bradley T. Animal physiology. 1963. Wiley. 409p. illus. $10.95.

A comprehensive treatment of basic physiological principles and their application to the entire animal kingdom, including domestic animals and man. Includes selected references for further reading.

A705 Schmidt-Neilsen, Knut. Animal physiology. 3d ed. 1970. Prentice-Hall. 145p. illus. $4.95; paper, $1.95.

A clearly written compilation of animal physiology fundamentals. Covers food and energy, oxygen, temperature, water, movements, etc. Has been recommended for supplementary reading use in introductory courses and for non-biologists interested in animal function systems.

A706 Scott, John P. Animal behavior. 1958. Chicago. 281p. illus. $5.75.

A clearly written introduction to animal behavior bringing together the major objectives, problems, and results of ethology. Primary attention focused on problems of adaptation of behavior to the environment.

A707 Storer, Tracy I., and Robert L. Usinger. General zoology. 4th ed. 1965. McGraw-Hill. 741p. illus. $9.95.

The latest edition of a textbook widely used in general elementary zoology courses. Begins with a general coverage of animal biology and follows with a complete description of the animal kingdom by phylum and class.

A708 Villee, Claude A., Warren F. Walker, and Frederick E. Smith. General zoology. 3d ed. 1968. Saunders. 844p. illus. $9.75.

An introductory general zoology textbook also useful for supplementary reading and reference. First part of book covers fundamental principles and second part deals with taxonomy.

A709 Walker, Ernest P., and others. Mammals of the world. 1964. Oxford. 3v. illus. v.1 and v.2 $25; v.3 $12.50.

A definitive work containing information on every genus of mammals. Has an encyclopedic organization which makes facts more readily available. Third volume comprised of a 50,000 entry bibliography arranged according to zoological groups, geographic areas, and general subjects. A major contribution to the study of mammalian biology of value to both layman and specialist.

A710 Wallace, George J. An introduction to ornithology. 2d ed. 1963. Macmillan. 491p. illus. $8.95.

An introductory study of the methodological, physical, and behavioral phases of bird biology. A useful reference for conservationists, wildlife managers, ornithologists, and amateur naturalists.

A711 Wasley, G. D., and J. W. May. Animal cell culture methods. 1970. Davis. 208p. illus. $9.

Designed for the student or laboratory worker with no previous experience in use of cell culture methods. Describes techniques and gives details of the preparation and understanding needed to begin this type of work.

A712 Watterson, Ray L., and Robert M. Sweeney. Laboratory studies of chick, pig, and frog embryos. 2d ed. 1970. Burgess. 179p. illus. paper, $6.50.

Designed for use in an introductory course in vertebrate embryology at the college level. This revision a result of numerous suggestions from undergraduates who used the first edition and from several graduate students who served as laboratory instructors.

A713 Wendt, Herbert. The sex life of the animals. tr. by Richard and Clara Winston. 1965. Simon & Schuster. 383p. illus. $7.95.

A comprehensive study of sex in animals. Begins with sex in the lowest forms of life and systematically proceeds to the large animals. Sex in man is not included. A useful reference book for students on both the high school and beginning college levels.

Animal Husbandry

A714* Acker, Duane C. Animal science and industry. 1963. Prentice-Hall. 502p. illus. $10.95.

Designed for use as a textbook in a beginning course in animal science and industry. Arrangement is by basic subjects important to the whole field of animal science rather than by species of livestock. Examples of subjects covered include nutrients, breeding, marketing, and meat technology. Graphs, tables, and pictures are used to help clarify the subject matter.

A715 American Kennel Club. The complete dog book. rev. ed. 1969. Doubleday. 580p. illus. $6.

This book's subtitle, *The history and standard of breeds admitted to AKC registration, and the training, feeding, care, and handling of pure-bred dogs,* defines both its scope and purpose.

A716 Anderson, Arthur L., and James J. Kiser. Introductory animal science. 4th ed. 1963. Macmillan. 800p. illus. $11.95.

A standard college reference and beginning textbook containing sections on the most important farm animals. Covers all aspects of livestock production and marketing.

A717 Asdell, Sydney A. Cattle fertility and sterility. 2d ed. 1968. Little, Brown. 276p. illus. $9.

A complete reference and guide to all aspects of cattle reproduction. Author shows how sound management practices can cut down on losses due to infertility. Of practical value to veterinary students and cattle breeders.

A718 ————. Dog breeding; reproduction and genetics. 1966. Little, Brown. 194p. illus. $6.50.

Molds into 1 volume most of the known facts about dog breeding. Discusses reproduction and genetics as related to the dog. Written primarily for dog breeders and dog owners.

A719* Beeson, William M., Roger E. Hunsley, and Julius E. Nordby. Livestock judging and evaluation. 1970. Interstate. 405p. illus. $9.75.

A practical presentation of information necessary for successful selection and improvement of livestock. Significant aspects are illustrated by pictures, tables, etc. Contains much material of value to breeders and livestock judges although primarily designed as a high school or college textbook.

A720 Bennett, Russell H. The compleat rancher. 2d ed. 1965. Denison. 253p. illus. $4.95.

A practical, down-to-earth book on modern ranching methods. Includes a number of author's personal experiences with cow-country philosophy and humor. Written for the be-

Animal Sciences

ginner seeking information on how to acquire and operate a western cattle ranch.

A721 Blount, William P., ed. Intensive livestock farming. 1968. Heinemann Med. 612p. illus. $13.50.

Contributions from 29 authors collected into 1 volume. Concerned with all phases of husbandry and disease control of food animals. Includes a chapter on fish farming and articles on world food problems, statistics, and computer use.

A722 Bogart, Ralph. Improvement of livestock. 1959. Macmillan. 436p. illus. $8.95.

Elementary information on breeding and genetics for students and farmers interested in livestock improvement. Discusses qualitative genetics, inheritance of abnormalities and color, animal breeding, and economic characteristics.

A723 Bowen, Edwin G., and **Ross W. Jenkins.** Chinchillas; history, husbandry, marketing. 1969. Bowen & Jenkins. 149p. illus. $14.75.

Presents an accurate picture of the chinchilla industry and covers the fundamentals necessary to the beginning rancher. Intended as an introduction and basic guide only. Mr. Bowen is currently president of the Empress Chinchilla Breeders Cooperative. A must for the beginning chinchilla rancher.

A724 Brady, Irene. America's horses and ponies. 1969. Houghton Mifflin. 202p. illus. $7.95.

A compilation of facts from many sources on the 38 most popular breeds of horses and ponies plus donkeys and mules. Illustrations are drawn to scale for easier comparison of animals. Does not include stable management, riding, or showing.

A725* Briggs, Hilton M. Modern breeds of livestock. 3d ed. 1969. Macmillan. 754p. illus. $12.95.

An introductory text for the student and prospective breeder. Impartially discusses the strengths, weaknesses, and developmental changes of various breeds of cattle, hogs, horses, and sheep. A useful reference manual reflecting current developments and research in the livestock industry.

A726 Buckett, M. Introduction to livestock husbandry. 1965. Pergamon. 186p. illus. $5.

A concise, well-organized text dealing with basic principles of livestock production and management. Covers sheep, pigs, and cows. Especially written for the British technical school student but also useful to American students.

A727* Bundy, Clarence E., and **Ronald V. Diggins.** Livestock and poultry production. 3d ed. 1968. Prentice-Hall. 723p. illus. $9.40.

A broad presentation of the basic factors essential to successful production, management, and marketing of livestock and poultry. Each chapter followed by a set of questions and a bibliography. Valuable as a high school or junior college text in elementary livestock production courses.

A728* ———, and ———. Swine production. 3d ed. 1970. Prentice-Hall. 342p. illus. $9.48.

A complete and up-to-date, introductory textbook which systematically handles the fundamentals of each phase of swine production and marketing. Reviews results of recent research in academic fields as well as in the meat packing industry. Many appropriate illustrations add to the book's effectiveness.

A729 Burns, Marca, and **Margaret N. Fraser.** Genetics of the dog. 2d ed. 1966. Lippincott. 230p. illus. $10.

A practical, well-documented account of the genetics and breeding of dogs, stressing actual breeding problems. Written in clear, easy-to-understand language. A very useful book for breeders, veterinarians, and other interested practitioners.

A730 Byerly, Theodore C. Livestock and livestock production. 1963. Prentice-Hall. 422p. illus. $10.50.

A concise but readable discussion of the common breeding, feeding, management, disease control, and marketing problems of the livestock industry. A useful reference for students on the vocational/technical school level.

A731* Campbell, John R., and **John F. Lasley.** The science of animals that serve mankind. 1969. McGraw-Hill. 771p. illus. $12.50.

Subject matter is presented from a biological point of view. Shows how biological science is applied to farm livestock production and uses examples to clarify the principles presented. Omits production and management practices. One of the few books of this type available and an excellent addition to technical agricultural school libraries.

A732* Cole, Harold H., ed. Introduction to livestock production, including dairy and poultry. 2d ed. 1966. Freeman. 827p. illus. $10.

Forty livestock authorities introduce the beginning student to scientific production of sheep, swine, dairy cattle, beef cattle, goats, horses, and poultry. Treats genetics, physiology, nutrition, medicine, etc., as related to livestock. Written for college level student but much of the text is not beyond high school understanding.

A733 ———, and **P. T. Cupps,** eds. Reproduction in domestic animals. 2d ed. 1969. Academic. 657p. illus. $19.50.

Reviews current knowledge in reproductive physiology as it pertains to domestic livestock and presents a great deal of essential background information. Useful for collateral reading and as a reference source.

A734 **Cook, Glen C.,** and **Elwood M. Juergenson.** Approved practices in swine production. 4th ed. 1962. Interstate. 329p. illus. $6.25; text ed., $4.75.

Consists of condensed yet thorough explanations of the major practices in swine production and management. Special features include a glossary of swine terms and a list of sources of useful information for swine producers.

A735 **Cooper, Malcolm M.,** and **Robert J. Thomas.** Profitable sheep farming. 1966. Farming Pr. 160p. illus. $4.20.

A practical book bringing together important facets of sheep husbandry. Covers selection, management, and disease control. Emphasizes the need to apply scientific knowledge to sheep farming. Will appeal to all concerned with sheep production.

A736 **Cunha, Tony J., Marvin Koger,** and **A. C. Warnick.** Crossbreeding beef cattle. 1963. Florida. 228p. illus. $8.50.

Consists of reports presented at the Beef Cattle Short Course on Crossbreeding held at the University of Florida in 1961. Provides a broad survey of crossbreeding with discussions of advantages and descriptions of particular breeding systems.

A737 ———, **A. C. Warnick,** and **Marvin Koger,** eds. Factors affecting calf crop. 1967. Florida. 376p. illus. $12.50.

A source manual and practical guide to factors affecting the calf crop in beef and dairy herds. Gives recommendations on steps to increase the calf crop and illustrates cattle raising practices in various parts of the world. Of value to both cattlemen and students.

A738* **Davis, Richard F.** Modern dairy cattle management. 1962. Prentice-Hall. 264p. illus. $7.95.

Presents the basic aspects of the dairy industry and the approved methods of dairy cattle management. Gives broad coverage to modern dairy methods and underlying principles, physiology of livestock, and milk production economics. A good source of reference for the dairyman and the student concerned with the dairy industry.

A739* **Diggins, Ronald V.,** and **Clarence E. Bundy.** Dairy production. 2d ed. 1961. Prentice-Hall. 341p. illus. $8.44.

A well-organized text containing basic information essential for practical dairy production. Illustrations are used to clarify the various aspects presented. Those who want more detailed background information will find the references useful. Includes career information.

A740* ———, and ———. Sheep production. 1958. Prentice-Hall. 369p. illus. $8.44.

Authoritative book on production and marketing of sheep written by experienced educators in agriculture. Topics treated include classes and breeds of sheep, selecting and establishing breeding flocks, selecting feeder lambs, feeds and feeding, housing and equipment, disease control, marketing of sheep and wool, and showing sheep.

A741* **Dukes, Henry H.** The physiology of domestic animals. 7th ed. 1955. Cornell. 1020p. illus. $19.50.

A comprehensive book covering all aspects of physiology as it pertains to domestic animals. Since the first edition appeared in 1933, it has become a must in all agricultural and veterinary colleges and has gained widespread acceptance by those active in the field. Considered one of the very best books available on the subject.

A742* **Ensminger, M. Eugene.** Animal science. 6th ed. 1969. Interstate. 1253p. illus. $19.75; text ed., $14.95.

A comprehensive study of the livestock industry designed to provide the most important features of a successful livestock program. Complete and technical enough for use as a text by students and practical enough for use as a ready reference by livestock farmers. Also discusses automation, environmental control, new breeds, agribusiness, and other new developments.

Animal Sciences

A743* ———. Beef cattle science. 4th ed. 1968. Interstate. 1020p. illus. $14.35; text ed., $10.75.

A modern text loaded with straight information on all areas of beef cattle production from breeding to feeding and management. Reflects recent technological advances and improved practices. An excellent text and reference for secondary, post-secondary and college classes studying practical scientific methods of beef cattle production.

A744* ———. Horses and horsemanship. 4th ed. 1969. Interstate. 907p. illus. $14.35; text ed., $10.75.

A valuable reference for everyone concerned with the care and management of horses. Covers a wide variety of topics such as history, anatomy, feeding, breeding, health, disease, judging, etc.

A745* ———. Sheep and wool science. 4th ed. 1970. Interstate. 948p. illus. $14.35; text ed., $10.75.

A comprehensive text covering all phases of sheep and wool production in a simple straightforward manner. Has outstanding chapter on feeding sheep and a very good discussion of diseases and parasites of sheep with emphasis on control and prevention. An excellent source for high school and technical school students needing information on the various aspects of sheep raising and wool production.

A746* ———. The stockman's handbook. 4th ed. 1970. Interstate. 957p. illus. $19.75.

An encyclopedia of basic information relative to beef, swine, sheep, and horses, with most emphasis in the area of beef production. Covers nearly 2,000 items from breeding to law. Most of information presented in tables and outlines for easy use. Excellent sourcebook for stockmen, farm managers, and students.

A747* ———. Swine science. 4th ed. 1970. Interstate. 881p. illus. $14.35; text ed., $10.75.

An up-to-date, basic text on swine husbandry and the hog industry in general. Deals with feeding, breeding, artificial insemination, confinement rearing, disease-free hogs, manure conservation, automation, marketing, and integration. Recommended as a text and reference for students and others interested in the swine industry.

A748 Farris, Edmond J., ed. The care and breeding of laboratory animals. 1950. Wiley. 515p. illus. $16.95.

A manual on the care and breeding of rats, mice, cats, dogs, fowl, fishes, and other species usually maintained for research purposes. Management, feeding, breeding, prevention of disease, and to a limited extent, treatment of disease, is discussed for each species. Intended for breeders, dealers, technicians, research workers, and others involved with laboratory animal care and breeding.

A749 Foust, Harry L., and Robert Getty. Anatomy of domestic animals. 3d ed. 1954. Iowa State. 105p. illus. paper, $8.75.

An atlas and dissection guide with brief but adequate directions given of method of procedure to follow. Includes instructions for dissecting and studying the horse, cow, pig, chicken, and dog. The illustrations are of excellent quality. Highly useful to practitioners as well as to students.

A750 Fowler, Stewart H. Beef production in the South. 1969. Interstate. 858p. illus. $13.75; text ed., $10.25.

Both a textbook for agricultural students and a reference book for southern cattlemen. Covers problems and procedures peculiar to the southern beef industry although not limited to that area in usefulness. Examples of subjects covered include historical development, breeds and breeding, and herd improvement.

A751* Frandson, R. D. Anatomy and physiology of farm animals. 1965. Lea & Febiger. 501p. illus. $12.50.

A well-planned and amply illustrated text describing the basics of anatomy and physiology common to all animals and those pertaining to species differences. Technical terms are clearly defined. Gives attention to cows, horses, swine, sheep, and dogs. Discusses the endocrine system, the digestive and reproductive systems, the mammary glands, and milk secretion. Written primarily for use in animal agriculture courses at the college level but also valuable to post-high school technical agriculture and young adult farmer classes.

A752 Fraser, Allan, and John T. Stamp. Sheep husbandry and diseases. 5th ed. 1968. Crosby Lockwood. 438p. illus. $6.

Contains most of the information required by the sheep farmer or student of sheep husbandry. Presentation is good and illustrations are clear and well-selected. Part 1, on husbandry, discusses breeds, management, reproduction, and nutrition. Part 2, on diseases, gives a clear and concise account of sheep diseases. Very few technical terms are used.

A753 Goodall, Daphne M. Horses of the world. 1965. Macmillan. 272p. illus. $10.95.

An illustrated survey of all breeds of horses and ponies known throughout the world. Contains more than 320 photographs, some in color, showing each breed in its native habitat. A good book for horse lovers.

A754 Greeley, R. Gordon. The art and science of horseshoeing. 1970. Lippincott. 176p. illus. $10.95.

A reliable and orderly account of horseshoeing. Begins with fundamentals and proceeds through proper fitting procedures to more advanced concepts involving foot pathology. Includes a general discussion on forge and anvil use and a guide to purchasing tools.

A755 Hafez, E. S. E. The behavior of domestic animals. 2d ed. 1969. Williams & Wilkins. 647p. illus. $22.50.

A collection of basic facts on animal behavior contributed by the editor and like-minded colleagues. Begins with the history and evolution of behavior, follows with explanations of behavior fundamentals, and finishes with discussions of the behavior of specific domestic animals and birds.

A756* ———, ed. Adaptation of domestic animals. 1968. Lea & Febiger. 417p. illus. $17.50.

A comprehensive text presenting material in a logical arrangement for livestock workers, farm building designers, and students in animal science, veterinary medicine, and environmental biology. Discusses all significant facets of environmental physiology as it relates to domestic livestock, poultry, and laboratory animals.

A757* ———, ed. Reproduction in farm animals. 2d ed. 1968. Lea & Febiger. 440p. illus. $15.

An up-to-date textbook covering both theoretical and practical aspects of animal reproduction. Nineteen world authorities present basic information on anatomy, physiology, genetics, etc., as they apply to reproduction in horses, cattle, sheep, swine, and chickens. A valuable text and reference for students of veterinary medicine and animal husbandry.

A758 Hagedoorn, Arend L. Animal breeding. ed. and annotated by Allen Frasser. 6th ed. 1962. Crosby Lockwood. 371p. illus. $3.

Not a textbook but an informal presentation of genetical theory and the role genetics plays in livestock improvement. Written for the ordinary livestock breeder and the beginning student in animal breeding.

A759 Hammond, John, and others. Farm animals: their growth, breeding, and inheritance. 4th ed. 1971. St. Martin's. 320p. illus. $17.50.

Outlines the practical application of scientific knowledge in livestock production for the student and the animal breeder. Deals with fertility and growth in farm animals, practical problems of inheritance in farm animal breeding, and problems in applying scientific techniques of selection and breeding.

A760 Harrison, James C., and others. Care and training of the trotter and pacer. 1968. U.S. Trotting Assn. 1054p. illus. $5.

One of the few practical handbooks available on the care, training, driving, shoeing, rigging, and balancing of the trotter and pacer. Written by leading trainers and drivers. Contains some general information also suitable for other breeds of horses.

A761* Hayes, Matthew. Points of the horse. 7th ed. 1969. Arco. 541p. illus. $12.50.

The subtitle, *A treatise on the conformation, movements, breeds, and evolution of the horse,* defines the book's scope. Has served as a standard work of reference for nearly 70 years. The author's thorough treatment of the anatomy of the horse is of outstanding value.

A762 ———. Stable management and exercise. 6th ed. rev. 1969. Arco. 369p. illus. $10.

Presents important aspects of stable management and exercise of the horse. Specifically written for the horse owner and student by a leading authority on all matters pertaining to horses.

A763 Henderson, Alexander E. Growing better wool. 1968. Reed. 108p. illus. $6.04.

A clear explanation of desirable wool characteristics and of different fiber problems, their causes, and results. Written primarily for New Zealanders and Australians but book's value not limited to these geographical areas.

A764 Hungate, Robert E. The rumen and its microbes. 1966. Academic. 533p. illus. $19.75.

A comprehensive account of rumen ecology for students and experienced workers in the animal sciences. Traces the developmental history of ruminants, discusses the rumen and its microbes, and explains the application of rumen ecology in agriculture.

A765* **Johansson, Ivar,** and **Jan Rendel.** Genetics and animal breeding. 1968. Freeman. 489p. illus. $17.50.

A general survey of the genetics of farm animals and their breeding for herd improvement. Provides methods of determining individual breeding values and clearly states the economic significance of selective breeding. A good introduction to animal breeding for agricultural and veterinary science students.

A766* **Judkins, Henry F.,** and **Harry A. Keener.** Milk production and processing. 1960. Wiley. 452p. illus. $10.95.

A thorough and specific treatment of milk production and processing. Deals with dairy farming essentials, dairy animal breeding and management, dairy operation processes, milk properties, milk testing, marketing of milk, and production of processed dairy products. Written for beginning college and advanced high school students.

A767* **Juergenson, Elwood M.** Approved practices in beef cattle production. 3d ed. 1964. Interstate. 353p. illus. $6.25; text ed., $4.75.

A comprehensive presentation of accepted practices in beef cattle production. Coverage ranges from herd management and feeding to disease control. Carefully written, in simple terminology, for both the agricultural student and the working beef producer.

A768* ———. Approved practices in sheep production. 2d ed. 1963. Interstate. 360p. illus. $6.25; text ed., $4.75.

A detailed examination of approved practices in sheep farming. Much of this information can readily be adapted and used in any geographic area. Selecting and breeding, feeding, shelter and equipment, disease control, and marketing are some of the subjects treated. A useful book for beginners in the sheep industry, students, and others interested in successful sheep production.

A769 **Kays, Donald J.** The horse. rev. ed. rev. by John M. Kays. 1969. Barnes. 439p. illus. $15.

An informative and practical book on horse production. Discusses breeding, feeding, management, judging, and marketing. Intended as a text for horse production courses at the college level but also useful as a reference for high school.

A770 **Kays, John M.** Basic animal husbandry. 1958. Prentice-Hall. 430p. illus. $10.

A comprehensive study of breeding, feeding, management, marketing, commercial uses, and other aspects of basic importance to the animal industry. Focuses attention on the most important domestic animals and stresses facets of production and marketing rather than show-ring characteristics.

A771 **Koehler, William R.** The Koehler method of guard dog training. 1967. Howell. 208p. illus. $7.95.

A simple, easy-to-understand guide to the selection, training, and maintenance of dogs for home protection, industrial security, police work, and military use. Written by an experienced dog trainer.

A772* **Krider, Jake,** and **William E. Carroll.** Swine production. 4th ed. 1970. McGraw-Hill. 528p. illus. $13.50.

Incorporates the latest research findings in this complete discussion of the swine enterprise. Considers breeding and reproduction, nutrition and feeding, housing and management, and marketing of feeder pigs, commercial hogs, and purebred swine. Designed for use in college swine production courses and for use by practicing hog farmers.

A773 **Lasley, John F.** Genetics of livestock improvement. 1963. Prentice-Hall. 342p. illus. $9.95.

Presents the principles of genetics and their role in livestock improvement through breeding. Covers selection, line breeding, crossbreeding, inbreeding, and outbreeding. Style of writing and organization of material makes the book suitable for both students and practical animal breeders.

A774 **Leonard, Albert H.** Modern mink management. 1966. Ralston Purina. 206p. illus. $4.

Not a Ralston Purina advertisement but an authoritative book written by the company's former manager of the mink department. Written in popular style using simple language and cartoon-like illustrations. Covers breeding, nutrition, health, and management of mink. Excellent book for the beginning mink rancher.

A775 **Lerner, I. Michael,** and **Hugh P. Donald.** Modern developments in animal breeding. 1966. Academic. 249p. illus. $10.50.

The authors, using a minimum of technical language, discuss animal breeding methods as they fit into modern agriculture. Artificial insemination, importation of new livestock, breed associations, and the attitudes of men to animals are examples of the subjects covered. A good reference for students concerned with animal breeding.

A776* **Lungwitz, Anton.** A textbook of horseshoeing for horseshoers and veterinarians. tr. by John W. Adams. 11th ed. 1966. Oregon State. 216p. illus. $6.

A scholarly approach to horseshoeing written by an outstanding authority in the field. Contains a thorough, detailed account of the anatomy of hooves and legs, explains the care of hooves, and illustrates how to prevent weaknesses and overcome defects. Although first published in 1884, still one of the best manuals on the subject.

A777 **McCoy, Joseph J.** The complete book of cat health and care. 1968. Putnam. 237p. illus. $5.95.

A practical, nontechnical guide for the cat owner or the worker in a pet home. Gives simple coverage to evolution and history of the cat, cat care, cat health, reproduction, and old-age care. Has detailed material on cause, symptoms, and treatment of various cat diseases but not intended to take the place of the veterinarian.

A778 **Mackenzie, David.** Goat husbandry. 3d ed. 1970. Transatlantic. 368p. illus. $15.75.

A standard work on the care and breeding of goats. Although primarily devoted to milk goat husbandry for British farmers, the book contains information of value to both the beginning and the experienced goat raiser.

A779 **Manwill, Marion C.** How to shoe a horse. 1968. Barnes. 109p. illus. $4.50.

Essentially a how-to book with many illustrations depicting the correct method of shoeing a horse. Written for use as a text in horseshoeing classes at the technical college level.

A780 **Miller, Malcolm E.,** and others. Anatomy of the dog. 1964. Saunders. 941p. illus. $20.

Contains a detailed treatment of each of the dog's organ systems with primary attention paid to the nervous system. Has outstanding illustrations. An extensive bibliography follows each chapter. An advanced book useful for anyone concerned with the anatomy of the dog.

A781 **Mosesson, Gloria R.,** and **Sheldon Scher.** Breeding laboratory animals. 1968. Sterling. 128p. illus. $5.95.

A practical manual for commercial breeders and an information source for students. Begins with a discussion of the use of animals in modern research. Follows with a chapter devoted to the breeding and care of each of the major laboratory animals. Special features include a glossary and tables of weights and measures.

A782 **Naether, Carl A.** The book of the domestic rabbit. 1967. McKay. 128p. illus. $4.95.

An elementary presentation of practical facts and theories concerning the selection, feeding, management, and marketing of domestic rabbits. Discusses and illustrates most of the common breeds and varieties.

A783* **Neumann, Alvin L.,** and **Roscoe R. Snapp.** Beef cattle. 6th ed. 1969. Wiley. 767p. illus. $12.95.

An up-to-date treatment of all phases of beef cattle production with emphasis on the financial risks and profit potentials of the various programs. Selection, breeding, feeding, reproduction, and marketing are examples of the aspects covered. Authors use scientific data to support the practices and procedures they recommend.

A784* **Nordby, Julius E., William M. Beeson,** and **David L. Fourt.** Livestock judging handbook. 9th ed. 1962. Interstate. 392p. illus. $7.75; text ed., $5.95.

A standard in the field. Author uses a practical, direct style of writing to present the subject of livestock judging and makes ample use of illustrations to clarify the principles presented. An important feature of the book is the extensive list of descriptive terms for each kind of livestock.

A785 **Nye, Nelson C.** The complete book of the quarter horse. 1964. Barnes. 471p. illus. $10.

A guide for breeders and owners and a reference for turfmen. Traces blood lines of individual, famous quarter horses, gives performance data, presents accepted breeding practices, describes uses of the quarter horse, etc.

A786* **Oppenheimer, Harold L.** Cowboy arithmetic. 1964. Interstate. 165p. illus. $6.95.

Not an arithmetic book or an accounting manual. The author, an investment counselor, gives practical advice on the economics of the cattle enterprise. Designed for use in any part of the United States where beef cattle are produced on a commercial scale. An excellent book for the beginner and for a text in a livestock production and marketing class.

A787 _____. Cowboy economics. 1966. Interstate. 285p. illus. $6.95.

A practical survey of the entire cattle industry paying particular attention to its transition into big business and the problems associated with this development. Written by the head of Oppenheimer Industries, Inc., an investment company specializing in the cattle business.

A788* _____, and **James D. Keast.** Cowboy litigation: cattle and the income tax. 1968. Interstate. 562p. illus. $8.95; text ed., $7.95.

A practical handbook covering the highlights of the legal and tax aspects of the livestock business. Covers agency and operational contracts, transactions in cattle and land, income taxes, and a summary of tax cases in the agricultural taxation field. Not intended as a substitute for competent legal counsel.

A789* **Patten, Bradley M.** Embryology of the pig. 3d ed. 1948. McGraw-Hill. 352p. illus. $7.50.

An introductory embryology text setting forth in simple terms the fundamentals of mammalian development. The author uses pig embryos to tell his story but stresses embryological phenomena involved rather than specific conditions existing in swine. An older book still valuable for reference use.

A790* **Perry, Enos J.,** ed. Artificial insemination of farm animals. 4th ed. rev. 1968. Rutgers. 473p. illus. $10.

An internationally known source of information on all aspects of artificial insemination in horses, cattle, sheep, goats, swine, dogs, poultry, and bees. Written in nontechnical language with an ample number of illustrations used to clarify the subject matter for students.

A791 **Porter, Arthur R., John A. Sims,** and **Charles F. Foreman.** Dairy cattle in American agriculture. 1965. Iowa State. 328p. illus. $5.50.

Authors use a practical and easy-to-understand style of writing in this dairy cattle performance text. Covers the history and development of modern dairy breeds, breeding programs, pedigree analysis, herd management, etc.

A792 **Porter, George,** and **William Lane-Petter,** eds. Notes for breeders of common laboratory animals. 1962. Academic. 208p. illus. $7.

Practical information on laboratory animal breeding presented in nontechnical language by knowledgable people in the field. Covers the breeding and keeping of all of the most frequently used laboratory animals.

A793 **Reaves, Paul M.,** and **Harry O. Henderson.** Dairy cattle feeding and management. 5th ed. 1963. Wiley. 448p. illus. $11.50.

A practical book examining significant aspects of dairy herd feeding, management, and development. Selection, breeding, genetics of milk production, nutrition, housing, and milking systems are examples of the subjects discussed. A useful reference for technicians, dairymen, and agribusinessmen who serve dairy farmers.

A794 _____, and **Calvin W. Pegram.** Southern dairy farming. 1956. Interstate. 598p. illus. $4.75.

A good textbook designed specifically for use in teaching the dairy business in the South. Covers herd management, milk production, and marketing.

A795* **Rice, Victor A.,** and others. Breeding and improvement of farm animals. 6th ed. 1967. McGraw-Hill. 362p. illus. $12.50.

A well-known and widely used textbook covering the entire field of animal breeding. Presents the basic principles of animal improvement and applies these principles to the major farm animals. Provides an excellent foundation in the modern methods of breeding for improvement of both type and productive performance of farm animals. Each chapter followed by a list of references.

A796 **Rouse, John E.** World cattle. 1970. Oklahoma. 2v. illus. $25.

A worthwhile reference on the breeds of cattle found in each of 85 major countries. Arrangement is by country with an index for breeds. Covers cattle breeds, management practices, cattle diseases, the relation of the government to the industry, and the economic outlook for the cattle business. The material is presented in an interesting manner that should appeal to the general reader.

A797 **Sainsbury, David.** Animal health and housing. 1967. Williams & Wilkins. 329p. illus. $10.75.

A clearly written account of animal health as affected by the housing environment. Deals concisely with site and materials, insulation and ventilation, and the different types of animal housing or systems. Dairy cattle needs are stressed with lesser attention given to beef cattle, sheep, pigs, and poultry. Recommended for agriculturists, veterinary students, and animal science students.

A798* **Salisbury, Glenn W.**, and **Noland L. Van Demark.** Physiology and reproduction and artificial insemination of cattle. 1961. Freeman. 639p. illus. $15.

A valuable reference source for students, technicians, cattlemen, veterinarians, and researchers. Reviews reproductive anatomy and physiology, explains the collection and processing of semen, describes artificial insemination techniques, and discusses reproduction problems and management.

A799 **Scott, Thomas C.** Obedience and security training for dogs. 1969. Arco. 171p. illus. $4.95.

A former dog trainer presents significant aspects of obedience and security training and the dog/handler relationship. Does not deal with only 1 specific breed of dog. An appendix contains the American Kennel Club Obedience Regulations, effective January 1, 1969.

A800 **Self, Margaret C.** The horseman's encyclopedia. 2d ed. 1963. Barnes. 428p. illus. $9.75.

An encyclopedia of facts, terms, definitions, and general information pertaining to horses. Does not cover treatment of diseases although does discuss some problems such as lameness or lacerations. Written in popular style for the general reader and not intended as a guide to breeding, feeding, and management.

A801* **Short, Douglas J.**, and **Dorothy P. Woodnott**, eds. The I.A.T. manual of laboratory animal practice and techniques. 2d ed. 1969. Thomas. 462p. illus. $14.

This Institute of Animal Technicians manual is intended as a textbook of laboratory animal husbandry for beginning students. Emphasizes practical aspects and provides the information needed for general management of an animal house. General topics include the animal house, its equipment, and animal care. Also discusses specific subjects such as the sexing of young animals and the production and use of germ-free animals.

A802 **Silvan, James C.** Raising laboratory animals: a handbook for biological and behavioral research. 1966. Nat. Hist. Pr. 225p. illus. $4.95; paper, $1.45.

Introduces the beginning student to laboratory animal breeding and care. Covers the whole range of animals from amoebas to opposums but omits rabbits. Provides accurate and documented information that will enable students to obtain and raise the various species described. An appendix lists supply sources for animals, cages, feeds, and biological materials.

A803 **Simpson, George G.** Horses, the story of the horse family in the modern world and through sixty million years of history. 1951. Oxford. 247p. illus. $9.

A detailed account of the evolutionary development of the horse. Includes a discussion of the general patterns and causes of evolution. Written on the beginning college level.

A804 **Smith, Vearl R.** Physiology of lactation. 5th ed. 1959. Iowa State. 291p. illus. $6.50.

Well-known book covering all phases of lactation and the factors involved. Deals basically with the cow but other animals are discussed when necessary to illustrate a point. A useful reference source for college students in dairy science.

A805 **Smythe, Reginald H.** The horse: structure and movement. 1967. Allen. 159p. illus. $4.95.

A general review of the anatomy of the horse. Discusses bones and joints, skin and surface contour of the body, and the functional relationship of certain parts of the horse in motion and at rest. Includes an extensive bibliography. A useful book for breeders, exhibitors, judges, and veterinary aids.

A806 **Spedding, C. R. W.** Sheep production and grazing management. 2d ed. 1965. Williams & Wilkins. 446p. illus. $15.25.

A timely and authoritative study of the agricultural ecology of the sheep. Covers husbandry, nutrition, pasture problems, health, disease and parasite problems, environmental effects on growth and productivity, and the biology of sheep production. Written for both agricultural and veterinary students.

A807 **Stoddart, Laurence A.**, and **Arthur D. Smith.** Range management. 2d ed. 1955. McGraw-Hill. 433p. illus. $10.50.

Reviews the history of range management, discusses basic principles, and explains the practical application of these principles to management problems. An older book but still useful.

Animal Sciences

A808 Swidler, David T. All about thorobred horse racing. 1967. Hialeah. 189p. illus. $10.

A book of general information concerning thoroughbred racing and the industry involved. Covers selection, training, equipment, supplies, racing rules, insurance, taxes, etc. Directed primarily at the amateur, but contains material of value to the more experienced.

A809 Templeton, George S. Domestic rabbit production. 4th ed. 1968. Interstate. 213p. illus. $6.25; text ed., $4.75.

A guide to nearly all aspects of rabbit production. Deals with such topics as selection, breeding, feeding, meat and wool production, housing, equipment, and marketing. A large number of illustrations are used to help clarify points difficult to get across by the written word.

A810 Trimberger, George W. Dairy cattle judging techniques. 1958. Prentice-Hall. 304p. illus. $9.95.

A comprehensive discussion of dairy cattle judging. Contains information for everyone from the beginner to the experienced judge. Contains separate sections on judging dairy bulls and dairy heifers.

A811 Von Bergen, Werner, ed. Wool handbook. 3d ed. 1963. Wiley. 2v. in 3. illus. v.1 $25; v.2, pt.1 $34.95; v.2, pt.2 $34.95.

A text and reference useful to the entire wool industry. Volume 1 deals with sheep raising and raw wool. Volume 2, part 1 covers the manufacturing processes of wool. Volume 2, part 2 discusses the basic processes in dyeing, finishing, and marketing.

A812 Wagnon, Kenneth A., Reuben Albaugh, and George H. Hart. Beef cattle production. 1960. Macmillan. 537p. illus. $9.95.

Covers the entire field of beef cattle production including breeding, feeding, management, and marketing. Emphasis is on western aspects although the book's usefulness is not limited to western areas. Important data are documented and both sides of controversial subjects are presented.

A813 Way, Robert F., and Donald G. Lee. The anatomy of the horse. 1965. Lippincott. 214p. illus. $12.50.

A pictorial approach is used to acquaint the reader with all facets of equine anatomy. Specifically intended for the horseman, the veterinary student, and the artist as an aid in visualizing the structure and function of the living horse.

A814 Williams, Stephen, and C. David Edgar. Planned beef production. 1966. Lockwood. 189p. illus. $4.20.

A practical account of using modern science and technology to develop the most profitable methods to suit the beef farmer's individual needs. Written by men experienced in both the technical and commercial branches of beef production.

A815 Wing, James M. Dairy cattle management: principles and applications. 1963. Van Nostrand. 394p. illus. $10.

Written especially for students of dairying and professional dairymen. A comprehensive treatment of feeding, breeding, lactation, health, reproduction, environmental effects, buildings, equipment, and business procedures. The 8 appendixes form a ready reference source of practical information.

A816 Winters, Laurence M. Animal breeding. 5th ed. 1954. Wiley. 420p. illus. $9.95.

A technical review of the principles of genetics and their application in livestock improvement programs. Primary attention is focused on inbreeding, crossbreeding, and selection.

A817 Wiseman, Robert F. The complete horseshoeing guide. 1968. Oklahoma. 238p. illus. $5.50.

A combination of basic principles and practical methods useful to both professional farriers and general horse owners. Briefly covers other topics such as tools, handling rough horses, diseases and injuries, metalwork, and special shoeing. Includes a short glossary of terms.

A818 Worden, A. N., and William Lane-Petter, eds. UFAW handbook on the care and management of laboratory animals. 3d ed. 1967. Williams & Wilkins. 1015p. illus. $23.

This Universities Federation for Animal Welfare handbook is a standard reference work not completely up-to-date because of the time spent collecting information from 10 different countries. Aimed principally at animal technicians and research workers unfamiliar with the various species of animals.

A819 Yapp, William W. Dairy cattle judging and selection. 1959. Wiley. 324p. illus. $7.95.

A college textbook covering the physical characteristics of various breeds of dairy cattle. Includes an assessment of type, a discussion of basic qualities, a summary of structure and appearance, and a review of the main dairy breeds.

A820 Yeates, Neil T. M. Modern aspects of animal production. 1965. Butterworth. 371p. illus. $17.50.

An unusual book basically devoted to beef and sheep production. Divided into 4 parts dealing with reproduction, influence of climate, meat, and wool. A practical text requiring some knowledge of biology and physiology.

A821* Youtz, H. G., and **A. C. Carlson.** Judging livestock, dairy cattle, poultry, and crops. 1970. Prentice-Hall. 195p. illus. $8.48.

Provides a sound approach to the judging of livestock, poultry, and crops. Author uses simple terminology to present both elementary and advanced information. Useful to students studying livestock, poultry, and plant production as well as to those learning to be judges.

A822* Zeuner, Frederick E. A history of domesticated animals. 1964. Harper. 560p. illus. $12.

A text and reference useful to both paleontologists and animal breeders. Gives general information about the origins of domestication and specific information on the uses of various domesticated animals throughout history. Covers mammals, birds, fishes, insects, reptiles, and amphibians.

Poultry Husbandry

A823* American standard of perfection. 5th ed. 1953, reprint 1966. Amer. Poultry Assn. 585p. illus. $8.50.

A detailed description of all recognized varieties of domesticated land and water fowl. Ample number of illustrations used to bring out points difficult to get across by the written word. This official book of standards is kept up-to-date with supplements.

A824 Bailey, Jackson W. Poultryman's manual: flock management and chicken diseases. 1957. Springer. 296p. illus. $4.95.

Flock management and the causes, symptoms, and treatment of health problems are discussed in simple terms for the poultry producer and the agricultural student.

A825 Biddle, George H., and **Elwood M. Juergenson.** Approved practices in poultry production. 3d ed. 1963. Interstate. 332p. illus. $6.25; text ed., $4.75.

Recommended poultry production practices are presented for the large-scale producer as well as the small operator or beginner. Topics covered include selection, feeding, housing, management, processing, and marketing.

A826 Bundy, Clarence E., and **Ronald V. Diggins.** Poultry production. 1960. Prentice-Hall. 370p. illus. $8.44.

A thorough treatment in simple language of the basic principles in the selection, production, and marketing of poultry and poultry products.

A827* Card, Leslie E., and **Malden C. Nesheim.** Poultry production. 10th ed. 1966. Lea & Febiger. 400p. illus. $8.50.

A comprehensive presentation of the basic principles underlying successful poultry and egg production. Covers all aspects from production to marketing including selection, breeding, feeding, nutrition, housing, equipment, disease control, and management. A valuable introduction to the poultry science field.

A828 Hale, Murray. The turkey stockman. 1964. Maclaren. 124p. illus. $4.50.

A small book packed with useful information. Covers all phases of turkey production including breeds, selection, housing, brooding, flock improvement, health, disease control, records, etc.

A829 Johnson, Alex A., and **William H. Payn.** Ornamental waterfowl: a guide to their care and breeding. 2d ed. rev. 1968. Witherby. 110p. illus. $4.20.

A British publication on the management, feeding, reproduction, and diseases of ducks, geese, and swans. Written in popular style with ample use of black and white illustrations and photographs. Animal health authorities gave their assistance in the section on ailments and diseases.

A830* Merck poultry serviceman's manual. 2d ed. 1967. Merck. 235p. paper, $1.45.

A practical manual for poultry servicemen, producers, and students. Describes poultry diseases, presents detailed instructions on diagnostic procedures and disease management, and gives tips on general management. Includes nutrition and husbandry charts and tables.

A831* Patten, Bradley M. Early embryology of the chick. 4th ed. 1957. McGraw-Hill. 244p. illus. $6.25.

A well-known embryology text designed primarily for the beginning student. Sets forth in brief and simple language the fundamentals of embryology. Presents development as illustrated by the chick because of its wide use as laboratory material in vertebrate embryology courses.

A832 Rice, James E., and Harold E. Botsford. Practical poultry management. 6th ed. 1956. Wiley. 449p. illus. $7.95.

Designed as a practical guide for students, poultrymen, and those involved in farm service businesses. Covers all phases of poultry production. An older book but still useful for reference.

Feeds and Feeding

A833 Abrams, John T. Animal nutrition and veterinary dietetics. 4th ed. 1961. Williams & Wilkins. 826p. illus. $16.

Brings together significant information on animal nutrition. Discusses the history of nutrition, characteristics of feedstuffs, environmental influences, malnutrition in domestic species, general principles for constructing rations, etc.

A834 Blaxter, Kenneth. The energy metabolism of ruminants. 2d ed. rev. 1967. Thomas. 329p. $14.50.

A brief, clear description of the fundamental principles of metabolism and the methods and techniques of meeting energy needs of ruminant livestock. An outstanding reference for those concerned with the nutrition of ruminants.

A835* Cassard, Daniel W., and Elwood M. Juergenson. Approved practices in feeds and feeding. 3d ed. 1963. Interstate. 362p. illus. $6.25; text ed., $4.75.

A ready reference book furnishing a comprehensive list of recommended practices. An explanation of how these approved procedures should be conducted is presented in practical terminology. Not only tells what to do but also what not to do. Should be of value in nearly any area with a livestock feeding problem.

A836* Crampton, Earle W., and Lorin E. Harris. Applied animal nutrition. 2d ed. 1969. Freeman. 753p. illus. $12.

A standard text and reference covering the theory of animal nutrition and its application to livestock feeding. Includes definitions of feedstuff terms and expressions, animal nutritional requirements, feed classification and ration formulation, and appendixes of essential tables. Excellent for use in post-high school programs related to animal nutrition and feeding.

A837 Cunha, Tony J. Swine feeding and nutrition. 1957. Wiley. 312p. illus. $9.95.

Provides abundant information on many aspects of swine feeding. Discusses the relative value of feeds, feeding pigs during various stages of growth, feeding the breeding herd, and swine management recommendations.

A838 Dent, John B., and Harold Casey. Linear programming and animal nutrition. 1967. Lippincott. 111p. illus. $7.50.

A nonmathematical approach to the techniques of linear programming and its practical use for least-cost ration formulation. Intended for those concerned with feeding farm animals.

A839 Feed additive compendium. 8th ed. 1970. Miller. 427p. $35. (includes 1-year subscription to updating service)

The feed additive encyclopedia for those who manufacture feed, mix feed, feed livestock, etc. A systematic arrangement of nearly all important facts concerning feed additives, including their uses, sources, and governmental restrictions. Kept up-to-date by monthly supplements.

A840 Halnan, Edward T., and Frank H. Garner. Principles and practice of feeding farm animals. 5th ed. rev. and enl. by Alfred Eden. 1966. Estates Gazette. 382p. illus. $9.12.

A British manual designed to provide practical information for livestock farmers. Contains only elementary facts concerning the physiology of digestion and metabolism. Covers dairy cattle, beef cattle, sheep, horses, pigs, and poultry.

A841 Jennings, Joseph B. Feeding, digestion, and assimilation in animals. 1965. Pergamon. 228p. illus. $6.50; paper, $4.95.

A general introduction to the study of animal nutrition using a zoological rather than a biochemical approach. Contents include essential components of the diet, animal feeding mechanisms, alimentary systems, digestion in general, and digestion in selected animal types.

A842* Maynard, Leonard A., and John K. Loosli. Animal nutrition. 6th ed. 1969. McGraw-Hill. 613p. illus. $12.50.

A scholarly presentation of technical information indispensable for all students studying animal nutrition beyond the most elementary level. Presents the basic principles of nutrition and their application in feeding practices. Specific topics discussed include nutritional processes, nutritional requirements of animals, measurement of body needs, etc.

A843* **Morrison, Frank.** Feeds and feeding. 22d ed. 1956. Morrison. 1165p. illus. $15.50; 9th abr. ed., $4.75.

An internationally known book. Considered an essential reference by everyone concerned with feeds and feeding. Consists of an account of nutritional fundamentals, an analysis of important feedstuffs, and a discussion of the feeding, care, and management of farm animals.

A844 **Nelson, Robert H.** An introduction to feeding farm livestock. 1964. Pergamon. 106p. paper, $5.

An introductory feeds and feeding text designed specifically for the study of agriculture on the vocational/technical level. Written for the British student but contains information of value to the American student.

A845* **Perry, Tilden W.** Feed formulations handbook. 1965. Interstate. 233p. illus. $6; text ed., $5.50.

A manual for small feed-mill operators, farmers, and stockmen who mix their own feed and students being trained for technical level jobs in feeds and feeding enterprises. Contains analyses of nutritional elements in feedstuffs, tables of nutrient requirements of farm livestock, instructions for regulating cost per pound of meat, etc. Includes a list of references for further study.

A846 **Schaible, Philip J.** Poultry: feeds and nutrition. 1970. AVI. 636p. illus. $22.50.

A comprehensive account of the principles of nutrition and their practical application in the poultry industry. Systematically presents every significant nutritional aspect of poultry feeds from the raw ingredients to the finished product.

A847 **Scott, Milton L., Malden C. Nesheim, and Robert J. Young.** Nutrition of the chicken. 1969. Scott. 511p. illus. $15.

Describes in detail the basic principles of nutrition and discusses the various known nutritional relationships and the scientific feeding of chickens on a commercial least-cost basis. Useful as a reference for the more advanced student.

A848 **Seiden, Rudolph, and W. H. Pfander.** The handbook of feed stuffs: production, formulation, medication. 1957. Springer. 591p. illus. $11.

A compilation of facts and figures on the grains, the minerals, and the plant and animal by-product feeds. For each listed gives information on production, utilization by farm animals, etc. An older ready reference for workers in the feedstuffs industry.

A849 **Simmons, Norman O.** Feed milling and associated subjects. 2d ed. 1963. Hill. 377p. illus. $11.40.

This British publication is the second edition of *Compound milling,* a standard reference book for many years. Gives a thorough account of the raw materials, the procedures, and the equipment used in the manufacture of mixed feeds.

A850* **Titus, Harry W.** The scientific feeding of chickens. 4th ed. 1961. Interstate. 297p. illus. $6.25; text ed., $4.75.

A concise but thorough presentation of the scientific feeding of poultry. Author makes full use of tables, graphs, and formulas to clarify his discussions of poultry nutrition essentials. Of practical value to the poultryman and feed manufacturer as well as the student.

A851 **Underwood, Eric J.** The mineral nutrition of livestock. 1966. Commonwealth Agricultural Bureaux (order from the British Information Services, 845 Third Ave., New York, N.Y. 10022). 237p. illus. $5.

A valuable collection of information by one of the foremost authorities in the field. Intended for students, teachers, progressive farmers, and research workers. Covers the causes, effects, diagnosis, and prevention of mineral excesses or deficiencies in the diet of various classes of livestock.

Veterinary Medicine

A852 **Adams, Ora R.** Lameness in horses. 2d ed. 1966. Lea & Febiger. 563p. illus. $12.50.

A widely accepted book describing practically every type of lameness in horses. Discusses symptoms, diagnosis, and treatment of lameness. Also includes information on conformation, corrective trimming and shoeing, and natural and artificial gaits. A valuable reference for veterinary students, practicing veterinarians, and laymen.

A853 **Andrewes, Sir Christopher H., and H. G. Pereira.** Viruses of vertebrates. 2d ed. 1967. Williams & Wilkins. 237p. illus. $12.25.

Deals with RNA viruses, DNA viruses, and unclassified viruses. A good reference on the properties of viruses and a starting point for further study. Includes an extensive bibliography.

A854 **Bailey, Jackson W.** Veterinary handbook for cattlemen. 3d ed. 1963. Springer. 439p. illus. $7.50.

A practical guide and handbook for agricultural students, technicians, and cattlemen. Covers the important cattle diseases and breeding problems, many of them concerning management as much as medicine.

A855 **Barnes, Charles D.,** and **Lorne G. Eltherington.** Drug dosage in laboratory animals. 1964. California. 302p. illus. paper, $8.

An unusual reference manual covering drug action in laboratory animals. Classifies drug responses by toxicity data, primary use, and secondary use. Briefly treats factors that modify drug response. Not concerned with treatment of diseases in laboratory animals.

A856* **Biester, Harry E.,** and **Louis H. Schwarte,** eds. Diseases of poultry. 5th ed. 1965. Iowa State. 1382p. illus. $18.

The most complete, up-to-date reference on poultry diseases currently available. Thirty-seven authorities have contributed discussions on poultry diseases and pests and the effect of various infections on breeding, feeding, management, and other phases of poultry production. A valuable source of information for everyone concerned with prevention and control of poultry diseases.

A857 **Blood, Douglas C.,** and **James A. Henderson.** Veterinary medicine. 3d ed. 1968. Williams & Wilkins. 927p. illus. $20.

An encyclopedic text on large animal medicine, of primary value to the student. Gives worldwide coverage with emphasis on cattle diseases. Made up of 2 major sections, the first concerned with the broad aspects of theory and methodology and the second dealing with individual diseases.

A858 **Boddie, George F.** Diagnostic methods in veterinary medicine. 6th ed. 1969. Lippincott. 447p. illus. $9.75.

A thorough, well-organized compilation of veterinary diagnostic procedures. Includes descriptions of the normal animal, physical examination of various systems, proper selection of specimens for laboratory diagnosis, and the interpretation of laboratory test results. Of more interest to the technician and general practitioner than to the researcher.

A859 **Coles, Embert H.** Veterinary clinical pathology. 1967. Saunders. 455p. illus. $15.

Includes chapters on hematology, blood chemistry, organ function, body fluids, blood coagulation, mastitis diagnosis, toxicology, etc. Gives detailed test procedures and information on equipment and supplies needed for a basic pathology laboratory.

A860* **Dunne, Howard W.,** ed. Diseases of swine. 3d ed. 1970. Iowa State. 1150p. illus. $23.

An exhaustive review of current knowledge concerning swine anatomy and physiology, disease, nutrition, and management. Presented in an easy-to-read form with ample use of illustrations. Indexing is cross-referenced to facilitate use. This valuable work represents the combined efforts of 56 authorities in the field.

A861 **Dykstra, Ralph.** Animal sanitation and disease control. 6th ed. 1961. Interstate. 858p. illus. $10; text ed., $8.

This manual, devoted to animal hygiene and disease control, is written specifically for animal husbandry students and livestock farmers. Covers such topics as animal health, disease, methods of disease control, decontamination, use of insecticides, etc.

A862* **Fox, M. W.,** ed. Abnormal behavior in animals. 1968. Saunders. 563p. illus. $19.50.

A compendium of current knowledge on the types, possible origins, and control of abnormal behavior in a wide variety of animal species. Includes information on farm animals, wild animals, fish, fowl, laboratory animals, and household pets. Valuable for veterinary science students, animal husbandry students, and those training for careers in zoo management.

A863 **Garner, Reuben J.** Veterinary toxicology. 3rd ed. ed. by Eustace G. C. Clarke and Myra L. Clarke. 1967. Williams & Wilkins. 477p. illus. $12.50.

An excellent text and ready reference on animal toxicology. Includes a general description of toxicology and a comprehensive treatment of toxic materials encountered by domestic livestock.

A864 **Gay, William I.,** ed. Methods of animal experimentation. 1965-68. Academic. 3v. illus. v.1 $14.50; v.2, $19.50; v.3 $9.50.

Discusses the use of animals in various fields of research, describes special techniques using laboratory animals, presents fundamental principles and methods of managing animals for experimental procedures, and reviews special animal care in relation to each experimental technique.

A865 Getty, Robert. Atlas for applied veterinary anatomy. 2d ed. 1964. Iowa State. 366p. illus. $7.50.

An atlas and laboratory guide intended for use with standard textbooks in veterinary medicine courses. Consists of photographs, sketches, and diagrams illustrating the anatomy of farm livestock and laboratory animals.

A866 Hagan, William A. Hagan's infectious diseases of domestic animals. 5th ed. ed. by Dorsey W. Bruner and James H. Gillespie. 1966. Cornell. 1105p. illus. $18.50.

An orderly presentation of all aspects of communicable diseases in animals. Comprehensively treats each infectious disease. A list of references presented following each disease making it easier to find additional information. Probably of more value to the teachers than the students in technical school.

A867* Hannah, Harold W., and Donald F. Storm. Law for the veterinarian and livestock owner. 2d ed. 1965. Interstate. 212p. illus. $7.50; text ed., $5.75.

A discussion of legal requirements concerning veterinary treatment of animals. Emphasis is on preventive rather than punitive aspects. Written for livestock producers and laymen who may have need of a veterinarian's assistance as well as for the veterinarian.

A868* Hayes, Matthew H. Veterinary notes for horse owners. 15th ed. ed. by John F. Tutt. 1964. Arco. 655p. illus. $12.50.

A horse owner's first-aid manual on the diagnosis and management of equine disorders. Intended for use while waiting for a veterinarian's assistance and in following his instructions. A good source of general information on horse diseases.

A869* Heath, J. S. Aids to veterinary nursing. 1970. Williams & Wilkins. 160p. illus. paper, $5.

A handbook for animal nursing technicians and trainees. A collection of useful information including descriptions of laboratory tests, a guide to tail docking, weights and measures, methods of determining strength of solutions, etc.

A870 Herbert, W. J. Veterinary immunology. 1970. Davis. 368p. illus. $15.

Systematically deals with fundamentals of immunity and immunological methods for diagnosis and prevention of disease. Should appeal to students of the veterinary and biological sciences.

A871 Herrick, John, and H. L. Self. Evaluation of fertility in the bull and the boar. 1962. Iowa State. 148p. illus. $6.

Designed to aid livestock breeders and technicians in selecting herd improvement sires. Discusses techniques for evaluating the reproductive potential of animals through observation and tests.

A872 Jensen, Rue, and Donald R. MacKey. Diseases of feedlot cattle. 1965. Lea & Febiger. 305p. illus. $13.50.

A compilation of disease syndromes grouped under virus, bacterial, fungal, protozoal, metazoan, parasitic, and miscellaneous causes. Documents specific cases and discusses both individual treatments and mass measures. Useful for cattle feeders, meat inspectors, students, and others.

A873* Jones, Bruce V., ed. Animal nursing. 1966. Pergamon. 2v. illus. $14; paper text ed., $12.

Designed as a text and reference for students training to become veterinary technologists or assistants. Basically devoted to animal care and nursing in a small animal hospital and concerned only with the assistant. Client relations and office practices are not considered. Specific subjects covered include feeding, hygiene, first aid, restraint, anesthesia, laboratory procedures, etc.

A874* Lapage, Geoffrey. Veterinary parasitology. 2d ed. 1968. Thomas. 1182p. illus. $27.50.

An encyclopedic work emphasizing the identification, life histories, and control of the common parasites of domestic animals. The material is well-organized with information on control and treatment following the description of each parasite. Useful as a general parasitological reference for nonspecialists.

A875 Merchant, Ival, and Ralph D. Barner. Outline of infectious diseases of domestic animals. 3d ed. 1964. Iowa State. 478p. illus. $7.95.

A comprehensive study of the important communicable diseases of domestic animals except those peculiar to small animals and poultry. A useful book for obtaining a quick review of a particular disease.

A876* Merck veterinary manual. 3d ed. ed. by O. H. Siegmund. 1967. Merck. 1674p. $11.25.

Provides a comprehensive source of current knowledge on the diagnosis, treatment, and prevention of the common animal diseases. Contains descriptions of diseases, metabolic disorders, and poisons. Coverage includes both large and small animals. An important reference work for all technical agriculture libraries.

A877 **Parker, William H.** Health and disease in farm animals. 1970. Pergamon. 301p. illus. $6.25.

A practical guide for those concerned with maintaining health in farm animals. Presents basic principles of disease prevention and discusses the important animal diseases. Written primarily for a British audience but much of the information is equally valuable to the U.S. reader.

A878 **Robinson, T. J.,** ed. Control of the ovarian cycle in the sheep. 1967. Sydney. 258p. illus. $9.95.

Concerned with increasing the efficiency of sheep breeding and sheep husbandry by regulation of heat and ovulation in the ewe. Not easily read because of the great amount of statistical details included. Of most value to advanced students and teachers.

A879* **Sarner, Harvey.** The business management of a small animal practice. 1967. Saunders. 248p. illus. $9.50.

The most comprehensive book available on modern business management problems. Although primarily intended for the professional veterinarian, it contains considerable information of value to the veterinary technologist. Covers such topics as accounting, bookkeeping, credit, and ethics.

A880* **Seiden, Rudolph.** Livestock health encyclopedia. 3d ed. ed. by W. James Gough. 1968. Springer. 628p. illus. $11.

An alphabetically arranged compendium of information on disease and parasite control in farm livestock. Uses simplified terminology for the benefit of nonspecialists. Content of each explanation based on up-to-date information from government publications and other authoritative sources.

A881 **Seneviratna, P.** Diseases of poultry (including cagebirds). 2d ed. 1969. Williams & Wilkins. 229p. illus. $11.50.

An account of the common poultry diseases, their causes, symptoms, diagnosis, treatment, and control. Briefly deals with allied subjects such as anatomy and physiology, poultry surgery, and poisons. Meets the needs of poultry breeders and students on the vocational/technical school level.

A882 **Soulsby, E. J. L.** Textbook of veterinary clinical parasitology. v.1 Helminths. 1965. Blackwell Scientific. 1120p. illus. $35.

Volume 1 is the only one published to date. It presents an account of pathological and clinical aspects of helminth infections in domestic animals. Intended for specialists and advanced students with a basic knowledge of helminth morphology.

A883 **Tavernor, W. D.,** ed. Nutrition and disease in experimental animals. 1970. Williams & Wilkins. 165p. illus. $7.95.

A collection of papers covering the care of laboratory animals and their nutritional and health requirements. Includes such topics as trace element deficiencies, vitamin requirements, formulation of diets, clinical approach to disease, and future trends in laboratory animal sciences.

A884 **Waterson, A. P.** Introduction to animal virology. 2d ed. 1968. Cambridge. 176p. illus. $5.50.

Provides the beginning student with a brief account of the major facts concerning viruses and virus diseases. Written in easy-to-read style and illustrated by electron micrographs of virus particles.

A885 **Westhues, Melchior,** and **Rudolph Fritsch.** Animal anaesthesia. 1965. Lippincott. 2v. illus. $18.

A comprehensive account of both local and general anesthesia in animals. Discusses principles of anesthesia, drugs used, techniques, and application to different animal species. Contains an extensive list of references for further reading.

Entomology

A886* **Anderson, Roger F.** Forest and shade tree entomology. 1960. Wiley. 428p. illus. $10.

A useful reference manual for forestry and entomology students. Contains a general discussion of the basic factors in insect life and a detailed treatment, including methods of control, of the more important forest and shade tree insects.

A887 **Andrewes, Sir Christopher H.** The lives of wasps and bees. 1969. Amer. Elsevier. 204p. illus. $5.75.

A discussion of the biology, behavior, and ecology of wasps and bees. Deals mainly with European and American species. The author, an entomologist, gives special emphasis to comparative ethology and its bearing on evolution.

A888* **Borror, Donald J.,** and **Dwight M. Delong.** An introduction to the study of insects. rev. ed. 1964. Holt. 819p. illus. $18.50.

Begins with a general discussion of insects and follows with a thorough treatment of each order. Also covers Arthropoda other than insects and gives detailed directions for collecting and preserving specimens. The excellent keys make it possible for beginning students to identify almost any insect. The complete coverage given each order makes it also useful to advanced students.

A889* ———, and **Richard E. White.** A field guide to the insects of America north of Mexico. 1970. Houghton Mifflin. 404p. illus. $5.95.

A comprehensive and up-to-date pocket guide to North American insects. Contains information on collecting and preserving insects, the study of living insects, their economic value, and other technical facts that beginners need to know. Describes nearly 600 families of insects plus other arthropods.

A890 **Brian, Michael V.** Social insect populations. 1966. Academic. 143p. illus. $6.

A succinct review of current knowledge on population studies of social insects. Covers ants, bees, termites, and wasps and includes such topics as reproduction, brood periodicity, competition, food supply, predators, and parasites.

A891 **Bursell, E.** An introduction to insect physiology. 1970. Academic. 276p. $10.

An easily read introductory work consisting of generalizations illustrated with up-to-date examples. Divided into 4 sections covering somatic physiology, neuromuscular sensory physiology, reproduction and development, and aspects of physiological ecology.

A892 **Chapman, R. F.** The insects: structure and function. 1969. Amer. Elsevier. 820p. illus. $13.75.

A comprehensive presentation of insect life. Relates the structure and function of the insect to its behavior under natural conditions. Divided into 6 sections: the head, ingestion and utilization of the food; the thorax and movement; the abdomen, reproduction and development; the cuticle, respiration and excretion; the nervous and sensory systems; the blood, hormones and pheromones.

A893 **Comstock, John H.,** and **Anna Comstock.** An introduction to entomology. 9th ed. 1940. Cornell. 1064p. illus. $17.50.

A classic in the field. Still useful to students of insects for the anatomical descriptions although the taxonomic sections may be out-dated. Much of the 1924 edition was reproduced without change in the 1940s.

A894 **Edwards, Clive A.,** and **Gordon W. Heath.** Principles of agricultural entomology. 1964. Thomas. 418p. illus. $16.

A British publication with material ranging from the academic to the practical. Organized in 3 parts: a condensed account of general principles; the descriptions, ecology, and control of pests; and, keys based on the type of damage done to each of 6 groups of crops.

A895 **Evans, Howard E.** Life on a little-known planet. 1968. Dutton. 318p. illus. $7.95.

A simple account of insects and their relatives which inhabit the soil, water, and air. Throughout his discussions the author stresses their ecological importance to man's future.

A896 **Gilmour, Darcy.** The metabolism of insects. 1965. Freeman. 195p. illus. paper, $2.50.

Presents the principles and theories essential for understanding insect metabolism. Contains a thorough treatment of energy metabolism, followed by carbohydrate and lipid metabolism, metabolisms of insecticides, amino acids, N-cyclic compounds, and proteins. Ends with a discussion on the control of metabolism. Some basic knowledge of organic chemistry in the reader is assumed.

A897 **Graham, Kenneth.** Concepts of forest entomology. 1963. Van Nostrand. 388p. illus. $10.

The author draws on experiences of his own and like-minded colleagues to present a blend of the older concepts of forest entomology with newer knowledge of ecological relationships and control methods. Should appeal to both students and practicing foresters.

A898 **Graham, Samuel A.,** and **Fred B. Knight.** Principles of forest entomology. 4th ed. 1965. McGraw-Hill. 417p. illus. $12.50.

A comprehensive analysis of principles that govern insect population. Includes history of forest entomology, ecological relationships, methods of control, and a discussion of representative species of insects that damage trees. A valuable reference for students of forestry, ecology, entomology, and horticulture.

Animal Sciences

A899 **Herms, William B.** Herms's medical entomology. 6th ed. ed. by Maurice T. James and Robert F. Harwood. 1969. Macmillan. 592p. illus. $15.

Book's purpose defined by its subtitle, *A textbook for use in schools and colleges, as well as a handbook for the use of physicians, veterinarians, and public health officials.* Gives complete and up-to-date information on the biology and control of all medically significant arthropods.

A900* **Hutchins, Ross E.** Insects. 1966. Prentice-Hall. 324p. illus. $7.95.

A renowned authority on insects surveys the whole field of insects. Begins with a general summary of the insect world and of characteristics common to all. Follows with descriptions of the most important ecological groups and specific accounts of their behavior. Written in popular style, using nontechnical terminology. An excellent book for anyone interested in insects.

A901 **Lanham, Urless N.** The insects. 1964. Columbia. 292p. illus. $8; paper, $2.95.

A survey of insect biology in general with some attention paid to characteristics of the important categories. Has an annotated bibliography and excellent illustrations. Useful as a supplementary text.

A902 **Little, Van A.** General and applied entomology. 2d ed. 1963. Harper. 543p. illus. $10.95.

Introduces beginning students to insects and their relationship to men. Presents the life history, habits, characteristics, and economic importance of representative insect groups. Also includes information on collecting and preserving insects.

A903* **Metcalf, Clell L.,** and **Wesley P. Flint.** Destructive and useful insects: their habits and control. 4th ed. rev. by R. L. Metcalf. 1962. McGraw-Hill. 1087p. illus. $20.

A standard beginning college textbook on applied entomology. Contains a section on anatomy, physiology, and classification of insects plus chapters on insects destructive to field crops, fruit, vegetables, stored products, domestic animals, etc. Useful for quick reference. Extensively revised and rewritten by R. L. Metcalf.

A904* **Pfadt, Robert E.** Fundamentals of applied entomology. 1962. Macmillan. 668p. illus. $11.50.

A useful text for agricultural students with little or no knowledge of entomology. Summarizes the history of entomology, characteristics common to all insects, methods of control, destructive insects, their hosts, and the injuries they cause. Section on chemical control and application not completely up-to-date.

A905 **Rolston, L. H.,** and **C. E. McCoy.** Introduction to applied entomology. 1967. Ronald. 208p. illus. $5.

An introduction to the management of insect populations. Includes discussions of principles having a broad application, economic and environmental management, chemical control, methods of application, and control programs. Useful as a textbook complementing a lecture course in general applied entomology.

A906 **Ross, Herbert H.** A textbook of entomology. 3d ed. 1965. Wiley. 539p. illus. $10.95.

An updated edition of a widely used, introductory textbook covering the fundamentals in all basic areas of entomology. Covers the same general territory as Comstock but more appealing to the beginning student.

A907 **Smith, Carroll N.,** ed. Insect colonization and mass production. 1966. Academic. 618p. illus. $27.

Deals with the insects injurious to plants, stored products, animals, and man, and explains laboratory mass production of insects for the sterile release techniques. Each group of arthropods discussed by an authority in that particular field.

A908 **Smith, Kenneth M.** Insect virology. 1967. Academic. 256p. illus. $11.50.

A thoroughly documented account of viruses which affect insects. Describes the different groups of viruses, the transmission and spread of insect viruses, the plant virus-insect vector relationships, and the use of viruses in biological control of insect pests.

A909 **Swain, Ralph B.** The insect guide. 1957. Doubleday. 261p. illus. $4.95.

A nontechnical guide providing ready identification of the major insect groups found in the United States and Canada. Has excellent illustrations, a good index, and a reading list for those who wish to read further.

A910 **Swan, Lester A.** Beneficial insects; nature's alternatives to chemical insecticides, animal predation, parasitism, disease organisms. 1964. Harper. 429p. illus. $7.95.

A well-written book concerned with biological control of destructive insects. Discusses control by other insects, by organisms causing insect disease, and by artificial biological methods. Will appeal to anyone interested in insect control.

A911 Urquhart, Frederick A. Introducing the insect. 1965. Warne. 258p. illus. $7.95.

Introduces the beginner and the would-be entomologist to insects. Gives a general discussion of entomology, and describes methods of collecting and preserving insects. Major portion of book devoted to identification keys to frequently found insects. Omits obscure families and orders. Keys highlighted with line drawings depicting aspects of insects' structure.

A912 Westcott, Cynthia. Gardener's bug book. 3d ed. 1964. Doubleday. 625p. illus. $9.95.

An excellent reference work for gardeners. Includes information on identification and control of nearly 1,900 garden pests, mostly insects, and descriptions of more than 700 species of host plants with a guide to their pests.

A913* Wigglesworth, Vincent B. Life of insects. 1964. World. 359p. illus. $12.50.

An excellent introduction for the beginning student in insect physiology. A clearly written, factual account of the natural history, physiology, and behavior of insects and their relationship to plants and to man. Includes a descriptive catalog of orders, new line drawings, and a good bibliography.

Apiculture

A914 Dadant, Camille P. First lessons in beekeeping. rev. ed. rev. and rewritten by James C. Dadant and Maurice G. Dadant. 1938, reprint 1967. Amer. Bee J. 127p. illus. $1.

One of the old standard handbooks on the keeping of bees. Provides facts essential for beekeeping for special purposes as well as for honey production.

A915* Eckert, John E., and **Frank R. Shaw.** Beekeeping. 1960. Macmillan. 536p. illus. $12.50.

Successor to *Beekeeping,* by E. F. Phillips, following much the same style and arrangement. Deals with all phases of beekeeping from buying equipment and bee colonies to the processing, marketing, and use of honey. Useful as a classroom text or as a reference for both beginners and commercial beekeepers.

A916 Frisch, Karl Von. The dancing bees. tr. from German by Dora Isle and Norman Walker. rev. ed. 1966. Harcourt. 198p. illus. $5.95; paper, $1.95.

An authoritative, nontechnical treatment of the honeybee. Explains how they communicate; describes the general life of the hive, their senses, intelligence, and behavior; and, discusses their enemies and diseases.

A917 Grout, Roy A., ed. The hive and the honeybee. rev. ed. 1949, reprint 1963. Dadant. 652p. illus. $5.75.

Successor to Longstroth's *The hive and the honeybee.* Designed especially for persons economically concerned with bees, including beekeepers, seedsmen, and fruitgrowers. Covers important aspects of bee life, basic beekeeping practices, and the industry's relation to plant agriculture.

A918 Hoyt, Murray. The world of bees. 1965. Coward. 254p. illus. $5.95.

Written on the young adult level and suitable for use as a classroom text. Discusses the usual topics found in books on bee life and bee culture. Topics of particular interest to commercial beekeepers include artificial insemination, hybrid crossing of queens, and multiple matings.

A919* Laidlaw, Harry H., and **John E. Eckert.** Queen rearing. 2d ed. 1962. California. 165p. illus. $4.95.

An important source of information on improved methods of queen rearing and the basic principles underlying these methods. Begins with the history of queen rearing and logically proceeds to bee breeding, artificial insemination techniques, disease control, etc. A valuable ready reference for anyone interested in honeybee culture.

A920* Root, Amos I. The ABC and XYZ of bee culture. 33d ed. rev. by E. R. Root. 1966. Root. 712p. illus. $4.95.

As its name implies, this book is an alphabetically arranged, complete encyclopedia on all major aspects of the scientific and practical culture of bees. Leading men in the industry have assisted in preparing this edition. Most of the scientific articles are accompanied by bibliographies. An important feature of the book is its glossary containing definitions of common beekeeping terms.

Pests and Pest Control

A921* DeBach, Paul, ed. Biological control of insect pests and weeds. 1964. Van Nostrand. 844p. illus. $25.

Leading authorities discuss all aspects of current theory, practice, and results of biological control of insect pests and weeds. Coverage ranges from the ecological basis of biological control to conservation of natural enemies, insect pathology, and the basic principles of weed control. Requires a basic knowledge of biology on the reader's part.

A922* **Frear, Donald E. H.**, ed. Pesticide handbook—Entoma. 22d ed. 1970. College Science. 284p. paper, $3.50.

A technical guide to pesticides covering insecticides, fungicides, rodenticides, etc. Pesticides and equipment are listed alphabetically according to use, and according to manufacturers. Also contains a list of poison control centers, information on antidotes, and general rules on the safe use of pesticides. Revised annually.

A923 **Gould, Robert F.**, ed. Organic pesticides in the environment. 1966. Amer. Chem. Soc. 309p. $10.50.

A collection of papers covering the use of pesticides, methods for detecting their presence, their movement in the environment, and their fate in air, water, soil, plants, and animals. A very useful reference book for both scientists and laymen interested in land and atmospheric pollution.

A924 **Gunther, Francis A.**, and **L. R. Jeppson.** Modern insecticides and world food production. 1960. Wiley. 284p. illus. $10.50.

Provides a general account of the taxonomy and structure of insects, the various methods of insect control, the many kinds of insecticides and acaricides, and the problems of their use and resulting from their use.

A925 **Hussey, Norman W., Wilfred H. Read,** and **John J. Hesling.** The pests of protected cultivation: the biology and control of glasshouse and mushroom pests. 1969. Amer. Elsevier. 404p. illus. $26.

An up-to-date treatment of greenhouse pest control and the pest problems of the mushroom industry. Covers materials, methods, economics, ecological aspects, and biological factors of pest control. Considerable attention is given to nematodes.

A926 **Kilgore, Wendell W.**, and **Richard L. Doutt**, eds. Pest control. 1967. Academic. 477p. illus. $19.50.

Largely concerned with control of insect and vertebrate pests and with those methods and chemical compounds more selective in their action and doing the least harm to the biological environment. Assumes many readers will have little background in biology. A valuable source of information for ecologists, food technologists, pest control operators, entomologists, etc.

A927 **Martin, Hubert.** The scientific principles of crop protection. 5th ed. 1964. St. Martin's. 376p. illus. $18.

Provides a wealth of information on nearly all aspects of crop protection. Covers host resistance, chemical pesticides, and biological control but omits some new developments such as resistant strains in pests. Also, chapters on pesticides not completely up-to-date because of new developments since book was published.

A928 **Munro, James W.** Pests of stored products. 1966. Hutchinson. 234p. illus. $5.50 (approx.)

Presents the history and practice of pest control in stored products. Includes a general introduction to entomology, descriptions of species that infest stored products, a report on the effect of infestation on international trade, and discussions on the prevention and control of pests of stored products.

A929 **Ordish, George.** Biological methods in crop pest control. 1967. Constable. 242p. illus. $3.60.

An informal account of what biologists are doing in the field of biological pest control. Author presents interesting stories and observations drawn from original research papers. Main purpose of the book is to acquaint the general public with the importance and promise of this applied science field.

A930* **Peairs, Leonard M.**, and **Ralph H. Davidson.** Insect pests of farm, garden, and orchard. 6th ed. 1966. Wiley. 675p. illus. $17.50.

Covers briefly but effectively the structure, function, development, classification, and control of major insect pests. Specific insects described in detail according to the stored products, plants, or animals associated with them. Various methods of control are discussed, including biological, chemical, cultural, and mechanical.

A931* **Pirone, Pascal P., Bernard O. Dodge,** and **Harold W. Rickett.** Diseases and pests of ornamental plants. 4th ed. 1970. Ronald. 546p. illus. $12.

A standard reference work widely used by both professional and amateur gardeners. Divided into 2 major parts: Part 1 discusses, by symptoms and by causes, diseases and pests and their control (does not recommend pesticides believed to be harmful to nature's ecological balance); part 2 is an alphabetical guide to nearly 800 genera of plants discussing diseases and pests to which they are susceptible, together with materials and methods of control.

A932 Seiden, Rudolph. Insect pests of livestock, poultry, and pets and their control. 1964. Springer. 161p. illus. $4.

A small reference book with 1 section devoted to insect pests and another to insecticides. Not completely up-to-date because of the many new developments since 1964. However, contains information still useful to students, livestock producers, farm suppliers, and others who use insecticides.

A933 Sweetman, Harvey L. Recognition of structural pests and their damage. 1965. Brown. 371p. illus. $7.75.

An illustrated, taxonomically arranged key to all pests found in and around structures. All terms are defined when first used. Can be used as a structural pest control text because of the short descriptions included. Also, a worthwhile aid in teaching pest controllers on the job.

A934 Wilson, George F. Horticultural pests, detection, and control. 3d ed. rev. by P. Becker. 1963. Chemical Pub. 240p. illus. $8.50.

Describes and illustrates the many symptoms produced in various plant parts as an aid in recognizing the particular pest causing the damage. Also, discusses the biological, chemical, cultural, and mechanical controls that can be employed.

Zoos and Zoo Management

A935* Crandall, Lee S. The management of wild mammals in captivity. 1964. Chicago. 761p. illus. $15.

A well-documented, exhaustive compilation of information for maintaining zoological parks. Covers common and technical names, general characteristics, breeding habits, care, and management of all kinds of mammals with captivity histories. Designed to appeal to the general reader as well as the specialist.

A936 _____. A zoo man's notebook. 1966. Chicago. 216p. illus. $4.95.

An abridged version of author's *Management of wild mammals in captivity*. Omits most of the minor details of the zoo business. Treats selected representative species and some little-known animals.

A937 Hahn, Emily. Animal gardens. 1967. Doubleday. 403p. illus. $6.95.

A lively presentation of zoos and zoo management. Describes the history of zoos and discusses general management aspects. Can be used as a starting point for further study.

A938 Kauffeld, Carl. Snakes: the keeper and the kept. 1969. Doubleday. 248p. illus. $5.95.

The director of Staten Island Zoo gives a brief and entertaining account of his experiences in keeping snakes. He includes details on collecting, housing, feeding, disease control, care, and general management of snakes. Intended for both the novice and the specialist.

A939 Kirchshofer, Rose, ed. The world of zoos. tr. by Hilda Morris. 1968. Viking. 327p. illus. $12.95.

A valuable collection of information concerning the world of zoos. Discusses modern zoo-keeping theory and philosophy, describes major aspects of zoo management, and surveys zoological gardens on a worldwide basis.

A940 Perry, John. The world's a zoo. 1969. Dodd. 308p. $6.95.

Concerned with the conservation of wildlife either in game preserves or in conventional-type zoos. Discusses collecting wild animals, their adaptation to captivity, breeding, care, and other problems facing zoo directors. Book's weakest feature is its lack of illustrations.

Fish and Marine Resources

A941* Bardach, John. Harvest of the sea. 1968. Harper. 301p. illus. $6.95.

Uses a minimum of technical terms to present current knowledge concerning the present and possible future uses of the sea. A valuable feature is the comprehensive discussion of agricultural techniques from planting of seed to harvesting. One of the best books available on marine resources.

A942 Bell, Frederick W., and **Jared E. Hazleton,** eds. Recent developments and research in fisheries economics. 1967. Oceania. 233p. illus. $10.75.

A collection of papers presented at a 1965 conference on fisheries economics. Topics covered include consumer demand for fish products, costs and the yield on capital, labor problems, resource and industry problems, the industry's present position, and its potential for future growth and development.

A943* **Bennett, George W.** Management of artificial lakes and ponds. 1962. Van Nostrand. 238p. illus. $10.50.

Scope of book confined to warm water fish species and their management in impounded waters. Specific subjects treated include fish management history, types of impounded waters, ecology of warm water fishes, fisheries management, and commercial aspects of sport fishing. Written specifically for the student and the professional fisheries worker.

A944 **Breder, Charles M.,** and **Donn Rosen.** Modes of reproduction in fishes. 1966. Nat. Hist. Pr. 957p. illus. $17.50.

A comprehensive classification, appraisal, and analysis of current knowledge of reproduction in fish. Includes breeding season and site, migration, sexual characteristics, courtship, etc., for several thousand known species of fish. A valuable reference book for libraries collecting fish and fishery literature.

A945* **Burgess, G. H. O.,** and others, eds. Fish handling and processing. 1967. Chemical Pub. 406p. illus. $10.

Treats in depth the problems of fish handling, processing, distribution, and storage and contains considerable information on new techniques and results of research. Written for persons with little or no science training who are actively engaged in the fish industry.

A945a **Carson, Rachel.** The sea around us. rev. ed. 1961. Oxford. 237p. illus. $6.50.

A popular classic written primarily for the general reader. Summarizes a wide variety of scientific information concerning the physical oceanography of the sea.

A945b **Davis, Herbert S.** Culture and diseases of game fishes. 1967. California. 332p. illus. $8.50.

Written primarily for the use of fish culturists with no special training in the field. Author makes no attempt to cover in detail all aspects of fish culture but gives a comprehensive account of methods most commonly used. Omits those diseases for which only fragmentary information is available.

A946 **Grant, Leonard J.,** ed. Wondrous world of fishes. 1965. Nat. Geographic. 367p. illus. $8.75.

A panoramic view of fishes and fisheries for the young adult. Begins with a general discussion of fishes and their habitat and proceeds to recreational fishing and commercial fishing.

A947 **Hickling, Charles F.** Fish culture. 1962. Faber. 295p. illus. $5.40.

Concerned only with the production of fish for food. Author treats fish as a form of stock-raising as well as a form of land and water usage. Contains information on fish farming management, feeding, breeding, disease control, and the various kinds of fish culture throughout the world.

A948 **Hoffman, Glenn L.** Parasites of North American freshwater fishes. 1967. California. 486p. illus. $15.

A compilation of keys to the identification of genera of North American freshwater fish parasites takes up the major portion of this book. Also, discusses fish parasitology, economic aspects, techniques of examining fish, and methods of parasite control.

A949 **Idyll, Clarence P.** The sea against hunger. 1970. Crowell. 221p. illus. $7.95.

An authority on marine biology investigates the possibility of harvesting the oceans to feed a hungry world. Author assesses man's food needs, examines potential food resources of the sea, foresees new techniques in fishing, and discusses pollution and other problems to be solved.

A950* **Iverson, Edwin S.** Farming the edge of the sea. 1968. Fishing News. 304p. illus. $10.20.

An American marine biologist discusses general aspects of cultivating marine animals, reports on current practices in the culture of seaweed, oysters, etc., and gives a general outline of diseases, predators, and pollution. Not a handbook but a practical account for persons considering fish farming as a career and for those with an interest in marine resources.

A951 **Jones, F. R. Harden.** Fish migration. 1968. St. Martin's. 325p. illus. $21.

An important addition to fish biology literature. Consists of a general introduction to fish and their migration, a critical review of available information on the subject, and a discussion of the basic mechanisms involved. A valuable reference for all students of fish biology.

A952 **Jones, John R. E.** Fish and river pollution. 1964. Butterworth. 203p. illus. $9.75.

Reviews what is known on toxicity to fish of a large variety of waste substances and discusses how pollution affects the growth rate of fish, their reproductive processes, etc.

A953 Lanham, Urless N. The fishes. 1962. Columbia. 116p. illus. paper, $2.25.

An easily read guide to the principal fish families of the world. Also covers the origin of fishes, their general structure, and selected biological information. Written for the general reader rather than the specialist.

A953a Mack, Jerry. Catfish farming handbook. 1971. Educator Books. 195p. illus. $12.95.

A handbook of information on the various phases of the catfish industry. Covers the general outlook for catfish farming, sources of stock, production and fielding, disease and treatment, harvesting and processing procedures, and marketing.

A954 McKee, Alexander. Farming the sea. 1969. Crowell. 198p. illus. $6.95.

An informal summary of what is known about this new industry. Reports on the increasing use of marine resources, the development of depth stations and underwater vehicles, and the possibilities of farming the sea to feed the growing world population.

A955* Marshall, Norman B. The life of fishes. 1966. World. 402p. illus. $12.50.

A masterly introduction to fish biology. Discusses the physiological and behavioral features of everyday life, the life histories of fishes, their adaptation to the environment, and their diversity. Black and white drawings are used to clarify significant aspects. A book for the professional as well as the serious-minded beginner.

A956 Netboy, Anthony. The Atlantic salmon: a vanishing species? 1968. Houghton Mifflin. 457p. illus. $6.95.

A nontechnical account of all aspects of salmon in all countries where it is important. Coverage ranges from the life history of salmon to current conservation programs designed to stop its extermination.

A957* Norman, John R. A history of fishes. 2d ed. rev. by Peter H. Greenwood. 1963. Hill & Wang. 398p. illus. $6.95.

A story of fish life in all its varied aspects, including structure, movement, feeding, breeding, development, distribution, behavior, and adaptation to environmental conditions. Contains a brief treatment of its economic uses and importance to man. Widely used as supplemental reading by those interested in or studying fishes and fisheries.

A958 Russell-Hunter, W. D. Aquatic productivity. 1970. Macmillan. 352p. (approx.). illus. paper, $4.95.

Book's scope is seen in its subtitle, *An introduction to some basic aspects of biological oceanography and limnology.* Stresses ocean's finite resources and methods for its future exploitation in food production. Written for beginning students in the field.

A959 Weyl, Peter K. Oceanography: an introduction to the marine environment. 1970. Wiley. 535p. illus. $12.50.

An excellent treatment of elementary oceanography. Discusses the history of the ocean; explains how marine organisms interact with each other and with the surrounding environment; and, shows the complex interactions between biological, chemical, geological, and physical processes.

PHYSICAL SCIENCES

General Works

A960* Avery, Thomas E. Interpretation of aerial photographs. 2d ed. 1968. Burgess. 324p. illus. $10.50.

An introductory textbook written from a do-it-yourself point of view. Applies the principles of the photogrammetry of aerial photographs to making useful measurements without the use of elaborate equipment. Written primarily for students in forestry, agriculture, and land economics.

A961* Buban, Peter, and **Marshall L. Schmitt.** Understanding electricity and electronics. 2d ed. 1969. McGraw-Hill. 438p. illus. $7.96.

A good beginning text presenting both theory and practical procedures. Contains some basic mathematics but requires no previous knowledge of electronics. Authors have focused their primary attention on materials, tools, electrical, and electronic equipment, and basic procedures and experiments.

A962 Carter, Robert C. Introduction to electrical circuit analysis. 1966. Holt. 500p. illus. $10.50.

Presents basic electrical circuit theory for the technical school student studying engineering. Requires the reader to have had basic courses in algebra, electricity, and plane trigonometry. Contains a glossary of terms and an appendix covering slide rule techniques.

A963 Crouse, William H. Automotive mechanics. 5th ed. 1965. McGraw-Hill. 616p. illus. $8.72; text ed., $7.28.

A comprehensive text covering all aspects of automotive mechanics in nontechnical language. Contains detailed information on automotive systems, their operation, maintenance, and repair. Significant points are emphasized by use of diagrams and photographs.

A964 Dezettel, Louis. ABC's of electrical soldering. 1968. Sams. 128p. $2.95.

A self-instruction manual for the beginner and a review source for the practicing technician. Explains what solder is, describes soldering tools, and tells how to make solder repairs.

A965 Draper, Alec. Electrical machines. 2d ed. 1967. Longmans. 384p. illus. $7.56.

Concerned with the principles and practical uses of semiconductor devices, induction motors, single-phase machines, etc. Written in a clear and simple style. Requires a basic background in electricity and electronics.

A966 Dwyer, James L. Contamination analysis and control. 1966. Van Nostrand. 343p. illus. $15.

A reference work presenting a broad picture of contamination control. Discusses sources of contamination, properties of aerosols and fluids, methods of analysis, and techniques of control.

A967 Edgar, Carroll. Fundamentals of manufacturing processes and materials. 1965. Addison-Wesley. 515p. illus. $11.75.

Designed for reference use in a beginning course in manufacturing processes at the technical school level. Gives extensive coverage to industrial shop processes and briefly treats basic principles underlying these processes. Assumes the student is familiar with machine tools and their operation.

A968 Ellison, Arthur J. Generalized electric machines. 1967. Harrap. 146p. illus. $2.40.

Designed for a laboratory course complementing a lecture course using the generalized approach to electric machines. With each experiment there is an introduction, a description of apparatus and procedures, an outline of significant theory and calculations, and a list of questions. Of most value to the inexperienced teacher.

A969 Feirer, John L. General metals. 3d ed. 1967. McGraw-Hill. 372p. illus. $7.36.

A basic, practical textbook on metalworking by an acknowledged expert. Covers art metal, bench metal, and sheet metal, metal working methods and techniques, and technology of metals. Each unit accompanied by questions and suggested projects.

A970 Frazee, Irving A. Automotive electrical systems. 3d ed. rev. by Walter E. Billiet and Leslie F. Goings. 1970. Amer. Tech. Soc. 393p. illus. $8.25.

Effectively and concisely explains the principles of electricity in simple language. Covers the newest as well as the older components and systems still in use. Gives complete details on construction, operation, servicing, and repairing.

A971 _____, and others. Automotive brakes and power transmission systems. 1956. Amer. Tech. Soc. 266p. $7.65.

A practical guide to troubleshooting, maintenance, and repair of both traditional types and the newer automotive systems. Intended for both beginning and experienced mechanics.

A972 _____, and others. Automotive fuel and ignition systems. 1953. Amer. Tech. Soc. 503p. illus. $7.65.

An elementary treatment of all aspects of automotive fuel and ignition systems arranged so that each section forms an independent unit. Useful as a text for the beginner and a review for the specialist.

A973 French, Thomas E., and **Charles J. Vierck.** Fundamentals of engineering drawing. 2d ed. 1966. McGraw-Hill. 447p. illus. $8.95.

A standard beginning textbook providing a thorough coverage of the fundamentals of mechanical drawing. Each unit accompanied by illustrations and exercises.

A974 Giachino, Joseph W., William R. Weeks, and **Elmer J. Brune.** Welding skills and practices. 3d ed. 1965. Amer. Tech. Soc. 352p. illus. $6.

Contains an accurate presentation of the various welding processes and provides practical instruction for developing essential welding skills.

Short-arc, pulsed-spray, vapor-shielded, laser beam, and ultrasonic welding are examples of the processes discussed.

A975 ———, ———, and **George S. Johnson.** Welding technology. 1968. Amer. Tech. Soc. 474p. illus. $7.50.

Consists of current information on industrial welding practices, metallurgy of welding, testing of weldments, strength of materials, welding symbols, safety measures, etc. Designed especially for training people who may eventually supervise production welding.

A976 Graham, Kennard C. Fundamentals of electricity. 5th ed. 1960. Amer. Tech. Soc. 312p. illus. $6.

A semitechnical book covering the traditional topics found in electricity textbooks. Begins with basic fundamentals and proceeds to the application of principles. Uses a minimum of mathematics. Can be used as a resource book for students in a technical agriculture program.

A977 ———. Understanding and servicing fractional horsepower motors. 1961. Amer. Tech. Soc. 256p. illus. $5.50.

An elementary text covering the principles of operation and repair of small electric motors, both AC and DC. Nonmathematical in approach, with review questions at the end of each chapter.

A978 Greenwood, Douglas. Mechanical power transmission. 1962. McGraw-Hill. 372p. illus. $13.50.

Intended as a basic reference for those involved with machinery that uses or transmits mechanical power. Contents include belt and chain drives, shafting and seals, bearings, couplings and clutches, gearing and speed reducers, and electric motor selection.

A979 Handley, William M., ed. Industrial safety handbook. 1969. McGraw-Hill. 475p. illus. $22.50.

A practical handbook which identifies and describes the main causes of industrial accidents, and deals with the proper design of equipment and buildings to provide safer working conditions. Also, covers the immediate and subsequent treatment of injuries.

A980 Jackson, Herbert W. Introduction to electric circuits. 2d ed. 1965. Prentice-Hall. 479p. illus. $9.95.

Designed for use as a textbook on electric circuits at the vocational/technical school level. Emphasis is on electrical theory rather than practical applications. Presupposes a basic knowledge of algebra.

A981* Jefferson, Ted B., and others. Welding encyclopedia. 1968. Welding Engineer Pub. 1067p. illus. $10.

A quick reference for basic data and current welding practices. Covers all aspects of welding plus data on specifications, normal weights, comparable products, etc. Also, contains a list of metal fabrication industry trade names, a buyers' manual, and an index.

A982 Kates, Edgar J. Diesel and high-compression gas engines. 2d ed. 1965. Amer. Tech. Soc. 448p. $7.95.

A standard text presenting the fundamentals of diesel and high-compression gas engines. Contains step-by-step instructions designed to develop proficiency in troubleshooting and repairing. Coverage includes new developments such as electronic ignitions, supercharging systems, etc.

A983 Kirk, Franklyn W., and **Nicholas R. Rimboi.** Instrumentation. 2d ed. 1966. Amer. Tech. Soc. 263p. illus. $6.95.

A concise, introductory treatment of the measuring instruments commonly used in the home and in industry. Authors use nontechnical language and simple diagrams to explain how these instruments are constructed and how they operate.

A984 Kirkpatrick, Elwood G. Quality control for managers and engineers. 1970. Wiley. 422p. illus. $11.95.

A discussion of the physical and economic facets of quality control. Covers a wide range of topics from product design to manufacturing and from planning to quality assurance. An instructor's manual is available.

A985 Lancaster, John F. The metallurgy of welding, brazing, and soldering. 1965. Amer. Elsevier. 291p. illus. $8.

Covers important aspects of the metal joint techniques of welding, brazing, and soldering. A large number of illustrations are used to help clarify the subject matter. Although primarily written for the engineering student, it can be used by any student beginning a study of the joining process.

A986 Lincoln Electric Company. How to read shop drawings, with special reference to welding symbols. 1961. Lincoln Electric. 187p. illus. $10.

A semitechnical work on shop drawing language with emphasis on welding symbols as standardized by the American Welding Society. Uses drawings and photographs to bring out points difficult to get across by the written word.

A987 Lytel, Allan. ABC's of electric motors and generators. 1964. Sams. 128p. illus. $2.95.

A basic introduction to electric motors and generators. Explains the principles of AC and DC motors, how they operate, their typical uses, and their repair.

A988 McNickle, L. S. Simplified hydraulics. 1966. McGraw-Hill. 196p. illus. $10.

A valuable, step-by-step presentation of hydraulics for the technician or the beginning student on the vocational/technical school level. Discusses and illustrates the underlying principles and the operation of hydraulic systems. Each chapter accompanied by practical exercises.

A989 Middleton, Robert G., and Milton Goldstein. Basic electricity for electronics. 1966. Holt. 694p. illus. $11.95.

Written especially to meet the needs of the beginning technology student. Begins with a brief history of electricity and proceeds with circuit analysis and electrical devices. Each chapter ends with a summary of fundamental concepts and a selection of practical questions and problems. Requires a basic background in algebra and trigonometry for full understanding.

A990 Nordell, Eskel. Water treatment for industrial and other uses. 2d ed. 1961. Van Nostrand. 598p. illus. $8.50.

A detailed account of water supply and treatment for both private and industrial uses. Covers allowable limits for impurities in water and the methods for removing harmful agents. A useful reference for knowledgable laymen as well as for those concerned with water supplies.

A991 Oldfield, R. L. The practical dictionary of electricity and electronics. 1959. Amer. Tech. Soc. 216p. illus. $5.50.

Provides concise, easily understood definitions and practical diagrams for basic terms and techniques in the fields of electricity, electronics, instrumentation, etc. A good ready reference for students on the technical school level.

A992 Parker, Marvin M. Farm welding. 3d ed. 1958. McGraw-Hill. 262p. illus. text ed., $6.20.

Covers arc and oxyacetylene welding and the equipment for both. Written in easy-to-understand style with an ample number of diagrams. A good reference book for farmers and agricultural students.

A993* Patton, W. J. The science and practice of welding. 1967. Prentice-Hall. 524p. illus. $12.95.

A thorough discussion of scientific principles and practical techniques of welding. Considers welding symbols, types of welds, methods of welding, testing of welds, etc. A good book for a technical school library welding collection.

A994 Pipe, Ted. Small gasoline engines training manual. 2d ed. 1969. Sams. 223p. illus. $5.25.

A comprehensive training manual for apprentices and students. Fully explains theory and design of small engines and covers troubleshooting and maintenance procedures step-by-step. Includes chain saws, stationary power units used to drive farm machinery, etc.

A995 Purvis, Judson A. All about small gas engines. 1963. Goodheart-Willcox. 304p. illus. $5.44.

Contains practical information on the construction, operating principles, and uses of small 2-cycle and 4-cycle gas engines. Also, covers troubleshooting, servicing, and repairing these engines.

A996 Rusinoff, Samuel E. Forging and forming metals. 1952, reprint 1964. Amer. Tech. Soc. 280p. illus. $6.25.

A text and manual covering the principles and practical applications of forging for the vocational/technical student. Explains the various forging techniques and the operations involved.

A997 ———. Manufacturing processes: materials and production. 3d ed. 1962. Amer. Tech. Soc. 753p. illus. $10.75.

Begins with the problem of selecting a manufacturing process, then covers machines and techniques used in the fabrication of plastics and metals. Quality control, inspection, and safety are also discussed.

A998 Sabersky, Rolf H., and Allen J. Acosta. Fluid flow. 1964. Macmillan. 393p. illus. $9.95.

A textbook for a beginning course in fluid mechanics. Presents the concepts of fluid flow and carries through to the practical problems of turbine engines, etc. Requires user to have a basic knowledge of mathematics.

A999 Sams, Howard W., and Co., Inc. Basic electricity and an introduction to electronics. 2d ed. 1967. Sams. 192p. illus. $4.25.

A practical introduction to electricity and electronics. Explains major concepts, provides practical projects, and includes useful information on cells and batteries, alternating current, heating, lighting, magnetism, etc.

A1000 Stephenson, George E. Small gas engines. 1964. Delmar. 165p. illus. $3.12.

A small textbook providing a concise but thorough explanation of small gas engines. Includes sections on maintenance, troubleshooting, and repair at the end of each teaching unit.

A1001 Strandberg, Carl H. Aerial discovery manual. 1967. Wiley. 249p. illus. $14.95.

Designed for use as a textbook in an air photo interpretation course or for individual study. The major portions of the book are devoted to aerial photographic interpretation, photo geology, and photo hydrology.

A1002 Town, Harold C. Automatic machine tools. 1968. Iliffe. 346p. illus. $9.

A lively presentation of information on automatic machinery and machine tools and the devices for machine control. A useful book for beginners in the field. Can be used as a source for ideas by people engaged in machine tool design.

A1003 Turner, Rufus P. Basic electricity. 2d ed. 1963. Holt. 412p. illus. $10.50.

Covers basic aspects of electricity and electronics with emphasis on fundamental principles. Uses a minimum of mathematics in the problems included. An authoritative book written particularly for the vocational student on the high school level.

A1004 Venk, Ernest A., and Walter E. Billiet. Automotive engines; maintenance and repair. 3d ed. 1964. Amer. Tech. Soc. 480p. illus. $7.15.

An authoritative book falling between the do-it-yourself manual and the technical text. Gives comprehensive coverage to the maintenance and repair of automotive engines including fuel and ignition systems.

A1005 _____, and _____. Automotive fundamentals. 3d ed. 1967. Amer. Tech. Soc. 566p. illus. $7.65.

A clearly written, detailed handbook on automotive basics. Will serve as a text for beginners and a review for the experienced mechanic. Disk brakes, alternators, transistorized ignitions, and air pollution control devices are examples of the topics considered.

A1006 _____, and **Edward D. Spicer.** Automotive maintenance and troubleshooting. 3d ed. 1963. Amer. Tech. Soc. 432p. illus. $7.65.

An aid in developing the student's ability to trace and diagnose logically and systematically any automotive difficulty that may arise. Covers electrical systems, power steering, power brakes, wheel alignment, safety devices, fuel systems, and other equipment and accessories.

A1007 Wetzel, Guy F. Automotive diagnosis and tune-up. 4th ed. 1965. McKnight & McKnight. 450p. illus. $8.96.

Presents fundamental concepts underlying automotive diagnosis, maintenance, and repair. Discusses analysis procedures, preventive methods, test instruments, and other important aspects in a practical manner with simple terminology.

Agricultural Engineering

A1008 Ashby, Wallace, J. Robert Dodge, and C. K. Shedd. Modern farm buildings. 1959. Prentice-Hall. 390p. illus. $9.32.

An older, but still useful book presenting practical information and detailed instructions accompanied by many photographs. Authors followed U.S. Dept. of Agriculture recommendations in their design of buildings for various farm purposes.

A1009 Bainer, Roy, and others. Principles of farm machinery. 1955. Wiley. 571p. illus. $12.95.

Another older book still valuable as a reference source. Covers machinery according to its uses. Includes some typical types and their utilization in tilling, planting, fertilizing, and harvesting. Introductory chapters cover such topics as cost analysis, hydraulic controls, power-take-off drives, and materials of construction.

A1010* Barger, Edgar L., and others. Tractors and their power units. 2d ed. 1963. Wiley. 524p. illus. $15.95.

Primarily intended for textbook use in college agricultural engineering courses. Emphasis is on the farm tractor and those features that distinguish it from other vehicles. A valuable source of information for technical level students needing to know engineering principles of tractors as part of their training in agricultural equipment technology.

A1011 **Brinker, Russell C.,** and **Warren C. Taylor.** Elementary surveying. 4th ed. 1961. International Textbook. 621p. illus. $10.

A concise but comprehensive presentation of fundamental aspects of surveying for college level students. Covers basic concepts and their practical application.

A1012* **Brown, Arlen D.,** and **Ivan G. Morrison.** Farm tractor maintenance. 3d ed. 1962. Interstate. 256p. illus. $5.50; text ed., $4.25.

Beginning chapters introduce the reader to tractors, diesel engines, fuels, oils, and lubricants. Section 2 emphasizes preventive maintenance of the various tractor parts and systems. The last section deals with the operation, repair, and storage of the tractor. Excellent diagrams show the function of each part of the tractor. Written for farmers, students, and teachers.

A1013 **Brown, Robert H.** Farm electrification. 1956. McGraw-Hill. 367p. illus. $10.50.

A complete general text on farm electrification designed for agricultural students who will take no other courses in electricity. Using a minimum of technical language, the author discusses electric wiring practices and the uses of electricity in agriculture.

A1014 **Cantor, Leonard M.** A world geography of irrigation. 1970. Praeger. 252p. illus. $8.50.

This book provides a broad, generalized, and nontechnical view of irrigation developments throughout the world. Useful as an introduction to the study of irrigation and as a reference for the nonspecialist.

A1015 **Clark, Colin.** The economics of irrigation. 1967. Pergamon. 116p. illus. $7.

A British author presents a discussion of elementary economic principles as applied to irrigation practices. Covers the water needs for various crops in different climatic regions. Intended for individuals involved in irrigation planning in any part of the world.

A1016* **Cook, Glen C.,** and others. Farm mechanics text and handbook. new ed. rev. by Lloyd J. Phipps and others. 1959. Interstate. 814p. illus. $10.75; text ed., $7.95.

A farm mechanics text designed for use by a wide variety of groups. Topics treated include using farm mechanics, farm shop work, farm buildings and conveniences, electrification, carpentry, soil and water management, fencing, sanitation, power machinery, etc. Of most value as a reference work.

A1017 **Dalzell, James R.,** and **Gilbert Townsend.** Concrete block construction for home and farm. 2d ed. 1957. Amer. Tech. Soc. 216p. illus. $4.15.

Filled with information on the use of concrete and concrete blocks in various types of construction. Mortar types, materials used in making concrete, mixing and formwork, constructing footings and foundations, and waterproofing are some of the subjects covered.

A1018 **Desrosier, Norman W.,** and **Henry M. Rosenstock.** Radiation technology in food, agriculture, and biology. 1960. AVI. 401p. illus. $13.50.

Important in the field of food science for its coverage of the utilization of atomic energy in food disinfection, stabilization, and distribution. Deals with radiation and its interaction with living organisms and then specifically with the nature of food and the role of radiation. A rather advanced subject for most technical students but useful in a food science library collection as an optional reference tool.

A1019 **Ehlers, Victor M.,** and **Ernest W. Steel.** Municipal and rural sanitation. 6th ed. 1965. McGraw-Hill. 663p. illus. $15.50.

A sanitary engineering text on the college level. Defines the field of sanitation, sets forth important problems confronting workers, and suggests solutions to these problems. Presents principles essential to all workers in urban and rural health.

A1020 **Eshelman, Philip V.** Tractors and crawlers. 2d ed. 1967. Amer. Tech. Soc. 274p. illus. $8.25.

A comparative study of power and tractor equipment. Covers types of power equipment, design features, components of tractors and crawlers, operating principles, maintenance, repair, and power mechanization terms. A good reference tool for technical school students training as mechanics or agricultural equipment salesmen.

A1021* **Farrall, Arthur W.,** and **Carl F. Albrecht,** eds. Agricultural engineering: a dictionary and handbook. 1965. Interstate. 434p. $11.25; text ed., $8.50.

A ready reference containing the answers to hundreds of frequent questions about all phases of agriculture. Part 1 defines terms relating to crop production and terms used in connection with agricultural equipment and procedures.

Part 2 is made up of charts and tables covering frequently used technical data and other reference material such as rates of seeding, densities of building materials, and storage properties of food crops.

A1022 Foss, Edward W. Construction and maintenance for farm and home. 1960. Wiley. 373p. illus. $8.95.

A clear, concise guide to farm and shop skills including farm mechanics, construction processes, woodworking, metalwork, etc. Necessary materials, tools, and equipment are discussed with each skill presented.

A1023 Fussell, George E. Farming techniques from prehistoric to modern times. 1966. Pergamon. 278p. illus. $7; paper, $4.95.

Begins with rudimentary techniques and equipment used in prehistoric farming and proceeds chronologically to the twentieth century discussing implements utilized and new methods developed. The only book devoted specifically to the history of farming techniques. A good reference source for background information and for illustrations of early tools and equipment.

A1024 Gray, Harold. Farm service buildings. 1955. McGraw-Hill. 458p. illus. $10.50.

Written for textbook use in a general course in farm structures. Covers development and planning of farm buildings, construction materials, structural requirements, livestock housing, storage buildings, and environmental control. Of most value as a reference for students, vocational/technical agriculture teachers, and farmers.

A1025 Gurnham, Fred C., ed. Industrial wastewater control. 1965. Academic. 476p. illus. $16.

A reference manual on wastewater management in American industries. Discusses the kinds of wastewater problems and their possible solutions. Important in the field of food technology because of the chapters devoted to wastewaters from food processing industries. Covers meatpacking, dairy processing, fruit and vegetable canning, beverage manufacture, and processing of sugar and starch. Remainder of book devoted to mining and mineral processing and manufacturing and other industries.

A1026 Hagan, Robert M., Howard R. Haise, and **Talcott W. Edminster,** eds. Irrigation of agricultural lands. 1967. Amer. Soc. of Agron. 1212p. illus. $22.50.

A collection of papers covering all aspects of irrigation including water sources, relation of water to plants and soil, methods and management of irrigation, and water conservation. An excellent source of information for students of irrigation, field workers, and organizations concerned with planning and operating irrigation projects.

A1027 Hamilton, James R. Using electricity on the farm. 1959. Prentice-Hall. 397p. illus. $8.84.

A simplified reference and how-to guide for agricultural students and farmers. Contains practical information on using electricity to improve farming efficiency and production. Includes practical illustrations and examples and instructions on how to plan and carry out each major job or project.

A1028 Harris, Anthony G., T. B. Muckle, and **J. A. Shaw.** Farm machinery. 1965. Oxford. 240p. illus. paper, $4.25.

A British publication written specifically for use as a textbook in a farm machinery course. A simple, well-illustrated presentation of the subject. Should have supplementary reading value for American students.

A1029 Howell, Ezra L., J. K. Coggin, and **G. W. Giles.** Building and equipping the farm shop. 1956. Interstate. 106p. illus. $2.95.

Pictures are used whenever possible to present ideas on building and equipping the farm shop. Contains very little written material. Covers the need, the type, the location, the structure, tools and equipment, interior arrangement, shop-made equipment, and shop safety.

A1030* Hunt, Donnell. Farm power and machinery management. 5th ed. 1968. Iowa State. 292p. illus. $5.95.

A compilation of current knowledge concerning profitable and efficient use of agricultural machinery. Combines basic management principles with machinery operating details and includes exercises applying principles to practice. Designed chiefly for students of vocational/technical agriculture and agricultural engineering. Should also be of value to the potential farm manager.

A1031* Israelsen, Orson W., and **Vaughn E. Hansen.** Irrigation principles and practices. 3d ed. 1962. Wiley. 447p. illus. $11.50.

A widely used textbook on irrigation and a comprehensive survey on a worldwide basis of irrigation problems and possible solutions. Outlines the need for irrigation; discusses water sources, storage, conveyance, and measurement;

explains the relationship of water to soil and plants; and, presents the types of irrigation, irrigation efficiencies, and the principles of irrigation design.

A1032* **Jones, Fred R.** Farm gas engines and tractors. 4th ed. 1963. McGraw-Hill. 318p. illus. $12.50.

Concerned with the construction, design, operation, servicing, maintenance, and economic importance of tractors and gas engines used in agriculture. Technical language kept to a minimum for the benefit of agriculture students who lack basic knowledge of the pure and applied sciences.

A1033 **Jones, Mack M.** Shopwork on the farm. 2d ed. 1955. McGraw-Hill. 626p. illus. $7.96.

Deals simply and clearly with materials, tools, processes, and operations rather than with specific shop projects or jobs. A useful source of information for agricultural students, teachers, and farmers.

A1034 **Kazmann, Raphael G.** Modern hydrology. 1965. Harper. 301p. illus. $11.95.

A refreshingly simple introduction to hydrology. Covers the historical development of the field, precipitation, evaporation, surface and ground water, water resource development, and man's effect on the hydrological cycle. A worthwhile reference source for anyone interested in water resource development and conservation.

A1035* **Kissam, Philip.** Surveying practice—the fundamentals of surveying. 1966. McGraw-Hill. 480p. illus. $9.95.

A successor to the author's earlier books on elementary surveying. Presents basic surveying procedures and the underlying principles. Ample use is made of illustrations to help the reader understand the steps involved. A good review of surveying fundamentals for technicians in the conservation field as well as a text and reference for students.

A1036 **Laverton, Sylvia.** Irrigation: its profitable use for agricultural and horticultural crops. 1964. Oxford. 166p. illus. $4.

A readable, comprehensive account of irrigation and its agricultural and horticultural uses. A book for the general reader and beginning agriculture student. Too elementary for the more advanced student of irrigation.

A1037 **Luthin, James N.** Drainage engineering. 1966. Wiley. 250p. illus. $13.50.

Emphasis is on methods for investigation of drainage problems and the designing of drainage systems. Discussion is largely limited to methods in common use today. Assumes some knowledge of soils and elementary fluid mechanics. Intended as a drainage engineering classroom text.

A1038* **McColly, Howard F.,** and **James W. Martin.** Introduction to agricultural engineering. 1955. McGraw-Hill. 553p. illus. $11.50.

A handy guide and an introductory text giving general coverage to the entire field of agricultural engineering. Covers engineering in agriculture, agricultural mechanics, farm power and machinery, electricity in agriculture, processing agricultural products, farm structures and conveniences, and soil and water conservation engineering.

A1039* **Neubauer, Loren W.,** and **Harry B. Walker.** Farm building design. 1961. Prentice-Hall. 611p. illus. $11.95.

A functional treatment of farm building design for the technical or college student seriously interested in the subject. Not designed for any particular geographical region. Includes cost estimation and other economic aspects. Covers farm houses, barns, sheds, silos, storage bins, animal shelters, etc.

A1040 **Olivier, Henry.** Irrigation and climate: new aids to engineering planning and development of water resources. 1961. St. Martin's. 250p. illus. $25.

Provides fairly simple methods for estimating basic crop water requirements in different localities. Methods are based upon use of standard climatic data. Intended for agriculturists and engineers involved in water resources development projects.

A1041* **Phipps, Lloyd J.** Mechanics in agriculture. 1967. Interstate. 808p. illus. $10.75; text ed., $7.95.

Highly recommended as a textbook for instruction in vocational/technical schools. Major sections of book devoted to agricultural shopwork, power and machinery, buildings and conveniences, rural electrification, and soil and water management. The detailed instructions and outstanding illustrations add to the book's value for the beginner.

A1042* **Promsberger, William J.,** and **Frank E. Bishop.** Modern farm power. 1962. Prentice-Hall. 280p. illus. $9.25.

A text and reference for vocational/technical classes in farm power. Discusses the history, principles, construction, operation, maintenance, housing, storage, selection, and uses of farm power machines. Also, provides facts pertaining to the importance of power in farming and safe practices in using mechanized equipment. Gives references, questions, and shop projects at the end of each chapter.

A1043 **Rayner, William H.,** and **Milton O. Schmidt.** Elementary surveying. 4th ed. 1963. Van Nostrand. 485p. illus. $7.50.

Includes all the basic aspects of surveying and mapping usually found in a beginning text. Intended to serve a wide student audience from engineering and geology to agriculture and forestry.

A1044* **Richey, C. B., Paul Jacobson,** and **Carl W. Hall,** eds. Agricultural engineers' handbook. 1961. McGraw-Hill. 880p. illus. $26.75.

Combines in 1 volume the basic principles and procedures for all areas of agricultural engineering and its application to agricultural production. Gives details on planting, harvesting, farm buildings, soil and water conservation, etc. Contains a minimum of definitions and other elementary information. Written for readers beyond the beginning level.

A1045* **Richter, Herbert P.** Practical electrical wiring: residential, farm, and industrial. 8th ed. 1970. McGraw-Hill. 664p. illus. $12.50.

A well-known instruction manual and text giving comprehensive coverage to electrical wiring based on the National Electrical Code. Covers electrical fundamentals and offers clear, concise answers to residential, farm, and industrial electrical wiring problems.

A1046 **Rubey, Harry.** Supplemental irrigation for eastern United States. 1954. Interstate. 209p. illus. $4.50; text ed., $3.

A simply written and well-illustrated summary of the possibilities of supplemental irrigation for eastern areas. Discusses planning, installing, and operating an efficient system, tells what to expect from it, and warns against pitfalls.

A1047* **Schwab, Glenn O.,** and others. Soil and water conservation engineering. 2d ed. 1966. Wiley. 683p. illus. $16.95.

An introductory text designed primarily for readers with little or no background in engineering. Considers agricultural, engineering, and economic aspects of soil and water conservation with emphasis on the engineering phases. Practices in surveying, mapping, erosion control, drainage, irrigation, land classification, wind erosion control, and simple hydrology are some of the topics covered.

A1048 **Shippen, John M.,** and **John C. Turner.** Basic farm machinery. 1966. Pergamon. 2v. illus. v.1 $4.95; v.2 $4.50.

Concerned with the general working principles of farm tractors and machinery. Deals individually with most of the major types of machinery. Intended primarily for use by British technical students and young agricultural mechanics. Worth considering as an addition to the agricultural mechanics library in the United States.

A1049 **Singleton, W. Ralph.** Nuclear radiation in food and agriculture. 1958. Van Nostrand. 379p. illus. $2.75.

A compilation of papers presented at a conference on the peaceful uses of atomic energy. Subjects covered range from agricultural and food processing research to basic research in radiobiology. Of special interest to agricultural researchers but can be used as a reference source for advanced students.

A1050* **Smith, Harris P.** Farm machinery and equipment. 5th ed. 1964. McGraw-Hill. 519p. illus. $12.50.

The design and operation of the many kinds of equipment necessary for efficient crop production are discussed in detail. Machinery and equipment for planting, minimum tillage, weed control, and harvesting of forage and field crops are examples of the areas treated. Written in simple terminology for the benefit of the beginning student yet sufficiently technical for the more advanced reader.

A1051* **Stone, Archie A.,** and **Harold E. Gulvin.** Machines for power farming. 2d ed. 1967. Wiley. 559p. illus. $11.95.

A complete guide to selecting, operating, and maintaining modern farm machines. Covers machines used for tillage, seeding, fertilizing, planting, weed control, insect and disease control, and harvesting. Emphasis is on the need for good technical and business management of farm machinery. Each chapter forms an independent unit, complete with review questions and projects.

A1052 **Ward, R. C.** Principles of hydrology. 1967. McGraw-Hill. 402p. illus. $12.75.

A nonmathematical and comprehensive treatment of the basic principles of precipitation, evaporation, soil moisture, runoff, and other aspects of hydrology. Illustrations and case histories mostly from Europe and Britain. Contains information of interest to the informed layman as well as those in various agricultural fields.

A1053 Wisler, Chester O., and **Ernest F. Brater.** Hydrology. 2d ed. 1959. Wiley. 408p. illus. $10.95.

A comprehensive review of principles and background information intended for the instructor and for both new and experienced workers in hydrology. Covers the topics generally found in hydrology texts such as precipitation, surface and ground water, runoff, evaporation, and water supply.

A1054* Zimmerman, Josef D. Irrigation. 1966. Wiley. 516p. illus. $18.95.

A creative approach to every phase of irrigation from planning and construction to operation and cost analysis. Emphasizes the need for thorough planning of irrigation projects, describes the techniques of modern irrigation methods, and discusses large scale networks and structures. Designed to appeal to a wide audience including engineers, farmers, students, and those engaged in water conservation and development.

Soils and Fertilizers

A1055 Alexander, Martin. Introduction to soil microbiology. 1961. Wiley. 472p. illus. $11.95.

Intended as a general introduction to the subject rather than an exhaustive treatment of any phase. Brings together a great deal of information from diverse disciplines such as biochemistry, biophysics, microbiology, etc. Microbial ecology, the carbon cycle, the nitrogen cycle, and mineral transportation are examples of the topics covered. Useful for both students and specialists concerned with soil bacteriology.

A1056* Archer, Sellers G. Soil conservation. 1956, reprint 1969. Oklahoma. 305p. illus. $5.95.

Material presented in simple terms easily understood by the general reader. Intended as a guide to practical soil conservation with emphasis on critical soil problems. Discusses soil conservation planning based on erosion control and crop and soil improvement. A widely used book now in its fourth printing.

A1057* Baver, Leonard D. Soil physics. 3d ed. 1956. Wiley. 489p. illus. $11.95.

A well-known source of information on the fundamental makeup of the soil and the physical properties of its various components. Specific topics include physical behavior of soil-water systems, soil air and water, principles of soil irrigation and drainage, soil temperature, and the physical properties of soil in relation to tillage and erosion.

A1058* Bear, Firman E. Soils in relationship to crop growth. 1965. Van Nostrand. 297p. illus. $14.95.

An authoritative, beginning text which presents the subject with a limited amount of scientific detail. Covers the practical aspects of soils in relation to their ability to produce high yields of superior quality crops. A large portion of the book devoted to the nutrient elements occurring naturally in soil.

A1059 _____, ed. Chemistry of the soil. 2d ed. 1964. Van Nostrand. 515p. illus. $22.50.

Mainly concerned with the nature of soil and with soil in relation to plant nutrition. Traces changes in the soil as a result of biological, geological, and meteorological agencies. Then, discusses soil properties, soil organic matter, trace elements, etc. Valuable to students and scientists as a reference source.

A1060 Bennett, Hugh H. Elements of soil conservation. 2d ed. 1955. McGraw-Hill. 358p. illus. $5.

A straightforward soil conservation guide and elementary text written by the father of the soil and water conservation movement. Covers the importance and the methods of soil conservation and the use and management of water in relation to the soil. Facts presented are based on research and on experience of the U.S. Soil Conservation Service.

A1061* Berger, Kermit C. Introductory soils. 1965. Macmillan. 371p. illus. $8.50.

Author uses nontechnical language to present the fundamentals of soil-plant science in relation to agriculture. Treats concisely, but fully, soil origin and classification, soil biology, chemistry and physics, the nutrient elements, soil testing, use of fertilizers, plant growth factors and soil conservation. Written primarily for use in introductory soils courses at the beginning college level.

A1062 Black, Charles A. Soil-plant relationships. 2d ed. 1968. Wiley. 792p. illus. $20.95.

A thorough analysis of the characteristics of soils which affect their ability to sustain plant growth. Gives an accurate account of current work in soil-plant relationships and the significance of measurements of soil properties. Intended for students who already have some knowledge of plant physiology and soils.

A1063 ———, ed. Methods of soil analysis. 1965. Amer. Soc. of Agron. 2v. illus. $30.

One of the best sources available concerning methods of soil analysis. Not a compilation of official procedures but a general discussion of the principles involved in specific measurements, limitations of procedures, and references to additional literature. Presupposes a good background in the sciences. Of most value to instructors of future soil technicians.

A1064 **Brewer, Roy.** Fabric and mineral analysis of soils. 1964. Wiley. 470p. illus. $16.

A comprehensive treatment of soil fabric and mineral analysis intended for students with a serious interest in soils. Provides a system of description and classification of soil phenomena and shows how to apply mineral analysis to studies of soil genesis.

A1065* **Buckman, Harry O.,** and **Nyle C. Brady.** Nature and properties of soils. 7th ed. rev. 1969. Macmillan. 653p. illus. $10.95.

A classic text on the nature and properties of soils. Reflects the latest findings on soils, enhanced by numerous tables and diagrams. Scope is seen in some representative topics: soil formation and classification, soil reaction and its control, nutrient elements of soil, and the use of fertilizers.

A1066 **Bunce, Arthur C.** Economics of soil conservation. 1942, reprint 1965. Nebraska. 227p. illus. paper, $1.30.

A consideration of the interrelationships between the economic and the technological problems of soil conservation. An older book but still valuable as a reference source for those concerned with soil conservation and use.

A1067 **Bunting, Brian T.** The geography of soil. rev. ed. 1965. Aldine. 213p. illus. $5; paper, $2.45.

A discussion of soil formation factors, soil description, classification, and nomenclature. Contains a list of references at the end of each chapter and a bibliography at the end of the book. Useful as a supplementary reference for students studying soil classification and geography.

A1068 **Childs, Ernest C.** An introduction to the physical basis of soil water phenomena. 1969. Wiley. 493p. illus. $17.50.

Begins with some elementary concepts and proceeds to a more advanced treatment of the subject. Primarily concerned with all kinds of soil water problems and with water movement in soil. A valuable source of ideas for soil science teachers in agriculture.

A1069 **Collings, Gilbeart H.** Commercial fertilizers, their sources and use. 5th ed. 1955. McGraw-Hill. 617p. illus. $14.50.

Intended for use as a text in commercial fertilizer courses on the college level. Covers sources, methods of manufacture, composition in relation to crop requirements, and the purchase and use of fertilizers. Contains an extensive bibliography for further reading. Useful as a reference in vocational/technical courses in fertilizers.

A1070 **Cook, Ray L.** Soil management for conservation and production. 1962. Wiley. 527p. illus. $11.95.

Treatment of the subject matter ranges from the very elementary to the highly technical. First part of book concerned with soil management problems in 6 climatic areas of the United States. The last part deals with the management of particular types of soils such as garden, greenhouse, turf, and forest. Requires a basic introductory knowledge of soils for complete understanding.

A1071 **Cooke, George W.** The control of soil fertility. 1967. Hafner. 526p. illus. $11.50.

Gives detailed coverage to the scientific basis for principles of crop nutrition and discusses results of recent experiments in soil fertility control. Topics covered include plant nutrients and soil fertility, plant nutrient cycles, use of fertilizers and manures, and soil productivity. A balanced treatment of the subject presented in a factual and logical style.

A1072* **Donahue, Roy L.** Our soils and their management. 2d ed. 1961. Interstate. 568p. illus. $10.75; text ed., $7.95.

An introduction to soil and water conservation designed as a high school or junior college text. Presents technical information in an easily understood style. Two-thirds of the book emphasizes soil fertility and fertilizers and the management of the more common types of soils, such as pasture, range, garden, lawn, orchard, and forest soils. The remaining third deals with special soil management problems.

Physical Sciences

A1073* ———. Soils: an introduction to soils and plant growth. 2d ed. 1965. Prentice-Hall. 363p. illus. $8.95.

An elementary text concerned with general soil science principles in relation to plant growth and their practical application to crop production problems. The extensive glossary containing definitions of hundreds of soil and plant terms is a feature of particular value to students.

A1074* Foster, Albert B. Approved practices in soil conservation. 3d ed. 1964. Interstate. 384p. illus. $6.95; text ed., $5.

A practical manual presenting the techniques of well-known soil conservation practices. Gives step-by-step instructions in the use of soil conservation instruments and in the application of important soil conservation practices. Written in readable language with a simple and direct style. Primarily intended for the agriculture student on the vocational/technical level.

A1075 Garrett, Stephen D. Soil fungi and soil fertility. 1963. Pergamon. 165p. illus. $5; paper, $2.95.

A study of the role of soil fungi in relation to soil fertility. Presents an account of fungi as members of the microbial community in the soil and discusses methods for studying fungi in the soil. Assumes a minimum of biological knowledge.

A1076* Hardy, Glenn W., ed. Soil testing and plant analysis. 1967. Soil Sci. Soc. of Amer. 2v. illus. paper, $2 each v.

A factual, informative, and up-to-date treatment of the methods, problems, interpretation, and utilization of soil tests and plant analyses. A valuable source of information for scientists, extension agents, and others concerned with crop production.

A1077 Held, R. Burnell, and Marion Clawson. Soil conservation in perspective. 1965. Johns Hopkins. 344p. illus. $9.

A realistic presentation of soil conservation rather than one designed to arouse public emotion. Discusses the history of the conservation movement, government agencies and their contributions, population and food supply problems, results of recent research, etc. An informative book for anyone interested in soil conservation.

A1078 Kellogg, Charles E. The soils that support us. 1941. Macmillan. 370p. illus. $6.95.

An informal account of the history of soil science, the origin and classification of soils, their basic components, their importance to man, etc. Author emphasizes ideas rather than data and facts. Written on an elementary level for the general reader.

A1079 Kevan, D. Keith McE. Soil animals. 1962. Witherby. 244p. illus. $4.20.

A general introduction to soil zoology. Discusses the wide variety of animals found in the soil, the life history and ecology of soil animals, methods of study, the effect of human activity on soil fauna, and the effect of soil animals on the soil. Will appeal not only to zoologists but to all interested in soil science.

A1080* Knuti, Leo L., Milton Korpi, and J. C. Hide. Profitable soil management. 2d ed. 1970. Prentice-Hall. 376p. illus. $8.36.

A standard introductory soil management text. Written especially for vocational/technical school use. Provides basic information on origin and classification of soils, principles and methods of soil improvement, irrigation and drainage, tillage practices, use of plant nutrients, and soil and water conservation.

A1081 Kohnke, Helmut. Soil physics. 1968. McGraw-Hill. 224p. illus. $10.50.

Basic essentials of soil physics are presented in this technically written textbook. Each chapter presents one important aspect of the subject. Useful to technical school students training as soil technicians, provided they have an elementary background in the sciences.

A1082 ———, and Anson R. Bertrand. Soil conservation. 1959. McGraw-Hill. 298p. illus. $10.50.

Gives a clear picture of the basic principles of soil conservation, methods and economics of soil improvement, and the importance of soil conservation to man's future welfare. Written in nontechnical language and amply illustrated.

A1083 Lamer, Mirko. The world fertilizer economy. 1957. Stanford. 715p. illus. $17.50.

Shows changes in the types and amount of commercial fertilizers provided and used by various countries during World War II. Contains an extensive bibliography and a wealth of statistical information. A valuable source of historical and background information for those interested in any phase of the fertilizer industry.

A1084* McVickar, Malcolm H. Using commercial fertilizers. 3d ed. 1970. Interstate. 354p. illus. $8.50.

The practical application of fertilizer knowledge and facts presented by an authority on fertilizer. Also covers specialized uses, chemical sources versus organic sources of plant nutrients, and the economics of fertilizer use. Intended for use in vocational/technical schools and as a reference manual for farmers and those in the fertilizer industry.

A1085 _____, **G. L. Bridger,** and **Lewis B. Nelson,** eds. Fertilizer technology and usage. 1963. Soil Sci. Soc. of Amer. 464p. illus. $8.

A compilation of papers containing a great deal of valuable information on all aspects of fertilizer technology and use. Written in literature review style with hundreds of references to other literature. Helpful to the advanced student in fertilizer technology.

A1086 Markham, Jesse W. The fertilizer industry, a study of an imperfect market. 1958. Greenwood. 249p. illus. $10.

A concisely written, well-organized study of the fertilizer industry. Describes the organization and structure of integrated fertilizer industries, briefly discusses specific types of industries, and assesses the imperfections of the fertilizer market.

A1087 Marshall, Charles E. The physical chemistry and mineralogy of soils. v.1 Soil materials. 1964. Wiley. 388p. illus. $12.95.

A source book rather than a textbook. Gives a general picture of the physical chemistry and mineralogy of soils and briefly covers a wide range of experimental work in the area. A wealth of references directs the reader to standard works for more detailed coverage of each aspect of the subject.

A1088 Millar, Charles E. Soil fertility. 1955. Wiley. 436p. illus. $8.95.

A well-developed reference book covering plant nutrition relationships, organic matter relationships, soil reaction, fertilizer and liming materials, and soil microorganisms. Although emphasis is on the plant, relevant aspects of soil chemistry, physics, and microbiology are discussed.

A1089 _____, **Lloyd M. Turk,** and **H. D. Foth.** Fundamentals of soil science. 4th ed. 1965. Wiley. 491p. illus. $11.95.

Written primarily as a textbook for elementary soils courses on the college level. The authors emphasize and explain in nontechnical terms the general principles of soil science, fundamental soil-plant relationships, and the principles underlying fertilizer treatments and practices. Proven soil management practices are briefly discussed. Can be used as a high school reference.

A1090 Pauli, Frederick W. Soil fertility: a biodynamical approach. 1967. Hilger. 204p. illus. $6.

Considers the soil ecosystem as a whole and relates recent research findings on soil organic matter to problems of soil fertility improvement. Of more value to teachers than to students.

A1091 Russell, Edward J. Soil conditions and plant growth. 9th ed. 1961. Wiley. 688p. $12.50.

A standard applied soil science reference book. Complements Cooke's *The control of soil fertility,* which bridges the gap between soil science and farm practices. Concerned with the interrelationship of soil conditions and plant growth. Soil biology, soil chemistry, nitrogen cycle in soils, humus, and soil phosphate are examples of the topics covered.

A1092 Sauchelli, Vincent, ed. Chemistry and technology of fertilizers. 1960. Van Nostrand. 692p. illus. $21.50.

A broad, detailed account of the fertilizer industry covering raw materials, fertilizer production and handling, and fertilizer plant design. A comprehensive reference for those involved in, or interested in, the production of chemical fertilizers.

A1093 _____, ed. Fertilizer nitrogen: its chemistry and technology. 1964. Van Nostrand. 424p. illus. $20.

A comprehensive review of various agricultural and technological phases of the nitrogen fertilizer industry. Authors have focused their primary attention on the industry as it relates to agriculture. Covers several aspects of vital importance to agribusiness.

A1094 Schaller, Friedrich. Soil animals. 1968. Michigan. 144p. illus. $5; paper, $1.95.

Begins with a brief description of soil fauna and methods for their collection and study. Then, follows with a discussion of the soil and an account of the life histories and behavior of typical soil animals.

A1095* Slack, Archie V. Chemistry and technology of fertilizers. 1967. Wiley. 142p. illus. paper, $6.95.

A revised reprint of the section on fertilizers in the second edition of the *Kirk-Othmer encyclopedia of chemical technology.* Outlines the modern fertilizer production technology with emphasis on the chemistry and technology of production of the nitrogen, phosphate, and potash fertilizers.

A1096 ———. Defense against famine: the role of the fertilizer industry. 1970. Doubleday. 232p. illus. $5.95.

A fertilizer specialist traces the history of the fertilizer industry and explains the role the fertilizer industry can play in preventing world famine. Included is a brief history of agriculture, a description of plant nutrition, and a discussion of the most important elements in fertilizers.

A1097 Stallings, James H. Soil conservation. 1957. Prentice-Hall. 575p. illus. $10.50.

An account of the soil and water conservation movement. Deals with the history of soil erosion, fundamental erosion problems, soil and water conservation practices, and farm and watershed planning. Useful as a supplemental reference for technical agriculture students.

A1098 ———. Soil use and improvement. 1957. Prentice-Hall. 403p. illus. $7.96.

An elementary text covering the technical aspects of soil erosion, conservation, and management. The book is divided into 3 parts concerned with historical aspects of erosion, factors involved in various types of erosion, and farming practices and techniques for soil improvement and protection.

A1099 Thompson, Louis M. Soils and soil fertility. 2d ed. 1957. McGraw-Hill. 451p. illus. $11.50.

Covers the usual material found in introductory textbooks for college freshmen soils courses. Author assumes student already has a basic understanding of botany and chemistry. Contains a great deal of worthwhile information for students in other agricultural fields as well as for students of soils.

A1100* Tisdale, Samuel L., and Werner L. Nelson. Soil fertility and fertilizers. 2d ed. 1966. Macmillan. 694p. illus. $12.95.

This book, written by authorities in the field, is designed for students who have already had a basic soils course. Presents basic concepts of plant nutrition and soil fertility, the properties of fertilizer and its manufacture, and the use of fertilizers in soil management programs. Each chapter ends with a summary and a list of questions.

A1101 Vanderford, Harvey B. Managing southern soils. 1957. Wiley. 378p. illus. $5.50.

A nontechnical approach to soil management written primarily for southern states. Deals with each phase of soil management in relation to the production of major southern crops. The use of illustrations and data from many states makes the book of value in other geographic areas.

A1102 Waksman, Selman A. Soil microbiology. 1952. Wiley. 356p. illus. $8.

A modern interpretation of microbiology as applied to soils. In addition to the study of microorganisms in soil, the book includes a brief history of soil work and 17 famous soil microbiologists who contributed to it. A bibliography is appended to each chapter. Will serve as a ready reference for all concerned with soil-plant relationships.

A1103 Wallwork, John A. Ecology of soil animals. 1970. McGraw-Hill. 283p. illus. $8.16.

A thorough and up-to-date coverage of the elements of pedology, systematic zoology, and quantitative ecology. Gives general treatment of major animal groups, decomposition processes, etc. Author's easy style of presentation should make the book especially appealing to students.

A1104 Welch, Charles D., and Gerald D. McCart. An introduction to soil science in the Southeast. 1963. North Carolina. 280p. illus. $6.

A vocational/technical level text covering the formation and classification of soils, soil water, soil testing, land use evaluation, plant nutrition and growth, soil fertility and fertilizers, and soil and crop records. Can be adapted for use in sections other than the Southeast.

A1105 Worthen, Edmund L., and Samuel R. Aldrich. Farm soils: their fertilization and management. 5th ed. 1956. Wiley. 439p. illus. $5.50.

A handy guidebook and manual presenting in as nontechnical a manner as possible, the fundamental principles of farm soil management and fertilization. Technical terminology not in common use is defined in the glossary.

Agricultural Chemistry and Physics

A1106* Farm chemicals handbook. 1970. Meister. 496p. illus. $16.

A comprehensive handbook kept up-to-date by annual revisions. Contains a dictionary of plant foods, a dictionary of pesticides, an encyclopedia of fertilizer trade names, and a buyers' guide. Useful as a reference for both teachers and students in farm chemicals classes.

A1107 Rose, C. W. Agricultural physics. 1966. Pergamon. 226p. illus. $6; paper, $4.50.

Deals primarily with the principles of physics as applied to agriculture, with the exception of agricultural mechanics. Intended to bridge the gap between high school and research levels of understanding. Useful as a reference source for vocational/technical students studying soils.

A1108* Sauchelli, Vincent. Trace elements in agriculture. 1969. Van Nostrand. 248p. illus. $15.

A survey for the student and general reader containing a wealth of current facts on micronutrients and their function in sustaining animal and plant life. Covers the trace elements essential to plants and animals, their importance to the feed and fertilizer industries, and practical information on the diagnosis and correction of element deficiencies. Contains references for further study.

A1109* Shirlaw, Douglas W. Gilchrest. A practical course in agricultural chemistry. 1967. Pergamon. 158p. illus. $6; paper, $4.50.

A textbook presenting practical methods in agricultural chemistry and omitting theory. Has chapters on the analysis of soils, fertilizers and manures, feeding stuffs, milk, and other subjects. Includes a chapter on the instruments and equipment used in agricultural chemistry. Suitable for college students not specializing in the subject and for those in technical agriculture schools.

A1110 Wijk, W. R. Van, ed. Physics of plant environment. 1966. Amer. Elsevier. 382p. illus. $12.95.

Concerned with physics as applied to soil-plant science. Temperature variations in soils, thermal properties of soils, atmospheric pollution, and air and soil physical characteristics which determine plant environment are some of the areas covered.

A1111 Zweig, Gunter. Analytical methods for pesticides, plant growth regulators, and food additives. 1963– . Academic. 6v. v.6 forthcoming. illus. v.1 $24; v.2 $23; v.3 $12; v.4 $12; v.5 $24; v.6 price not set.

Brings together current information on analytical methods for laboratory workers responsible for studying pesticides, plant growth regulators, and food additives. The first 4 volumes deal with principles, methods, and general applications; insecticides; fungicides, nematocides, rodenticides, and food and feed additives; and herbicides. Volume 5 complements volume 1 and updates volumes 2 through 4. Volume 6 not yet published.

Climatology and Meteorology

A1112 Battan, Louis J. Cloud physics and cloud seeding. 1962. Doubleday. 144p. illus. paper, $1.25.

A simple, brief explanation of the mechanics of cloud formation, the different forms of precipitation, and man's efforts to artificially control rainfall. Written for young adults with little or no science background.

A1113 ———. Harvesting the clouds: advances in weather modification. 1969. Doubleday. 148p. illus. $4.95.

An introduction to weather control for the student and layman. Reviews rain formation processes and cloud-seeding techniques, gives an account of ways man has attempted to modify the weather, and describes possible effects of weather modification on the earth's ecology.

A1114 Cantzlaar, George L. Your guide to the weather. 1964. Barnes & Noble. 242p. illus. $4.50; paper, $1.50.

An orderly arranged treatment of the weather, using a minimum of technical language and presupposing no background in the subject. Deals with all aspects ranging from atmospheric fundamentals to weather observation, weather forecasting, and research projects.

A1115* Chang, Jen-hu. Climate and agriculture: an ecological survey. 1968. Aldine. 304p. illus. $9.75.

A broad survey of the relationship between crop plants and climate, emphasizing general principles rather than experimental details. Intended for use as an agricultural climatology classroom text or as a reference manual for research workers.

A1116 Critchfield, Howard J. General climatology. 2d ed. 1966. Prentice-Hall. 465p. illus. $10.50.

A nontechnical approach to the physical, regional, and applied phases of climatology. Explains the fundamentals of weather, describes the climates of the world, and discusses the effects of climate on plant and animal life.

A1117 Day, John A. The science of weather. 1966. Addison-Wesley. 214p. illus. $8.25.

A simplified, authoritative book covering all major aspects of weather science. Intended as an elementary meteorology textbook for the nonscience major on the advanced high school or community college level. Amply illustrated with pictures and diagrams. Requires some knowledge of algebra for full understanding.

A1118 Lowry, William P. Weather and life: an introduction to biometeorology. 1969. Academic. 306p. illus. $5.95.

A timely account of the biological effects of weather and climate on plants and animals, including man. Written from a meteorological rather than a biological viewpoint.

A1119 McIntosh, Douglas H., and **A. S. Thom.** Essentials of meteorology. 1969. Springer-Verlag. 238p. illus. paper, $2.80.

An authoritative presentation of current meteorological knowledge. Gives a concise introduction, on a beginning college level, to the major fundamentals of modern meteorology. Assumes the reader has a basic background in mathematical science.

A1120 Petterssen, Sverre. Introduction to meteorology. 3d ed. 1968. McGraw-Hill. 333p. illus. $11.50.

A general introduction to meteorology presented on an elementary level for students of various interests. Subjects discussed include types of clouds and precipitation, laws of motion, widespread weather phenomena, world climates, etc. Little attention given to the science of weather forecasting. Appendixes cover instruments, symbols, technical terms, and tables of equivalents. Requires little mathematical background.

A1121 Rumney, George R. Climatology and the world's climates. 1968. Macmillan. 656p. illus. $12.95.

Introduces the reader to general climatology and the distribution and nature of the various climates of the world. The major portion of the book is devoted to descriptions of the world's climatic regions. Classified according to the principal plant formations. Will appeal to general readers as well as students of meteorology, forestry, ecology, and plant geography.

A1122 Taylor, James A., ed. Weather and agriculture. 1967. Pergamon. 225p. illus. $13.50.

A collection of papers concerned with weather and its relationship to agriculture, particularly in England and Wales. Covers weather and climate in respect to the growing season, shelter problems, land usage, soil, plant and animal disease, agricultural productivity, and hazards.

A1123 Thompson, Phillip D., Robert O'Brien, and the editors of *Life*. Weather. 1965. Time. 200p. illus. $3.95.

The story of weather in word and picture intended mainly for the young adult and the nonspecialist. It depicts weather phenomena and its atmospheric causes and discusses the past, present, and future of meteorology.

A1124* Waggoner, Paul E., ed. Agricultural meteorology. 1965. Amer. Met. Soc. 188p. illus. $18.

Current information on climate in relation to agriculture is presented in a concise and orderly manner. Considers the response of animals and plants to heat, cold, and light, plant-soil-water-atmosphere relationships, and the use of atmospheric relationships in making management decisions.

FOOD SCIENCES

General Works

A1125* Bender, Arnold E. Dictionary of nutrition and food technology. 3d ed. 1965. Shoe String. 228p. $9.50.

An essential reference tool for anyone interested in food science and for any library collection in this field. Not only supplies food and nutrition terms but also gives initials and abbreviations, such as A.F.D. (accelerated freeze drying) and H.T.S.T. (high-temperature, short-time pasteurization). Some of the listed terms and their accompanying descriptions are for product names, herbs, cereals, nutritional diseases, processes, and special techniques used in the food industry. A special feature is the extra references listed with almost every entry.

A1126 Borgstrom, Georg. Principles of food science. 1968. Macmillan. 2v. illus. $13.95 each v.

Each volume of this current set deals with food from a different viewpoint. The first considers human food, production and preservation of it, particular foods, and wastage and ruination of such products. The second volume covers the various phases of sanitation and disposal of wastes in food processing. Of interest to agriculturists, future crop growers, biochemists, and those involved with food production and packaging.

A1127 Peterson, Martin S., and Donald K. Tressler. Food technology the world over. 1963-65. AVI. 2v. illus. $15 each v.

A universal picture of food technology and nutrition. Not an essential work for the vocational/technical school library but an informative background text written by well-known specialists from various areas in the food science field. Volume 1 covers the Western Hemisphere and Australia, and the second volume describes food production in South America, Africa, the Near and Far East, and the requirements for running a modern food industry.

A1128* Potter, Norman N. Food science. 1968. AVI. 653p. illus. $15.

A comprehensive introduction to food science presenting all aspects of the field and written essentially for the inexperienced. Treats control methods, unit operations, and specific commodities, as well as world food needs, definitions of food science, characteristics of the food trade, and properties, significance, and nutritive aspects of foods. The style is simplified and explanatory, and the book is helpfully illustrated with diagrams, photographs, and charts.

A1129 Rietz, Carl A., and Jeremiah J. Wanderstock. A guide to the selection, combination, and cooking of foods. 1961-65. AVI. 2v. illus. $17 each v.

Of interest to those learning manufacturing and preserving of foods. Discusses how to select raw products qualitatively, how to combine foods to most enhance their flavor, and how to cook foods properly. Includes understanding and appreciating foods, consumer appeal, the history of food cookery, a discussion of separate foods, how to select and combine them, and special sections on such topics as relative sourness of foods, flavor appraisal procedures in package cookery, quality control, and starch.

A1130 Smith, Donald B., and Arthur H. Walters. Introductory food science. 1967. Classic Pubns. 164p. illus. $3 (approx.).

A general account of the fundamentals of food science written in conversational style which also examines the function of the foods specialist in promoting human health and nutrition. May be used as a career guide for this field because it describes available occupational opportunities. Some of the subjects covered include microbiology and food spoilage, chemistry and biology of food elements, food raw materials, and processing methods.

A1131* Woollen, Anthony, ed. Food industries manual. 20th ed. 1969. Chemical Pub. 524p. illus. $15.

This manual, a reprint of the twentieth British edition, serves as a dictionary-glossary of ingredients, food manufacturing products and equipment, technological terms, and an encyclopedia describing the more frequently used techniques in food processing. This reference text belongs in any food science collection because, though the material is written in technical language, the author includes adequate explanations for readers who are not knowledgable in food science.

Food Composition and Analysis

A1132 Altschul, Aaron M. Proteins: their chemistry and politics. 1965. Basic. 337p. illus. $7.50.

Important for its information on the new sources for proteins and for its presentation of background ideas and facts which are helpful for the high school and first year college student. Discusses protein supply and increased protein production in relation to the world's nutritional needs and its demand for food.

A1133* American Association of Cereal Chemists. Cereal laboratory methods. 7th ed. 1962. Amer. Assn. of Cereal Chemists. 800p. illus. $13.50 (approx.)

A laboratory manual, a quick reference tool, and a review guide which includes 282 procedures used in feed and cereal laboratory science. Gives the scope, definition, apparatus, reagents, procedures, and references for each method and leaves space for new methods developed. Looseleaf for easy revision.

A1134 Amerine, Maynard A., Rose M. Pangborn, and Edward B. Roessler. Principles of sensory evaluation of food. 1965. Academic. 602p. illus. $19.50.

This monograph, a good laboratory reference book, is written in fairly simplified language understandable by junior college students as well as researchers in the field. Prepared from a University of California course on the analysis of foods by sensory methods. Chapters are written on the senses, such as olfaction, taste, vision, etc., and on laboratory studies, test procedures, factors influencing sensory measurements, consumer studies, and other similar topics.

A1135 Bolton, Edward R., and **Cecil Revis.** Oils, fats, and fatty foods; their practical examination. 4th ed. ed. by Kenneth A. Williams. 1966. Amer. Elsevier. 488p. illus. $18.50.

Written by a British author but useful to American as well as British students and researchers because most of the procedures utilized in the 2 countries are similar. Information on laboratory analysis of fats and fatty foods is written in a practical how-to-do-it style. Whenever alternative analytic procedures are discussed, the best methods and the advantages and disadvantages of each are noted.

A1136 CRC handbook of food additives. 1968. Chemical Rubber. 771p. illus. $26.

Fourteen subject areas are covered in this extensively used, widely known manual for the food technologist. These consist of color additives, enzymes, vitamins and amino acids, antimicrobials, antioxidants, acidulants, sequestrants, gums, starches, surfactants, polyhydric alcohols, flavorings, potentators, and nonnutritive sweeteners. A legal approach to food additives is given along with the fundamental characteristics and uses for them. Valuable because it provides in 1 volume information which before was scattered throughout various forms of literature.

A1137* Cox, Henry E., and **David Pearson.** Chemical analysis of foods. 1962. Chemical Pub. 479p. illus. $10.

This is the first American edition of Cox's work which is taken from the fifth British edition. Surveys the methods for testing every type of food and food product for analytical chemists in the food business, teachers on all levels, and beginning as well as advanced students. Includes requirements specified by the U.S. Food, Drug, and Cosmetic Act and supporting bibliographies for further reading. Discusses each type of food separately. A comprehensive work, whose price is low enough for any library. The author, a man of world renown in food analysis, bases the material presented on his long experience. Emphasizes adulteration detection rather than procedures for research in quality control of processing or standards of identification in foods.

A1138 Dixon, Malcolm, and **Edwin C. Webb.** Enzymes. 2d ed. 1964. Academic. 950p. illus. $18.

A treatise specifically written for biochemists, enzymologists, and all those in fields related to enzyme study or research. Includes details on enzyme isolation, techniques, kinetics, specificity, classification, mechanisms, inhibitors, cofactors, structure, formation, systems, and biology. Additional features are an extensive list of references, an atlas of crystalline enzymes, and a lengthy table of most all of the enzymes discovered by 1962.

A1139 Dyke, Stanley F. The chemistry of the vitamins. 1965. Wiley. 363p. illus. $13.95.

Gives essential practices and principles of vitamin chemistry. Thoroughly discusses the 5 vitamin groups, the 8 separate vitamins, and the marginal substances. A useful text in food analysis.

A1140 Goodwin, Raymond W. Chemical additives in food. 1967. Williams & Wilkins. 128p. illus. $10.

A collection of 7 papers given at a symposium which provide a brief survey of chemical additives in foods. The papers describe a new additive to be developed, antioxidants, food packaging aspects, flavors and colors of foods, public health, and legislative actions. Good for almost anyone interested in food science.

A1141* Jacobs, Morris B. The chemical analysis of foods and food products. 3d ed. 1958. Van Nostrand. 970p. illus. $23.50.

Spans the entire subject of food analysis and tells how parts of different foods should be identified and measured. Designed to serve as a reference manual, a guide to U.S. food laws, and a review of specific topics such as pesticide residues and radiochemical determinations. Indispensable for food chemists, quality control directors, and others interested in the subject.

A1142 Jenness, Robert and **Stuart Patton.** Principles of dairy chemistry. 1959. Wiley. 446p. illus. $12.50.

In this work, the author assumes that the reader has a fundamental knowledge of chemistry and biochemistry. The text is an introduction to the field dealing with the basic characteristics of milk, the effect of the combination of various components of milk in different situations, and the significance of these factors in dairy product processing and storage. Gives only the general aspects of dairy chemistry and does not claim to provide comprehensive treatment of each subject reviewed.

A1143 Joslyn, Maynard A., ed. Methods in food analysis; physical, chemical, and instrumental methods of analysis. 2d ed. 1970. Academic. 845p. illus. $35.

Number 9 of Food Science and Technology, a series of monographs. Devotes special attention to the general methods of food analysis and only explains specific food and food product identification and examination to illustrate these. Text also describes how to practically apply the general analytic procedures to separate foods.

A1144 Kertesz, Zoltan I. The pectic substances. 1961. Wiley. 628p. illus. $24.50.

Suitable text for scientific and nonscientific readers because it acquaints them with the major aspects of the pectin substances, including their chemistry, botany, biochemistry, function, manufacture and applications, in a style which is authoritative and, at the same time, understandable. Written from an objective point of view and completely indexed.

A1145 Lowe, Belle. Experimental cookery from the chemical and physical standpoint. 4th ed. 1955. Wiley. 573p. illus. $9.50.

Summarizes the latest research in experimental cookery and concentrates on such specific topics as the relation of cooking to colloid chemistry, sugar cookery, fruits and vegetables, pectic substances, jelly, jam, gelatin, meat and poultry, and eggs. May be used as a reference tool or a text for college courses. Deals with sensory methods and food acceptability as a relatively new area of interest.

A1146* Mackinney, Gordon, and **Angela C. Little.** Color of foods. 1962. AVI. 308p. illus. $12.70.

A book written on an extremely important subject because proper color in food products is a valuable indicator of crop maturity in raw materials and because it affects consumer appeal, quality of food processing, and stability of food in storage. Chapters cover such factors of color determination as vision and perception of color, evaluation of color differences, color tolerances, instrumentation, the natural coloring matters, and color specifications for foods. A section on current problems is a helpful feature. A must for the basic collection in food technology and nutrition.

A1147* Matz, Samuel A. Food texture. 1962. AVI. 286p. illus. $12.50.

Another essential reference for the basic collection, this volume deals with the fundamentals of food texture, the history, methods for obtaining and preserving quality texture, and the reasons for changes in texture during processing or storage. Extensive bibliographic supplements to chapters and a clear, concise, well-organized writing style make this book especially good for the student, teacher, or administrator in the vocational/technical food science program.

A1148* ———. Water in foods. 1965. AVI. 275p. illus. $14.50.

An extremely significant work because of the place of water and its quality in food technology and the failure of other texts on hydrology, water supply, engineering, etc., to emphasize water science from the food technological point of view. Because H_2O is the universal ingredient in all foodstuffs, its volume, form, and position in foods determines the overall quality of the product. Discusses water itself, its nature and properties, the interaction of water with other components, conversion of salt water, wastage, its treatment, and its disposal.

A1149* Merory, Joseph. Food flavorings: composition, manufacture, and use. 2d ed. 1968. AVI. 478p. illus. $19.

This text is founded on the underlying premise that flavor technology is based on proper use of food flavorings, man-made products. Thus, a knowledge of food flavor chemistry as well as the characteristics of the biological substances in foodstuffs is necessary for the flavor chemist, technologist, or other food specialist. Contains 350 formulas for flavor composition, manufacturing techniques, and the latest FDA regulations for production.

A1150 Mohsenin, Nuri N. Physical properties of plant and animal material. 1970– . Gordon & Breach. 2v. v.2 forthcoming. illus. $45.

The analysis of plant and animal material used as food. Compares the physical and mechanical characteristics of these substances in relation to their quality evaluation and control, and their application in processes, machines, structures, and other uses. Encompasses raw materials (meat and fruit, for example), organic fluids, and food products that have been processed.

A1151 Neurath, Hans, ed. The proteins: composition, structure, and function. 2d ed.

1963– . Academic. 6v. v.6 forthcoming. illus. $21 (approx.) each v.

A fairly advanced set for the technical/vocational school level but valuable for reference because it is one of the few unified comprehensive texts on the proteins. Volume 1 discusses analytical methods for studying amino acids, and the later works examine the nature, behavior, structure, and interaction of proteins.

A1152 **Pigman, William W.,** ed. The carbohydrates: biochemistry, physiology. 2d ed. 1970. Academic. 469p. illus. $22.50.

A good reference tool for students involved with the chemistry of carbohydrates. Gives a complete, well-organized account of each major kind of carbohydrate, its structure, physiology, and function. Especially valuable in food analysis.

A1153 **Scheraga, Harold A.** Protein structure. 1961. Academic. 305p. illus. $10.

Even though intended for the graduate student or researcher, this text, which is based on lectures given by Scheraga, contains helpful details about the properties of proteins. Concentrates on the internal structure of protein molecules and other related subjects.

A1154 **Schultz, Harold W.,** ed. Food enzymes. 1960. AVI. 144p. illus. $7.50.

Some of the areas reviewed in this treatise on developments in enzymology are enzymatic reactions of carbohydrates, those concerned with polysaccharide synthesis and breakdown, and those relating to energy production from oxidation of carbohydrates. Emphasizes the practical application of the described enzymological concepts in food technology.

A1155 _____, ed. Symposium on foods: the chemistry and physiology of flavors. 1967. AVI. 552p. illus. $3.

The chemistry and physiology of flavor prepared by over 20 renowned scientists. This book consists of 4 parts: (1) physiological aspects of olfaction and gustation; (2) advances in analytical methodology; (3) flavors of foods; and (4) the sources of food flavor. Useful for many readers including the researcher, flavor chemist, and the student of food analysis.

A1156 _____, and **A. F. Anglemier,** eds. Symposium on foods: proteins and their reactions. 1964. AVI. 472p. illus. $3.

Purpose of book is to stress the relationship between the biochemists in foods and those in the other sciences. Portrays recent methods, protein characteristics and morphology, interaction and breakdown of proteins, specific protein systems, and biological effects of protein interactions. The final chapter, presented in panel discussion format, relates and summarizes all of the facts given in earlier sections for easy review.

A1157 _____, **Robert F. Cain,** and **R. E. Wrolstad,** eds. Carbohydrates and their roles. 1969. AVI. 458p. illus. $3.

Important to the food chemistry and technology student for its appraisal of the role of carbohydrates in nutrition and in the food industry. An advanced work but a useful reference tool for a food technology vocational program. Offers extensive discussion of techniques for determining the fat content of carbohydrates, the reactions of carbohydrates with other food contents, and the chemical and physical structures of cellulose, starch, pectins, etc.

A1158 _____, **E. A. Day,** and **R. O. Sinnhuber,** eds. Symposium on foods: lipids and their oxidation. 1962. AVI. 442p. illus. $3.

Twenty-one papers comprise this text which reviews the nature and processes of lipid oxidation degradation. Because they were originally prepared for a foods symposium, the chapters are written in fairly complex language. Valuable for the technical or professional food analyst, pathologist, and nutritionist because it is one of the few sources which provides modern ideas on the subject in well-organized form.

A1159* **Triebold, Howard O.,** and **Leonard W. Aurand.** Food composition and analysis. 1963. Van Nostrand. 497p. illus. $16.50.

A broad survey of food analysis and composition which acquaints the student and the working food technologist with the general principles and details of the composition, deterioration, and preservation of food. Highlights procedures used in common food analysis and the relevant food laws and regulations. Indispensable as a ready reference source in the food laboratory.

A1160 **Webb, Byron H.,** and **Arnold H. Johnson,** eds. Fundamentals of dairy chemistry. 1965. AVI. 827p. illus. $25.75.

A volume written as successor to *Fundamentals of dairy science* which may serve as a complete reference work in dairy chemistry for students undergoing vocational/technical training in food technology. Begins with a discussion of dairy product composition, then describes product components, and finally covers the equilibria of the fat and protein systems which are influenced by forces applied to them.

Food Products

A1161 Altschul, Aaron M., ed. Processed plant protein foodstuffs. 1958. Academic. 955p. illus. $28.50.

Deals with the plant protein aspect of food production and is directed toward nutritionists, consumers, product producers, and students involved with the study of proteins. Presents the problems and methods for production and utilization of this type of food from both the general and detailed points of view. Somewhat dated but its vast accumulation of established factual information is very helpful.

A1162 American Meat Institute Foundation. The science of meat and meat products. 1960. Freeman. 438p. illus. $8.50.

Outlines meat science and meat processing in an easily understood style. Effectively illustrated to support points made in the text. Discusses the application of the different scientific fields in meat experimentation. Not only of interest to general readers but also executives and administrators in meat and related industries, universities and agencies with programs concerned with meat science, and students learning to process and preserve meats.

A1163 Amerine, Maynard A., Harold W. Berg, and **William V. Cruess.** Technology in winemaking. 2d ed. 1967. AVI. 784p. illus. $27.

An extensive account of wines and wine production all over the world, which delves into raw product composition, the types of wine and their components, molds and yeasts used, and the chemistry of fermentation, winery operations, production processes, spoilage, by-products, and the legal limitations on winemaking. An extremely specialized, useful book which is comprehensive in every way.

A1164 Andersen, Aage J. C., and **Percy N. Williams.** Margarine. 2d ed. 1965. Pergamon. 420p. illus. $17.50.

An enlarged version of its earlier edition containing additional sections on the newest developments in margarine production as well as the fundamental concepts and facts established by Andersen in the first work. A well-developed, clear text easily read by the technical school student.

A1165 Arbuckle, Wendell S. Ice cream. 2d ed. 1966. AVI. 403p. illus. $11.50.

Priced to fit the budget of the small as well as the large technical school library, this volume, like other books on specialized subjects, is a good collection builder when specialization in various areas of food technology is desired. Includes the basic introductory material on manufacturing frozen dairy foods and also recent findings in ice cream processing.

A1166 Austin, Cedric. The science of wine. 1968. Amer. Elsevier. 216p. illus. $6.75.

Assumes that the reader has background knowledge in organic chemistry and biochemistry. Austin gives more technically descriptive information than that available in Amerine's book, *Technology in winemaking*. Examines the basic components of wine, its chemistry, the problems of purifying and preserving it, and several other aspects of wine science.

A1167 Binsted, Raymond H., and **James D. Devey.** Soup manufacture: canning, dehydration, and quick freezing. 3d ed. 1970. Food Trade Pr. 260p. illus. $10.

A comprehensive treatise on soup processing published in Great Britain but applicable to U.S. soup manufacturing as well. Gives a brief history of soups, and then discusses soup stock, the soup-making plant, canning, condensing, processing, and freezing of soups, legal aspects, and other phases of the industry. Presentation is simplified enough for 2-year technical school students and yet detailed enough for the professional.

A1168* Brandly, Paul J., George Migaki, and **Kenneth E. Taylor.** Meat hygiene. 3d ed. 1966. Lea & Febiger. 789p. illus. $15.

A moderately technical work that is well-organized and written in easily understood language. Some of the areas described in the text are: slaughtering; sanitation in plant operation; preparation of meat and meat products; meat grading, and statistics on meat production, consumption, and composition. A helpful guide for anyone working with or studying the meat industry from the meat inspector to the student of animal or food science.

A1169* Bull, Sleeter. Meat for the table. 1951. McGraw-Hill. 240p. illus. $8.50.

Bull writes in simple language because his book is intended for home economics and agriculture students on the high school and college level. Describes cuts and grades of meat of all kinds with information on proper cooking for each. Also discusses the physical structure of the meat-packing industry and the nutritive value of meat.

A1170* Cruess, William V. Commercial fruit and vegetable products. 4th ed. 1958. McGraw-Hill. 884p. illus. $18.50.

A valuable contribution to food product literature. Cruess devotes equal attention to processing methods and the reasons for them. May be used as a reference volume by those in canning, freezing, juice producing, preserving, and other related phases of the food industry. The author begins by discussing the general principles of canning, and then considers the different types of fruits and vegetables, preserve and syrup making, tomato products, and dehydration. Dwells more on fruits than on vegetables.

A1171 **Davis, John G.** Cheese. 1965. Amer. Elsevier. 4v. v.3 and v.4 forthcoming. illus. $16 (approx.) each v.

Covers cheese and its manufacture in a general and technological manner. In volumes 1 and 2, details are included on the history and origin of cheese, milk, the tools and materials used in cheese manufacture, the packaging, storing, and ripening techniques of modern industry, and a comprehensive bibliography on cheese. When published, volumes 3 and 4 will cover the practical economic and the scientific aspects of cheese making. Quite useful to the junior college food technician for reference.

A1172 **Glicksman, Martin.** Gum technology in the food industry. 1969. Academic. 590p. $27.50.

An interesting approach to gum technology including terminology and classification, function in food products, hydrophilic gums such as starch, pectin, and gelatin, and practical applications in various food industries. A highly technical book but written in a style not difficult to read and understand. Complete and well organized for quick access of information.

A1173 **Goose, Peter G.,** and **Raymond Binsted.** Tomato paste, puree, juice, and powder. 1964. Food Trade Pr. 151p. illus. $9.

A well-illustrated, general text offering an informative discussion of the processing of tomato products. Not claiming to be comprehensive in all areas, this book begins with growing and selecting the raw material, and then covers processing, packaging, plant operation, and equipment used. A good source for the novice in the field as well as for those working in tomato product manufacture, quality control, and factory supervision.

A1174 **Hall, Carl W.,** and **Theodore I. Hedrick.** Drying of milk and milk products. 1966. AVI. 320p. illus. $16.50.

This work is devoted to the history of the dry milk products industry, all types of drying equipment, production of dry and fluid products, quality control, use of by-products, plant sanitation, and chemical and physical characteristics of milk in dried form. Meant to be a student's textbook as well as a technologist's handbook. Also of interest to those in marketing and sales.

A1175 _____, and **G. Malcolm Trout.** Milk pasteurization. 1968. AVI. 234p. illus. $14.50.

A comprehensive exposition written on the level of the layman and the technologist. It describes the history, effects of heat treatment, equipment used for pasteurization, milk plant automation, and the economics of, and the plant requirements for, pasteurization. Emphasizes the heat treatment of milk and includes the latest innovations in processing.

A1176 **Hummel, Charles.** Macaroni products —manufacture, processing, and packing. 2d ed. 1966. Food Trade Pr. 287p. illus. $20.

Deals with the importance of macaroni products, manufacture of macaroni, ingredients used, different processes applied, typical extrusion presses, drying of products, quality control, and storing and packaging. Profusely illustrated in color and black and white, this text is adequate for students and technologists needing information on all phases of macaroni production.

A1177 **Jacobs, Morris B.** Carbonated beverages. 1959. Chemical Pub. 333p. illus. $11.

Chapters in this book cover the manufacture and evaluation of all types of carbonated, non-alcoholic beverages, including the raw materials, steps in processing, and composition of the final products. Suggests ways to prevent spoilage, methods for analyzing the products, and gives flow diagrams and flavor and color formula charts for the reader. Uses simplified, concise wordage and a clear, organizational style.

A1178 **Kent, Norman L.** Technology of cereals with special reference to wheat. 1966. Pergamon. 262p. illus. $8.

An orientation to the technology of the cereal grains. Outlines the actual processing of wheat, barley, oats, rye, rice, and maize from raw material production to derivation of the final food. Also views cereal technology from the nutritional angle. Universal in scope and readable for students, engineers, and others seeking an introduction to the field.

A1179 **Kent-Jones, Douglas W.,** and **Arthur J. Amos.** Modern cereal chemistry. 6th ed. 1967. Food Trade Pr. 730p. illus. $30.

Combines details on cereal chemistry, production, and preservation with those on the nutritional aspects of cereals to form a single, complete text on cereals for the professional, technologist, student, or teacher in cereal chemistry, technology, or microbiology. Discusses the various types of cereals, their characteristics and processing, and then delves into subjects such as nutritive value, infestation by insects and mites, and moisture in cereals.

A1180 Knight, James W. The starch industry. 1969. Pergamon. 189p. illus. $3.

Designed to provide a realistic image of the starch industry for students, teachers, and industrial employers or employees. Organized in 10 chapters which examine the history, structure, manufacture, and other characteristics of starch as well as the future trends of the industry. One of its most notable assets is the currency of its information.

A1181 Kosikowski, Frank V. Cheese and fermented milk foods. 1966. Edwards Bros. 429p. illus. $16.

A college text which may also be used in technical schools because it is presented from an applied instead of a theoretical point of view. The author is experienced in teaching and research and also in making the foods which he discusses. The background procedures are used in many countries but a majority of them are practiced in America. Helpful illustrations support the text and explain processes more clearly.

A1182 Lampert, Lincoln M. Modern dairy products. 1970. Chemical Pub. 407p. illus. $15.

Presents the latest information on dairy product composition, processing, nutritive value, chemistry, and microbiology in a nontechnical style for those interested directly or indirectly in dairy technology. Of special interest to teachers, students, and others in agricultural and vocational schools for its insight into the processes involved in the manufacture of these products.

A1183* Lawrie, Ralston A. Meat science. 1966. Pergamon. 368p. illus. $9; paper, $5.95.

An extensive account of the origin and nature of meat in relation to the historical development of the muscle. This treatise is based on the idea that meat is a product derived from the muscle of dead animals. A thoroughly documented work meant to serve the undergraduate student of meat science. It is adequately indexed, and it acquaints the reader with the biological, chemical, and biochemical factors which influence meat quality and production.

A1184* Levie, Albert. The meat handbook. 3d ed. 1970. AVI. 350p. illus. $12.

An up-to-date, comprehensive manual on meat. Covers all aspects of production from livestock to prepared meat that is ready to cook. New additions to the revised volume include discussions of the Wholesale Meat Act of 1967 and its effects, the newly revised Federal Grading Standards and Labeling Rules, and more information on meat quality, cooking and palatability. Written in lay language, it is understandable by the housewife, food service operator, amateur chef, and the student.

A1185 Matz, Samuel A. Cereal science. 1969. AVI. 248p. illus. $13.

This book is based on the first 8 chapters of the *Chemistry and technology of cereals as food and feed,* by Matz, published in 1959 and since out of print. A basic review of the cereal grains, their composition, nutritional value, quality criteria, production, and trade data. For the student and the food scientist. Chapters in this work are of high merit, written by research specialists in the field.

A1186 ———. Cereal technology. 1970. AVI. 388p. illus. $21.25.

Derived from the chapters on processing in the now out of print *Chemistry and technology of cereals as food and feed,* by Matz. Each of the sections on such topics as milling, baking, malting, and macaroni products is written by an expert American cereal technologist. The information in the text is enhanced by the accompanying illustrations, charts, and diagrams. The material is presented in clear, concise, and well-organized form.

A1187 ———. Cookie and cracker technology. 1968. AVI. 320p. illus. $15.

One of the few complete sources of information on cookie, cracker, and snack food production. Although the text presupposes that the reader has a basic background in physical science, it is nevertheless understandable in most parts by the completely inexperienced student. Mechanized production is the emphasized method in the text because small unit production techniques are dealt with in most other references. Covers ingredients, formulas, and procedures and other phases of the industry.

A1188 Minifie, Bernard W. Chocolate, cocoa, and confectionery: science and technology. 1970. AVI. 624p. illus. $23.50.

Not only includes common descriptions of chocolate, cocoa, and confectionery technology, such as production of jellies and jams, wrapping chocolate bars, nut processing, and fruit dehydration, but also examines microbiological aspects, shelf life, and storage problems, which prior to the publication of this book were subjects hardly mentioned in the food literature. Intended for those with general or specific interest in this science whether they are students of quality control, factory managers, food scientists, or public consultants.

A1189* Mountney, George J. Poultry products technology. 1966. AVI. 264p. illus. $15.

This book omits all materials not directly associated with poultry products and by-products and deals with the processing, quality control, handling, marketing, and preparation of the poultry products for the food industry. Written and organized in a simple manner to enhance its value to beginners in addition to food technologists needing a quick reference tool. A special feature is the summary of the latest feed information for producing large, healthy chickens, turkeys, and ducks from the young.

A1190 Parry, John W. Spices. 1969. Chemical Pub. 2v. illus. $8.75 each v.

In these 2 volumes, the whole story of spices is depicted beginning with a historical account on each spice and a comprehensive review of the spice industry, and progressing into a discussion of the spices' morphology, histology, and chemistry. A guidebook to identification, utilization, evaluation, and analysis of all types of spices for the technician, student, and the general reader. Reasonably priced for such a complete set though too specialized for the ordinary basic technical school collection.

A1191 Rosengarten, Frederic. Book of spices. 1969. Livingston. 489p. illus. $20.

Rosengarten's book is eminently valuable for its colorful, large illustrations of 35 different types of spices. The text is composed of general interest material including the origin, historical development, and use of each spice and recipes for cooking with them. Gives miscellaneous details for every type, such as their names in several languages, their characteristics, and location.

A1192 Sivetz, Michael, and **H. Elliot Foote.** Coffee processing technology. 1963. AVI. 2v. illus. v.1 $17.25; v.2 $12.50.

A 2-volume work which describes every facet of coffee processing from growing and harvesting to packaging. It offers historical, practical, and theoretical information on coffee technology. A major purpose of this set is to train new personnel, and thus, it is good for the food technology student specializing in this large industry.

A1193 Spencer, Guilford L., and **George P. Meade.** Cane sugar handbook. 9th ed. 1963. Wiley. 845p. illus. $25.50.

In this handbook, Spencer and Meade present an updated, detailed account of sugar, its manufacture, and refining. A well-known and often used guide for the industry with chapters written by specialists in the field. Complete in content. References to further reading are numerous, and the illustrations greatly enhance the text.

A1194 Stansby, Maurice E. Fish oils; their chemistry, technology, stability, nutritional properties, and uses. 1967. AVI. 440p. illus. $19.

Another book treating a highly specialized subject in foods and food technology, this volume deals with the composition and analysis, production methods and utilization, stability and deterioration, nutritional properties, and geographical aspects of fish oils. A timely representation of the newest research advances and techniques. Very useful as a reference tool, but the factors of high price, technical language, and subject specialization should be carefully considered before purchase.

A1195 _____. Industrial fishery technology. 1963. Van Nostrand. 393p. illus. $12.75.

Specifically written to describe the fisheries and fishery production of the United States because there is a dearth of available fishery material on this geographical area. Intended for a course textbook, and, in general, is well written, fairly comprehensive, well organized, and helpfully illustrated. Statistics are often dated but the overall value of the work is high because it coordinates the scattered information on the subject.

A1196 Sultan, William J. Practical baking. 2d ed. 1969. AVI. 492p. illus. $15.

An excellent guide to commercial baking and other food preparation techniques for the student. Includes bread and rolls, sweet yeast dough products, biscuits and muffins, fried products, pies and pastries, cakes and cake specialties, cookies, and detailed recipes. Basic principles, techniques, and problems are discussed.

A1197 Talburt, William F., and **Ora Smith.** Potato processing. 2d ed. 1967. AVI. 588p. illus. $16.

One of the most complete, up-to-date books on the potato industry. Describes the many techniques and procedures for processing as well as the various methods of raw material selection and storage, the storage diseases, and evaluation of potato varieties. Meant to be a quick reference manual as well as a college text. The new chapter on water disposal makes this work especially relevant to the modern world which is presently concerned with industrial pollution.

A1198 Thorner, M. E., and R. J. Herzberg. Food beverage service handbook. 1970. AVI. 332p. $18.

A practical manual covering all kinds of nonalcoholic beverages. Separate chapters are devoted to each product and include origin of the raw material, processing, and dispensing by the food service industry. Quality control methods and the importance of sanitation and water quality are emphasized. The glossary and the simple terminology used make this an excellent guide for the junior college student in food technology or nutrition.

A1199 Tressler, Donald K., and Maynard A. Joslyn. Fruit and vegetable juice processing technology. 1961. AVI. 1028p. illus. $25.

A detailed reference which examines the processing of fruit and vegetable juices. This work has added scope because the manufacturing methods used in America, Europe, and elsewhere are discussed. Spans the whole field from the history to the quality control, processing methods, and the individual types of juices.

A1200 Webb, Byron H., and Earle O. Whittier, eds. Byproducts from milk. 2d ed. 1970. AVI. 430p. illus. $18.

A review of the information in government bulletins, patents, and other sources on the commercial procedures for using dairy by-products. Although the material included is technical, background discussions are provided to enable students and others having minimal preliminary knowledge of the field to understand the manufacturing processes. Chapters are written by dairy research scientists on such subjects as fermentation products from skim milk, the by-products of milk, sterilized products, and disposal of dairy wastes.

A1201 Weiss, Theodore J. Food oils and their uses. 1970. AVI. 224p. illus. $16.50.

Serves the student, food processor, manufacturer of equipment, plant operator, refinery salesman, and the researcher. Portrays the science and technology of food oils including their chemical and physical nature, processing, composition, packaging, uses, and the problems commonly encountered in using these oils. Presents material in a nontechnical style.

A1202* Ziegler, Percival T. The meat we eat. 1966. Interstate. 547p. illus. $9.75; text ed., $7.50.

One of the best, most comprehensive texts written on meat. Acquaints the reader with the full process of meat preparation from the animal in the field to the meat on the dinner plate. Valuable for vocational/technical school students because it examines such subjects as meat inspection, slaughtering, preservation, storage, and grading. Also includes poultry and game preparation, and the nutritional, chemical, physical, and morphological characteristics of meat.

Food Processing and Preservation

A1203 The Almanac of the canning, freezing, preserving industries. Annual. 55th ed. 1970. Judge. 246p. illus. $8.

Supplies the latest facts on the food and drug standards, the U.S. Dept. of Agriculture grading standards, labeling laws and legislation, good manufacturing regulations, the FOCUS Frozen Food Code, World Buyers and International Trade Statistics, canned food prices, and other similar information as well as the standard policies and procedures of these industries. A practical guide for anyone studying or working in this field.

A1204 Charm, Stanley E. Fundamentals of food engineering. 2d ed. 1970. AVI. 629p. illus. $29.

A mathematical, quantitative approach to food engineering which is considered by the U.S. Office of Education to be on the technical school level although the book assumes that the reader has a background knowledge of calculus. (Charm's work is included in the Office of Education's . . . *Suggested 2-year post high school curriculum* guides for vocational education programs.) Gives the fundamental concepts, their application, and the problems which develop in this area of food science. The principles and problems are illustrated and further explained by good examples.

A1205 Copson, David A. Microwave heating in freeze drying, electronic ovens, and other applications. 3d ed. 1962. AVI. 433p. illus. $16.

Copson begins by discussing the theory of microwave heating, and then, he covers its practical application in freeze-drying, electric ovens, and other areas. Surveys the nutritional and microbiological aspects, the heating, materials used, accessory devices, and components in engineering of the heating process. Useful in the technical school for reference because it is one of the few simplified but comprehensive works written on the subject from the food science point of view. It is a timely work because of the new interest in freeze-drying and microwave heating.

A1206* **Desrosier, Norman W.** The technology of food preservation. 3d ed. 1970. AVI. 488p. illus. $12.

Considers the fundamental techniques of food preservation including salting, canning, drying, pickling, fermentation, freezing, chemical additives, and radiation in detailed content in fairly simple style. Data are given to enable proper application of these principles to food processing. Emphasizes quality in foods, public health in food handling, the use of additives to prevent spoilage, and food-packaging advantages and disadvantages. A college textbook which is quite useful for the vocational school student learning food processing and technology. An up-to-date text with good supportive bibliographies following the chapters.

A1207 **Earle, R. L.** Unit operations in food processing. 1966. Pergamon. 342p. illus. $9.50.

A broad view of food process engineering that gives general coverage of this science. Each process is dealt with separately and details on their applications are omitted. Written for the student and the engineer involved with unit operations of food production.

A1208 **Farrall, Arthur W.** Engineering for dairy and food products. 1963. Wiley. 674p. illus. $17.95.

Encompasses the many aspects of dairy and food product engineering. Supplies information on sanitary engineering, systematic and individual processing units, automation, power application, steam generation and utilization, freezing, and other related topics. The first part of the text examines fundamental principles of mechanical engineering, and latter parts describe plants and equipment. Gives better coverage of the engineering in dairy science than in food processing.

A1209 **Goldblith, Samuel A., Maynard A. Joslyn,** and **J. T. Nicherson.** Introduction to thermal processing of foods. 1961. AVI. 1128p. illus. $16.50.

One of the original works forming the basis of food science and technology literature. Contains papers written on microbiological factors, heat transfer, process calculation, and adaptation to commercial conditions. Not only provides a vital introduction to the field but in doing so, outlines the history of food-processing research. A good background book but not essential for a vocational school collection.

A1210* **Heid, John L.,** and **Maynard A. Joslyn.** Fundamentals of food processing operations. 1967. AVI. 740p. illus. $15.

A condensation of the 3 volumes entitled *Food processing operations,* published in 1963–64, which summarizes their mass of material and produces a more simplified, less expensive textbook or reference work for smaller colleges and universities. Deals with food ingredients, methods of food preservation, and packaging. Authoritatively describes the basic principles for food processing in well-organized format, and offers extensive bibliographies.

A1211 Laboratory manual for food canners and processors. 1968. Nat. Canners Assn. 2v. illus. v.1 $12.50; v.2 $14.50.

For those working with, or studying, freezing or dehydrating of foods, developing new and improved food products, or striving for food quality in canning. Text includes microbiological methods, heat penetration and processing times, processing of canned foods, basic statistical calculations, sanitation, and other similar subjects. Helpfully illustrated and written in practical, handbook terminology.

A1212 **Lock, Arthur.** Practical canning. 3d ed. 1969. Food Trade Pr. 428p. illus. $9.

The author, a consultant to the canning industry, writes with expertise. He acquaints the reader with the latest machinery and equipment used in Great Britain, the Continent, and the United States as well as the techniques used in processing. Gives directions and practical advice for the cannery supervisor and the student planning a career in the canning business. Includes discussions on starting and organizing a cannery, can sealing, canning of different foods, retort operation, and general labeling, storage, hydrogen swells, and quality.

A1213 **Marsh, R. Warren,** and **C. Thomas Olivo.** Principles of refrigeration. 1966. Delmar. 374p. illus. $7.60.

All aspects of refrigeration and refrigerating machinery are discussed in this volume. Includes commercial and industrial as well as domestic systems and gives the fundamental concepts and applications as well as supporting illustrations for each.

A1214 Matz, Samuel A. Bakery technology and engineering. 1960. AVI. 669p. illus. $15.

A new approach to bakery technology which treats the engineering aspects as well as the fundamental and scientific principles of baking. Covers large-scale production and automation techniques and supplies lengthy descriptions of the chemical and technological characteristics of most raw materials. The text has appeal for those interested in cereal chemistry, bakery engineering, supplying raw materials for bakery production, and marketing and sales.

A1215 Reed, Gerald. Enzymes in food processing. 1966. Academic. 483p. illus. $18.50.

Adapted for use more as a textbook than a quick reference tool because specific descriptions are lacking in this general exposition. Adequately documented and comprehensive. This is one of the few books written on the use of enzymes in food production, and it is divided into 2 sections, enzyme chemistry and enzyme technology.

A1216* Sacharow, Stanley, and Roger C. Griffin. Food packaging. 1970. AVI. 412p. illus. $16.

Intended as a text for the inexperienced worker and the student in elementary food technology who is interested in producing, marketing, or utilizing food packages. The only book to cover the whole spectrum of packaging in such a comprehensive, well-organized, and easily understandable manner. Discusses the origins of food packaging and the basic food processes, and then deals with all types of foods separately, their packaging in all kinds of materials, and advice on the best packaging methods. Concludes with a chapter on regulations for the packaging industry. Helpfully defines terms and illustrates points and processes mentioned in the text. Another must for the vocational/technical school library and teacher or administrator.

A1217 Tressler, Donald K., Wallace B. Van Arsdel, and Michael J. Copley. Freezing preservation of foods. 4th ed. 1968. AVI. 4v. illus. v.1 $14.; v.2 $17.; v.3 $19.; v.4 $22.

Volume 1 is a comprehensive survey of refrigeration and refrigeration equipment, which begins with its history and principles and ends with frozen food transport, and the second volume is comprised of the factors affecting quality in frozen foods. The third deals with commercial food freezing operations for fresh foods and discusses the selection and processing of each important kind of food. The fourth covers freezing of precooked and prepared foods. Fairly extensive bibliographies follow each chapter, and even though the language is rather technical, the set is valuable for its practical how-to and its problem-solving information.

A1218 Van Arsdel, Wallace B., and Michael J. Copley, eds. Food dehydration. 1963–64. AVI. 2v. illus. v.1 $10; v.2 $23.50.

A 2-volume work in which the first volume covers the principles of food dehydration and the second, the processes used and products derived. Conditions affecting the drying process such as humidity, temperature, the characteristics of food materials, and the behavior of certain types of dryers are among the details provided by the authors in this set. Written in fairly technical style but includes definitions and explanations for the novice.

A1219 Woolrich, Willis R. Handbook of refrigerating engineering. 4th ed. 1965–66. AVI. 2v. illus. v.1 $10; v.2 $13.50.

Comprehensively presents available information on the fundamentals of refrigeration and food processing and on storing, transporting, and preserving products. Charts, graphs, formulas, tables, and other types of handbook illustrations clarify the detailed text. Important for its coverage of the equipment and facilities utilized for refrigeration. More useful for a reference book than a textbook on the technical school level.

A1220 _____, and **Elliot R. Hallowell.** Cold and freezer storage manual. 1970. AVI. 352p. illus. $21.50.

Written by 2 experts in the refrigeration field, this manual is for cold storage and freezer operators and those in the warehouse business. Subjects included are construction, equipment and materials used, problems in converting caves to warehouses, and warehouse management. The glossary of terms and the conversion factors increase the already certain value of this book as a good reference for those interested in food preservation as a profession.

Food Microbiology and Toxicology

A1221 Ayres, John C., and others, eds. Chemical and biological hazards in foods. 1969. Iowa State. 383p. $9.50.

Adapted from a symposium on food protection held at Iowa State University. Issues discussed were: the advantages and disadvantages of food additives; intentional and incidental additives; pathogenic organisms; and microbial toxins. A clear, concisely worded report evaluating recent progress and knowledge. This work has considerable scope and should be especially useful for those involved with chemical hazards and processing contamination of foods.

A1222   Dack, Gail M. Food poisoning. 3d ed. 1956. Chicago. 251p. $6.95.

Dack, director of the Food Research Institute of the University of Chicago, has produced an indispensable food-poisoning reference for doctors, health officers, and sanitarians based on his extensive research in foods. Necessary revisions have been made in such areas as bacterial food poisoning and the problems evolving from new methods of processing food. Discusses Salmonella, poisonous plants and animals, botulism, and other forms of toxicity.

A1223* Foster, Edwin M., and others. Dairy microbiology. 1957. Prentice-Hall. 492p. illus. $9.95.

Many readers should profit from the practical procedures and details outlined in this exposition on dairy microbiology. Written well and organized to facilitate quick use. Suitable for students as well as laboratory technicians and dairy fieldmen.

A1224   Frazier, William C. Food microbiology. 2d ed. 1967. McGraw-Hill. 537p. illus. $13.50.

A digest of the food microbiology literature which delineates the principles fundamental to this science. Gives examples which further explain the basic concepts and recent developments. The text is in 6 parts: (1) a description of molds, yeasts, and bacteria, especially those responsible for food contamination; (2) the microbiological factors of food preservation; (3) examples of microbial spoilage in food products; (4) the importance of enzyme chemistry to food microbiology; (5) foods in relation to disease; and (6) food sanitation control and inspection.

A1225   ———, E. H. Marth, and R. H. Deibel. Laboratory manual for food microbiology. 4th ed. 1968. Burgess. 122p. $4.

The underlying purpose of this manual is to acquaint the student with procedures for identification and analysis of bacteria and with the general aspects of food microbiology. For any laboratory course concentrating on this subject.

A1226   Graham-Rack, Barry, and Raymond Binsted. Hygiene in food manufacture and handling. 1964. Food Trade Pr. 148p. illus. $9.

A good general book concerned with food spoilage and handling. Although it is published in Great Britain, the general principles and methods described are applicable anywhere. The authors offer helpful answers to questions which the food manufacturer and processor ask about sanitation, as well as know-how information for the student. Some of the chapters cover these topics: food pests and their control; construction and layout of plants and equipment; types of food poisoning; and bacteria, molds, and yeasts.

A1227   Hersom, A. C., and E. D. Hulland. Canned foods, an introduction to their microbiology. 5th ed. 1964. Chemical Pub. 291p. illus. $10.

A very comprehensive text on the microbiological aspects of canned foods for the student, teacher, and researcher in bacteriology, biochemistry, microbiology, and other related fields. Several topics are new in this edition, and they include antibiotics, glass packs, revised testing programs, and ionizing radiation. Extensive reference sections following each chapter are a valuable feature of the book.

A1228   Jay, James M. Modern food microbiology. 1970. Van Nostrand. 328p. illus. $10.25.

In this microbiology book, Jay reviews the history of microorganisms in food, food spoilage, food preservation, food poisoning, and the molecular biology of some important diseases transferred in food. Readers should have a good knowledge of biology and chemistry to benefit from this work, and a background in organic chemistry and biochemistry would enable them to fully comprehend all parts of the text.

A1229   Liener, Irvin E., ed. Toxic constituents of plant foodstuffs. 1969. Academic. 504p. illus. $20.

Discusses the chemical composition and physical properties, the processing techniques which reduce or eliminate the harmful components, and the dietary effects of toxic substances in plant food materials. Deals with the primary sources of these toxins and their place of appearance in the food chain. All of the articles which comprise this book are written by well-known specialists.

A1230   Riemann, Hans, ed. Food-borne infections and intoxications. 1969. Academic. 700p. illus. $28.

Volume 5 of the Food Science and Technology series. Articles are contributed by experts in their fields and by the editor, and thus, the work is highly authoritative. Designed to present the latest developments in food infections and toxic constituents. Even though comprehensive, this is not good for a textbook because the material lacks unification. Excellent for reference.

A1231* **Scharf, John H.,** ed. Recommended methods for the microbiological examination of foods. 2d. ed. 1966. Amer. Public Health Assn. 205p. illus. $6.

Another guidebook to laboratory work dealing with the study of microbiological aspects of foods. In addition to describing the usual methods of examination, it outlines special procedures to be used. Arranges most of the information in chapters on particular types of foods and is only lacking in detailed explanations of how to interpret results. This omission is a result of limited space in the volume.

A1232 **Stumbo, Charles R.** Thermobacteriology in food processing. 1966. Academic. 236p. illus. $10.50.

Stumbo is concerned with clearly representing the bacteriological phases of food thermoprocessing in this book. A very simplified but adequately comprehensive account of heat treatment in foods for the prevention of microorganismic contamination. Consists of 2 major parts, microbiology in the canning industry and the principles and methods of thermoprocessing.

A1233 **Thatcher, F. S.,** and **D. S. Clark.** Microorganisms in foods. 1969. Toronto. 234p. $12.50.

Designed to advance the international standardization of techniques for microbiological examination of foods. These methods were suggested and agreed upon by renowned microbiologists from 11 countries. An essential reference volume for information on universally used techniques.

A1234* **Weiser, Harry H.** Practical food microbiology and technology. 1962. AVI. 345p. illus. $11.

A basic work in food microbiology literature that treats the fundamental relationships between food and microorganisms, the problem of loss from spoilage and how to remedy it, and how to improve food quality. The author assumes that the reader has an elementary knowledge of bacteriology but the layman or beginning student looking for mere practical knowhow information may benefit as much from the book.

Food Quality Control

A1235* **American Public Health Association, Inc.** Standard methods for the examination of dairy products. 12th ed. 1967. Amer. Public Health Assn. 319p. illus. $10.

The twelfth edition of a work that has been in use for 68 years and which has since become an essential reference for anyone interested in testing dairy products. A collection of various proven procedures for examination in the laboratory. Emphasizes 2 basic methodologies, the reference procedure applied in official control, and the faster, more simplified, and less costly method used in milk and other dairy product control.

A1236 **Ayres, John C.,** and others, eds. The safety of foods. 1968. AVI. 367p. illus. $15.

Consists of 6 significant sections which contain information on laws for control of foods in the Americas, the importance of developing universal standards, natural toxicants and those produced by other means, and other relevant topics. Considers several issues of current interest. and thus it is a timely work. The material included was first presented at an international symposium on the Safety and Importance of Foods in the Western Hemisphere.

A1237 **Christensen, Clyde M.,** and **Kaufmann, Henry H.** Grain storage: the role of fungi in quality loss. 1969. Minnesota. 153p. illus. $6.50.

Christensen and Kaufmann provide detailed instruction on how to prevent the deterioration of grains and seeds in storage. Condenses the research findings on the subject and presents them in a style understandable by any reader who is involved with buying, selling, storing, and processing grains or seeds. Concentrates on deterioration from fungus growth in stored products.

A1238 **Gunderson, Frank L., Helen W. Gunderson,** and **Egbert R. Ferguson, Jr.** Food standards and definitions in the United States. 1963. Academic. 269p. $10.

One of the few complete manuals outlining the foundation of the food standards in America. Portrays the role of every governmental agency that is in any way responsible for defining or regulating foods. Meant to serve the lay consumer as well as the food industry employee and the food lawyer with answers to the problems of food regulation. An excellent guide that is well organized, clear, and accurate.

A1239 Herschdoerfer, S. M., ed. Quality control in the food industry. 1967– . Academic. 3v. v.3 forthcoming. illus. v.1 $16; v.2 $17.50; v.3 price not set.

Combines all aspects of quality control from the different areas of food science into 1 3-volume set, which not only presents essays by researchers and officials in the food industry but also provides a philosophy of control and then discusses those factors affecting its application. Specifically, the first volume covers the purposes for control and the ways to effect it, and volumes 2 and 3 cover control of quality in the industrial branches (with a separate discussion of each branch). Intended for the food product distributor, the public health official, the teacher, and the college student.

A1240 Kramer, Amihud, and **Bernard A. Twigg.** Quality control for the food industry. 1970. AVI. 2v. v.2 forthcoming. illus. v.1 $16.50; v.2 price not set.

Although the second volume is not yet published, the first is available and is indicative of the quality and completeness of the whole set. Together, these books will examine quality control in the food industry by describing the basic principles and their applications. The first volume, which deals with the fundamental concepts generally, discusses measurement, reporting, and decision-making practices.

A1241* Nelson, John A., and **G. Malcolm Trout.** Judging dairy products. 4th ed. 1964. Olsen. 453p. illus. $15.

A standard manual for judging and grading dairy products. The usual procedures for grading and scoring all types of milk, cheese, ice cream, and other dairy derivatives are presented. Special techniques are also included for detecting defects and other deficiencies. An important feature is the examples of forms used for grading that are provided in the text. A must for technical school trainees who are learning to rate these products.

A1242 Newlander, John A., and **Henry V. Atherton.** Chemistry and testing of dairy products. 1964. Olsen. 365p. illus. $9.95.

Complete, updated survey of the composition and characteristics of dairy products. Stresses the chemistry of the products and the testing methods for determining the quality. Valuable for the student and the technologist because of its detailed descriptions of chemical properties, laboratory tests, and quality control. Well organized, helpfully illustrated to explain points made in the text, and written in a simplified style understandable by the novice.

A1243* Thornton, Horace. Textbook of meat inspection. 5th ed. 1968. Williams & Wilkins. 596p. illus. $11.75.

Thornton's book is universal in scope because the meat inspection problems of Great Britain have become common throughout the world as a result of increased international food trade. The author is concerned with new methods in plant sanitation and problems such as parasitic diseases and salmonellosis as well as the standard procedures and concepts for examination of meat products.

A1244 Van Arsdel, Wallace B., Michael J. Copley, and **Robert L. Olson,** eds. Quality and stability of frozen foods: time-temperature tolerance and its significance. 1969. Wiley. 384p. illus. $19.95.

Gathers all information available on achieving quality in frozen food products and summarizes it in this single volume for workers in allied industries, food technologists who may or may not be experts in frozen foods, and for students. Founded on the research done by the U.S. Dept. of Agriculture on various commercial frozen products and the derived results which reported the best methods for sustaining high quality food from the raw product to the grocery frozen food shelf. A valuable textbook-guide.

Food Distribution and Trade

A1245 Cross, Jennifer. The supermarket trap. 1970. Indiana. 258p. illus. $5.95.

Reveals the exploitation of the consumer by the supermarket and the food industry due to the deceptions in advertising, packaging, placement of products on the shelves, and other methods for encouraging consumer buying. Cross writes from the citizen's point of view. Stresses the food industry's knowledge of consumer psychology as the basic reason for its success in deceiving customers. Should benefit the producers, buyers, and sellers in the food business as well as the housewife or other citizen who needs deeper insight into the "service" given the customer.

A1246* Darrah, Lawrence B. Food marketing. 1967. Ronald. 358p. illus. $7.50.

An excellent introductory text for agricultural students desiring a complete, detailed book on marketing. Concentrates on the retailing aspects of food distribution including the causes for increased public demand for products. An easily understandable work that omits most technical material.

A1247 Rogers, John L. Automatic vending—merchandising, catering. 1958. Food Trade Review. 105p. illus. $4.

Of interest to those manufacturing foods and food products for the vending industry as well as to those learning the business of supplying the industry and those working directly in the business. Treats all areas of automatic vending and catering including the commercial possibilities, machines and equipment, cafe service vending, area operators and their roles, and goods and materials used. Has some dated sections on equipment and machinery, but overall the book is of high value because of the scarcity of texts available on this subject.

A1248 Williams, Sheldon W., and others. Organization and competition in the Midwest dairy industries. 1970. Iowa State. 339p. illus. $12.50.

Based on research done by the North Central Regional Dairy Marketing Committee on the Midwest Dairy Industries. Written in a professional tone, and gives a comprehensive view of a dynamic field. Care should be taken in evaluating the material because much of the information is impression, not true research analysis. Chapters are on specific areas of the industry such as the fluid milk and ice cream industry, the creamery butter industry, and others.

Nutrition

A1249 Beaton, George H. and **Earle W. McHenry,** eds. Nutrition: a comprehensive treatise. 1964–66. Academic. 3v. illus. $18 (approx.) each v.

An up-to-date approach to the entire field of nutrition in 3 volumes. The first surveys recent knowledge of nutrients and nutrient requirements at different stages of life in the human. The second and third volumes review nutritional adaptation, dietary needs, factors affecting food choice, malnutrition, and other important subjects. Quite a detailed work. It may be too advanced in some parts for the technical school student but it is nevertheless a valuable reference volume for the larger vocational/technical school library.

A1250 Bender, Arnold E. Dietetic foods. 1968. Chemical Pub. 286p. illus. $13.50.

Chapters cover foods for health, dietetic foods, nutritional requirements, and programs to follow. Considers normal people and people afflicted with diseases or conditions which require special dietary regulations. Diet plans are based on the National Research Council regulations for formulas, and special sections are devoted to low calorie products.

A1251 Goldblith, Samuel A., and **Maynard A. Joslyn.** Milestones in nutrition. 1964. AVI. 800p. illus. $15.

An important document in food science and nutrition because it presents, under 1 cover, the most valuable nutrition articles published in the last 150 years. It not only offers the articles as they appeared originally but it also introduces each one by explaining its historical and scientific significance. Some of the chapters are on minerals, the Vitamin B complex, pellagra, the essential fatty acids, and nutrition in the future. The great variety of subjects included makes this a work of very wide scope.

A1252 Lawrie, Ralston A., ed. Proteins as human food. 1970. AVI. 525p. illus. $20.50.

Written by expert British nutritionists, this book deals with the world's supply of protein and the significance of proteins in human nutrition. Special attention is given to the quality of proteins and the effect which certain processing techniques have on them. The protein contents of different kinds of food and raw products are discussed and compared in various ways. Good for students, teachers, and others involved with the biochemistry of foods and nutrition as well as allied fields.

A1253 Munro, Hamish N., and **James B. Allison,** eds. Mammalian protein metabolism. 1964. Academic. 4v. illus. $22 (approx.) each v.

A complex 4-volume treatise on protein metabolism in mammals written by knowledgable specialists. In most cases, this is too advanced for a textbook but it is useful for reference on the vocational/technical school level. For students of biochemistry, nutrition, agriculture, food technology, and clinical science. A work of high quality. (Listed in the U.S. Office of Education's . . . *Suggested 2-year post high school curriculum* guides)

A1254 Pyke, Magnus. Food and society. 1968. Transatlantic. 178p. $6.95.

Outlines the development of nutritional science and the unusual aspects of nutrition that are omitted in most standard texts on the subject. Deals with the world's foods and the specific diets of various regions. Presents this information from a sociological point of view. An easy-to-read book which has a highly valuable final section on today's principles of nutrition and their applications.

A1255* Taylor, Clara M., and Orrea F. Pye. Foundations of nutrition. 6th ed. 1966. Macmillan. 564p. illus. $8.95.

The sixth edition of one of the oldest nutrition texts available today. It has been completely updated to include all of the significant developments in modern nutritional science. A good text, reference tool, or supplementary reading source for the post-high school student.

A1256 Underwood, Eric J. Trace elements in human and animal nutrition. 2d ed. 1962. Academic. 429p. illus. $13.50.

Introduces the reader to the function of trace elements and to their biochemical and pathological role in animals. Oriented toward the general importance and effects of trace elements rather than the specific. Separate sections are written on copper, zinc, iodine, cobalt, and others. Also contains a discussion of soil, plant, and animal interaction and interdependency in relation to these elements.

A1257 Wilson, Eva D., Katherine H. Fisher, and Mary E. Fuqua. Principles of nutrition. 2d ed. 1965. Wiley. 596p. illus. $8.95.

Attempts to provide a simplified nutritional text for the beginning college student or for anyone with enough background knowledge to understand the material. Presents the fundamental nutritional concepts and their applications in nontechnical style. Arranged by broad subject headings, and throughout the work the practical aspects are emphasized.

Food Supply

A1258 Borgstrom, Georg. The hungry planet: the modern world at the edge of famine. 1965. Macmillan. 487p. illus. $7.95.

An influential book on the world food problems resulting from the population explosion and the failure of food producers to meet the new demands. Originally Borgstrom's material was presented as a series of broadcasts. His book was chosen by the American Library Association as one of the 50 most significant books published in 1965. Examines marine resources, synthetic nutrients, water needs, air pollution, and many other factors affecting food supply.

A1259 Brown, Lester R. Seeds of change: the green revolution and development in the 1970's. 1970. Praeger. 205p. illus. $6.95; paper, $2.50.

A critical view of the potential development of agriculture and its ability to satisfy the food needs of the developed as well as the underdeveloped countries. Gives suggestions for definite agricultural policies to be followed and emphasizes the integral part which agricultural improvement plays in the whole developmental process.

A1260 Cochrane, Willard W. The world food problem: a guardedly optimistic view. 1969. Crowell. 331p. $7.95.

An up-to-date treatment of the food supply for the world's population written from the economic point of view. The author, experienced in agriculture as a researcher, administrator, and consultant, supplies current data on food needs and the volume of food that should be produced. Presents the problem realistically, but in a fairly optimistic tone.

A1261 Desrosier, Norman W. Attack on starvation. 1961. AVI. 312p. illus. $7.75.

Completely covers the food supply needs of the world and makes good suggestions for meeting them. Nutrition is given special emphasis because it is the basis of the food supply problem. Using a nutritional study as background, Desrosier then presents a history of our present food systems, their inability to supply today's population, and what must be done for future generations to insure against world starvation. His style is easily understandable for students on the first and second year college level.

A1262 Hardin, Clifford M., ed. Overcoming world hunger. 1969. Prentice-Hall. 177p. $4.95; paper, $1.95.

Intended as background reading for the general reader and anyone working on the twenty-first century hunger crisis. The papers which comprise this volume were first prepared for a National Conference of the American Assembly at Columbia University. They give an excellent review of the important issues and the ideas of noted specialists concerning them.

A1263 Hutchinson, Sir Joseph B., ed. Population and food supply. 1969. Cambridge. 144p. illus. $4.95.

Deals with human fertility, population explosion, birth control, food per capita, space per person, and the other pertinent food and population problems. An excellent introduction and source for information not easily available in most other references. It should provide helpful background knowledge for those in the well-fed countries who will be responsible for the starving nations in the future.

A1264 Mittleider, Jacob R., and **Andrew N. Nelson.** Food for everyone. 1970. College. Pr. 608p. illus. $15.95.

Food for everyone describes special methods for changing infertile land into productive acreage and for improving crop yields from fertile land, and then shows how this Mittleider method of farming can be used to end food scarcity. The authors have written a useful introductory text for plant and food science students and a helpful guide for food producers in countries whose people are already suffering from shortage of food. Chapters cover such subjects as Mittleider gardening, plants and their characteristics, soils and their properties, marketing, costs, and crops and crop care.

A1265 National Farm Institute. Farmers and a hungry world. 1967. Iowa State. 136p. illus. $3.95.

This conference report offers insight into the role of the United States in meeting the world's food shortage. The participants of the meeting represented all regions of America, and these delegates came to 4 major conclusions: (1) farmers of the United States cannot feed the population of the world, and thus American technical assistance in agriculture should be encouraged; (2) trade is preferable to aid; (3) multilateral aid is preferable to bilateral; and (4) it will take 15 years for birth control to substantially reduce population growth. These and other more minor conclusions are discussed fully in the text.

A1266 Pyke, Magnus. Man and food. 1970. Weidenfield. 256p. illus. $4.20.

Designed to acquaint the reader with the problems of food supply and the possible answers to them. Explains that there are 3 factors essential to adequately providing food for people. These include a knowledge of the quantity of necessary components for a healthy diet, how these can be obtained, and methods for proper food distribution. Along with these suggestions, Pyke also discusses population, synthetic foods, leaf protein extraction, food for the elderly, and other items of importance in this area.

A1267 Slater, Sir William. Man must eat. 1964. Chicago. 112p. $3.75.

A highly comprehensive, authoritative account on survival and the international food situation, at present and in the future. Based on 4 public lectures given at the University of Chicago, which included information on nuclear science and agriculture, how this can be applied to the food problem, and a history of the reasons for today's food shortage. A very clear and rational approach to the subject.

NATURAL RESOURCES

General Conservation

A1268* Allen, Shirley W., and **Justin W. Leonard.** Conserving natural resources. 3d ed. 1966. McGraw-Hill. 432p. illus. $10.50.

This textbook for conservation courses covers the complete field of natural resources including soil, water, air, plants, animals, and human resources. Defines conservation and discusses governmental agencies and their programs. Written on the high school and junior college level.

A1269 Bates, Marston. The forest and the sea. 1960. Random. 277p. $6.95.

An examination of the interactions of organisms with animate and inanimate environment for the purpose of demonstrating man's place in nature. More a book of biological and philosophical ideas than a book of facts.

A1270 Black, John D. The management and conservation of biological resources. 1968. Davis. 339p. illus. $7.50.

A teaching manual emphasizing wildlife conservation although it does include chapters on soil, water, forest, and grassland conservation. Useful in training future teachers and for young people interested in conservation careers.

A1271 Borgstrom, Georg. Too many: the biological limitations of our earth. 1969. Macmillan. 384p. illus. $7.75.

Excellent supplementary reading in food economics and food supply courses at the college level. Discusses the biological limits of world natural resources and man's effort to exploit these resources as solutions to the world food problem.

A1272 Burton, Ian, and Robert W. Kates, eds. Readings in resource management and conservation. 1965. Chicago. 609p. illus. $8.50.

An anthology of writings on conservation and resource management rather than a standard type textbook on the subject. A broad range of selections. Covers resources and economic development, scientific and technological change, managerial experiences and alternative approaches. Valuable as a reference book or as supplementary reading on the advanced student and the professional levels.

A1273 Callison, Charles H., ed. America's natural resources. rev. ed. 1967. Ronald. 220p. $6.

An informal approach to the problems of conservation sponsored by the Natural Resources Council of America. Brings together in brief, easily understandable form, basic background information on each of our major natural resources.

A1274 Carr, Donald E. Death of the sweet waters. 1966. Norton. 257p. illus. $5.95.

A candid treatment of water supplies and resources, water use, and misuse in relation to pollution problems and water management. Covers the geographical areas of Africa, Europe, and Asia but devotes the major portion of the book to water resources and problems in the United States.

A1275* Carson, Rachel L. Silent spring. 1962. Houghton Mifflin. 368p. illus. $5.95.

Highly controversial study of harmful effects of chemical pesticides, industrial poisons, etc., on the balance of nature. This classic book by an eminent biologist was influential in arousing the general public and the federal government to concern for our natural resources. No natural resources library collection is complete without a copy.

A1276 Clepper, Henry. Careers in conservation. 1963. Ronald. 141p. illus. $4.50.

A guide to career opportunities in the various branches of conservation. Types of jobs available are listed and specific educational requirements are discussed for each. Covers such areas as soil conservation, wildlife and fishery management, range and watershed management, park development, and forestry. Not completely up-to-date because of advances in research and development since date book was published.

A1277* ———, ed. Origins of American conservation. 1966. Ronald. 193p. illus. $5.

This accurate, readable account of the history and evolution of conservation in the United States was sponsored by the Natural Resources Council of America. Each major field has its own chapter written by a specialist. Includes wildlife, forests, aquatic resources, soil, water, range and forage, parks, and wilderness. List of references for further reading at end of each chapter. Should appeal to a wide audience, including students, laymen, and specialists.

A1278 Cooley, Richard A., and Geoffrey Wandesforde-Smith, eds. Congress and the environment. 1970. Washington. 277p. illus. $8.95.

Case histories of congressional action on environmental and conservation issues and proposals are reported in this book. The authors raise questions but offer no answers in the handling of these issues, in an effort to increase public concern and awareness.

A1279* Darling, Frank F., and John P. Milton, eds. Future environments of North America. 1966. Nat. Hist. Pr. 767p. $12.50.

Man's responsibility to himself and to future generations is the central theme of this collection of papers. Concerned with America's natural resources, man's influence on the environment, regional planning and development, and proposals for modifying existing patterns of environmental change. A valuable reference for students in the fields of conservation and ecology.

A1280 Dasmann, Raymond F. A different kind of country. 1968. Macmillan. 276p. illus. $5.95.

The goal of this book is to present the need for planning to preserve the irreplaceable natural resources and to prevent population growth from exceeding the limit the land can support. Author suggests management programs for preserving natural diversity of the environment. A good reference for those interested in human ecology and preservation of wilderness areas.

A1281 ———. Environmental conservation. 2d ed. 1968. Wiley. 375p. illus. $8.95; paper, $4.95.

An interesting, nontechnical look at the human environment from the viewpoint of conservation. Discusses all the major natural resources with emphasis on those that are renewable. Author ties together the political, economic, and social problems involved in the conservation of these resources.

A1282   Davies, J. Clarence. The politics of pollution. 1970. Pegasus. 231p. $6.

Federal efforts to control air and water pollution is the major concern of this book. It deals with the growth of interest in pollution problems, discusses each of the major water and air pollution acts passed by Congress, describes the part played by various forces in shaping pollution policy, and examines the major pollution control policy processes. Designed to inform rather than to arouse the reader.

A1283*   Ehrlich, Paul R., and Anne H. Ehrlich. Population, resources, environment. 1970. Freeman. 383p. $8.95.

A comprehensive and thoroughly documented analysis of the overpopulation crisis and its effect on the world's environment and resources. Offers constructive proposals to brighten man's future prospects. A timely textbook for environmental science courses and a sourcebook for interested laymen. The food production chapter will be of particular interest to agricultural students.

A1284   Eisenberg, David S., and Walter Kauzmann. The structure and properties of water. 1969. Oxford. 296p. illus. $10; paper, $4.50.

A clear, concise treatment of water, relating its properties to its structure. Summarizes literature on water, provides some major water data, and presents theories effective in correlating these data. Requires only beginning knowledge in chemistry and physics for understanding.

A1285   Graham, Frank. Since silent spring. 1970. Houghton Mifflin. 333p. $6.95.

Begins with a discussion of Rachel Carson, her book, and her reaction to the controversy it started. Then, proceeds to give a factual account of each aspect of environmental pollution by chemical pesticides in the years since *Silent spring* was published. Includes an extensive list of primary sources for readers who wish to study further.

A1286   Helfman, Elizabeth S. Rivers and watersheds in America's future. 1965. McKay. 244p. illus. $4.95.

Not an in-depth study of water resource problems of the United States but a well-written, semipopular presentation of information concerning rivers, dams, watersheds, water management, and conservation. Discusses water resources development and its economic importance to surrounding areas.

A1287   Herbert, Frederick W. Careers in natural resource conservation. 1965. Walck. 110p. illus. $4.50.

A brief guide to careers in conservation of renewable natural resources. In addition to professional careers it describes aid and technician jobs requiring less than a college degree.

A1288*   Jackson, Nora, and Philip Penn. Dictionary of natural resources and their principal use. 2d ed. 1966. Pergamon. 128p. illus. $5.50; paper, $2.95.

A concise, easy-to-use dictionary of the world's natural resources and their principal uses. Describes raw materials and commercial products obtained from these resources, such as alpaca, cork, mica, sesame, wool, and yams.

A1289   Joffe, Joyce. Conservation. 1969. Doubleday. 188p. illus. $6.95.

A short nontechnical account of the fundamental part man plays in his environment. Discusses the damage man has done to the environment and what is being done to repair some of that damage, to preserve the earth's natural beauty, and to more wisely use its resources.

A1290   Lauwerys, Joseph A. Man's impact on nature. 1970. Nat. Hist. Pr. 188p. illus. $6.95.

Briefly discusses man's rise to dominance in nature and how he has influenced its forms and altered its function. The last chapter deals with the critical problem of overpopulation. Written to appeal to the general public.

A1291   McClellan, Grant S. Protecting our environment. 1970. Wilson. 218p. $4.

A compilation of reprints of journal articles, excerpts from books, and addresses on the environment. Covers global aspects of pollution, air and noise pollution, water resources and land pollution, and finally, what can be done about it.

A1292*   McHarg, Ian R. Design with nature. 1969. Nat. Hist. Pr. 197p. illus. $19.95.

A book on ecological planning written by a practicing landscape architect. He emphasizes building and developing without destroying nature and demonstrates with concrete examples. Agriculturally relevant because of the discussions on land use.

A1293   McNall, Preston E. Our natural resources. 1964. Interstate. 280p. illus. $4.95.

Covers all major natural resources in relation to human needs. Gives extensive treatment to soil conservation but covers other natural resources in much less detail. Ignores water and air pollution problems. Valuable as a soil conservation teaching aid on the vocational/technical school level. Can be used as supplementary reading or for reference purposes in other areas of conservation.

A1294 Mellanby, Kenneth. Pesticides and pollution. 1967. Collins. 221p. illus. $6.

Purpose of the book is to create public awareness of the whole problem of environmental pollution. Shows how generalizations about pollutants may not apply in all geographic areas. Discusses present and future dangers, sources of pollution, and legal action against pollutants and polluters. Primarily concerned with Great Britain but not limited to one country in scope.

A1295 Neal, Harry E. Nature's guardians: your career in conservation. 1963. Messner. 191p. illus. $3.64.

Gives basic information on the field of natural resources, chiefly in wildlife management, fisheries, and animal biology. Outlines types of work available and educational requirements, mainly on the professional level but does include technicians such as aids or laboratory assistants.

A1296 Nikolaieff, George A. The water crisis. 1967. Wilson. 192p. $3.50.

Consists of reprints from other sources. Deals with America's water situation, pollution and other factors that limit man's water supply, how the total water supply can be increased, and the question of who pays.

A1297* Osborn, Fairfield. Our plundered planet. 1948. Little, Brown. 217p. $4.50; paper, $1.95.

Demonstrates what man has done and is doing to his environment through the misuse of soil and water resources. Shows how man's conflict with nature is threatening his very existence. Out-of-date in many respects but a valuable source of background information on farming practices. A still popular classic.

A1298 _____, ed. Our crowded planet. 1962. Doubleday. 240p. $4.50.

A collection of essays concerned with overpopulation in relation to the economic, political, moral, and religious life of man and its effect on the natural environment. List of authors includes outstanding men in science, religion, etc. Sponsored by the Conservation Foundation.

A1299 Parson, Ruben L. Conserving American resources. 2d ed. 1964. Prentice-Hall. 521p. illus. $10.25.

A lively presentation of a serious subject. Covers the whole range of natural resources and ends with a chapter on man's prospects and responsibility for intelligent use of natural resources. A reliable aid for the classroom teacher and an interesting account for the general layman.

A1300 Pitts, James N., and Robert L. Metcalf, eds. Advances in environmental sciences. 1969. Interscience. 358p. illus. $15.95.

The first volume of a planned series of multi-authored books devoted to the environment, its quality, and the technology of its conservation. The chapters of volume 1 are reviews more than reports of advances and are written by experts in their respective disciplines. Useful for guiding students toward careers in this important field.

A1301 Roosevelt, Nicholas. Conservation: now or never. 1970. Dodd. 238p. $5.95.

The conservation of scenic resources is the central theme of this book. Using a popular style of writing, the author presents the story of the struggles made in the past to conserve the wilderness areas. Useful as collateral reading for conservation courses at the high school or junior college level.

A1302 Smith, Frank E. The politics of conservation. 1966. Pantheon. 338p. $7.95.

An account of the political history of conservation and development of natural resources in the United States. Covers the complete range of natural resources from forestry to reclamation of arid lands. Written in an easy-to-read style using few technical terms.

A1303* Smith, Guy-Harold, ed. Conservation of natural resources. 4th ed. 1971. Wiley. 685p. illus. $11.95.

A widely used text and reference presenting factual, unbiased information on natural resources conservation. Devotes most space to the conservation of land and water resources. Each chapter ends with a list of references for further reading. The author's treatment of soils, forests, fisheries, and irrigation agriculture will be of particular interest to agriculture students.

A1304 Udall, Stewart L. The quiet crisis. 1963. Holt. 209p. illus. $5.

Vividly describes the history of the use and misuse of America's natural resources. Discusses contributions men such as Muir and Pinchot

A1305   Van Dyne, George M., ed. The ecosystem concept in natural resource management. 1969. Academic. 383p. illus. $16.50.

A collection of papers reviewing and discussing the ecosystem concept in relation to natural resource management. Covers the evolution, development, and importance of the ecosystem concept; its application in natural resource research and management; and the use of the ecosystem concept in natural resource sciences training. Contains mostly biological and agricultural applications.

made to the conservation movement. Interprets man's relation to the land and the present trends in conservation policies. Important for background reading in conservation classes.

A1306   Veatch, Jethro O., and **Clifford R. Humphreys.** Water and water use terminology. 1966. Thomas Print. 375p. illus. $12.95.

An authoritative guide to the language of water and water use. Terms are listed alphabetically, defined, and discussed. A large number of illustrations are used to aid in clarifying the subject matter. Recommended for all who have a special interest in water, whether land and water use specialists, students, conservationists, or legislators.

A1307   Watt, Kenneth E. F. Ecology and resource management. 1968. McGraw-Hill. 450p. illus. $14.50.

Covers the problem, theory, principles, and methods of resource management. Major portion of the book devoted to application of the techniques of operations analysis in resource management. Calculus and basic statistics are essential prerequisites for understanding.

A1308   White, Gilbert F. Strategies of American water management. 1969. Michigan. 155p. illus. $5.95.

An appraisal of American experience in water use and control. Explains the major strategies practiced in the United States for managing water and their social, environmental, and economic effect on American citizens. Of special interest to agriculturists is the section on rural water-supply management.

A1309* Whitten, Jamie L. That we may live. 1966. Van Nostrand. 251p. $6.95.

Presents the opposite side of the story from that presented in Carson's *Silent spring.* Gives a fairly balanced account of the benefits of pesticides and the associated dangers. Book's control theme is that pesticides used widely constitute no real threat to man. Anyone who reads *Silent spring* should also read this book.

A1310   Wilber, Charles G. The biological aspects of water pollution. 1969. Thomas. 296p. illus. $23.75.

An overview of the biological influence of pollution on plants, animals, and man. Author attempts to reflect the urgency of the situation without preaching panic. Treatment of subject matter ranges from elementary to technical. Not a textbook but a source of information for all who are involved in water pollution problems.

A1311   Wright, James C. The coming water famine. 1966. Coward. 255p. $6.95.

A description of the increasing abuse of our water supply, what is being done about it, and what should be done. At times, the author attempts to appeal to the reader's emotions as well as his intellect. Can be used as supplementary reading for students studying water conservation.

Wildlife Management

A1312   Allen, Durward L. Our wildlife legacy. 1962. Funk & Wagnalls. 422p. illus. $7.95.

Shows how our wildlife interests can be managed successfully in relation to the often conflicting interests of public officials, administrators, and sportsmen. Written by a man with many years of experience in federal and state game management. Primarily for the layman, but also of interest to all in this field.

A1313* Dasmann, Raymond F. Wildlife biology. 1964. Wiley. 231p. illus. $7.95.

An introductory textbook stressing the ecological principles underlying the management of mammal and bird resources. Discusses the history of wildlife conservation and its aesthetic, ethical, and practical values; wildlife communities; principles and techniques of wildlife management; methods of studying wildlife populations; and wildlife populations and their regulation.

A1314   Douglas, William O. A wilderness bill of rights. 1965. Little, Brown. 192p. illus. $6.75; paper, $1.95.

One of Justice Douglas' best-known books in his extracurricular field of interest—nature and its conservation. Here he reviews the practices and major problems of land use by federal, commercial, and private interests in relation to a rapidly growing population and advances a

12-point proposal for a bill of rights designed to preserve the wilderness through changing administrations of government and shifts of government policies. Useful for alerting people to the need for balance between the various interests in land use.

A1315 Ehrenfeld, David W. Biological conservation. 1970. Holt. 228p. illus. paper, $3.50.

Reviews the history of conservation in the United States, discusses factors that threaten natural communities and individual species, and gives methods for their preservation. One of the few introductory textbooks on this subject beginning to be considered as an interdisciplinary one involving biology, the social sciences, and the humanities.

A1316* Fisher, James, and others. Wildlife in danger. 1969. Viking. 368p. illus. $12.95.

A standard reference work for all libraries. Encyclopedic in style, with high quality illustrations. For each endangered species, the biology and status of extinction is given. Covers only living forms of mammals, birds, reptiles, amphibians, fishes, and plants.

A1317 Gabrielson, Ira N. Wildlife conservation. 2d ed. 1959. Macmillan. 244p. illus. $5.95.

Revised edition of an early textbook on wildlife conservation by a noted authority and former director of the U.S. Fish and Wildlife Service. Puts into simple language the basic facts in this field and emphasizes that the various programs for the conservation of wildlife, forests, water, and soil are closely interwoven. Divided into 2 parts—one showing the interdependence of conservation programs, and the other dealing with specific problems of certain groups of wildlife.

A1318 Hewitt, Oliver H., ed. The wild turkey and its management. 1967. Wildlife Soc. 589p. illus. $6.

The story of progress in restoring the wild turkey, written by professional biologists in a factual, scientifically documented manner for those in wildlife management and for the general reader. Covers historical background, taxonomy, distribution, and present status of the wild turkey; biological and ecological material; management history of the various turkey races by geographic areas; turkey hunting, and a look to the future.

A1319 Laycock, George. America's endangered wildlife. 1969. Norton. 226p. illus. $4.95.

Has chapters devoted to individual representative animal species threatened with extinction. Describes their present status, what is being done, and what should be done to protect them. An outstanding feature of this book is its appendix with an annotated list of prominent species of rare and endangered American wildlife, followed by a list of concerned organizations. A valuable reference for students and laymen concerned with wildlife conservation.

A1320* Leopold, Aldo C. Game management. 1933. Scribner. 481p. illus. $8.95.

First comprehensive attempt to adapt the art of game management to biological principles and to American conditions and traditions, and to develop and define terms in the field. Still the acknowledged bible for the science, and its author is known as the father of modern wildlife management. Intended as a college textbook and also as an explanation of how game management relates to their field of interest for the sportsman, biologist, agricultural expert, and forester.

A1321 McClung, Robert M. Lost wild America: the story of our extinct and vanishing wildlife. 1969. Morrow. 240p. illus. $5.95.

Narrative of how Americans have treated wildlife since colonial days, followed by brief accounts of some endangered species of animals, including man himself. Covers the extinct, the nearly extinct, and the possible victims of the future. Gives excellent information on current problems in the control of our natural environment and suggests a conservation program for the future.

A1322* Martin, Alexander C., Herbert S. Zim, and **Arnold L. Nelson.** American wildlife and plants: a guide to wildlife food habits. 1951. Dover. 500p. illus. paper, $3.

A standard reference guide, based on research by the U.S. Fish and Wildlife Service, covering the food habits of mammals, birds, reptiles, amphibians, and fishes and the plants they feed upon. Also discusses geographical distribution of wildlife and their migratory patterns. Good for teaching principles of plant-animal relationships, although not a textbook.

A1323 Matthiessen, Peter. Wildlife in America. 1964. Viking. 304p. illus. $1.95.

A history of man's effect on American wildlife from colonial times to the present and a record of the conservation movement as applied to wildlife. Well documented, and written with distinction.

A1324 Murphy, Robert. Wildlife sanctuaries: our national wildlife refuges—a heritage restored. 1968. Dutton. 288p. illus. $22.50.

Attractive book describing a selection of the nation's wildlife refuges and some of the animals which inhabit them. Introductory chapters give the history of wildlife conservation and information on migratory bird pathways. Main body organized by regions of the country, with an appendix giving brief annotations of refuges by state. Copiously illustrated with color and black and white photographs and maps. For pleasure reading and for reference.

A1325 Schorger, Arlie W. The wild turkey: its history and domestication. 1966. Oklahoma. 625p. illus. $10.

Deals with the history, biology, and management of the wild turkey. Also has chapters on the turkey in art, on hunting, and on utilization of the turkey. Includes a lengthy bibliography. A well-illustrated book for the student or specialist, not the general reader.

A1326 Shomon, Joseph J., and others. Wildlife habitat improvement. 1966. Nat. Audubon Society. 96p. illus. $2.50.

A practical handbook giving suggestions for improvement of habitats of many kinds of wild birds and mammals in the United States. Numerous photographs, diagrams, and drawings help to explain the textual material. Useful for vocational/technical students as well as landowners, community leaders, and managers of nature preserves.

A1327* Trippensee, Reuben E. Wildlife management. v.1 1948, v.2 1953. McGraw-Hill. 2v. v.1 $9.50; v.2 $11.50.

Contains sound practical information based on proven research which would be invaluable for solving problems in wildlife management. Volume 1 covers general principles of wildlife management and the life histories, ecology, and management of upland game species; volume 2 deals with water conservation, the management of swamp and marsh habitats, and the life histories, ecology, and management of the animals which inhabit them.

A1328 Van Dersal, William R. Wildlife for America: the story of wildlife conservation. rev. ed. 1970. Walck. 160p. illus. $6.

Describes how America has endangered its wildlife, what has already been done to protect it, and measures which can be taken in the future to prevent other species from becoming extinct. Simply written, readable presentation for the student and layman.

A1329 Ziswiler, Vinzenz. Extinct and vanishing animals: a biology of extinction and survival. 1967. Springer-Verlag. 133p. illus. paper, $3.40.

A translation and revision of a German book which gives a reasonably complete report on how world wildlife is being destroyed, why protection of wildlife is desirable, and what can and is being done about its preservation. Perhaps overdoes the horrors, according to some reviewers.

Parks and Recreation

A1330* Brockman, Christian Frank. Recreational use of wild land. 1959. McGraw-Hill. 346p. illus. $11.50.

The first book to treat this subject in 1 volume and to coordinate the various points of view on it. Considered by many as a solid and lasting contribution to the field. Useful not only as a classroom text, but also to administrators of wild lands and zoning.

A1331 Clawson, Marion, and **Jack L. Knetsch.** Economics of outdoor recreation. 1966. Johns Hopkins. 328p. illus. $8.50.

Deals with important operational and policy decisions which must be met by workers in the field and provides the background information necessary for understanding the nature of outdoor recreation. Raises issues of public policy on this subject, but does not try to solve them. Written by economists, but aimed at park and recreation workers and at students.

A1332* Doell, Charles E. Elements of park and recreation administration. 2d ed. 1968. Burgess. 334p. illus. $7.

Presents the basic knowledge necessary for general park administration. Discusses all classes and types of parks and the various policies for their acquisition and administration. Valuable as textbook for students and quick reference for practitioner. Written by an expert in the field, both as an administrator and as an educator.

A1333 Douglass, Robert W. Forest recreation. 1969. Pergamon. 335p. illus. $14.

Brings together in 1 volume for the first time the specific research findings on how to plan, develop, and administer forest recreation areas, the problems that confront forest recreation administrators, and the desires of the using public. Recounts actual problems and offers practical solutions. Written for the forest land manager and for those doing the job on the technician level, but also useful as a student textbook.

A1334 **Jensen, Clayne R.** Outdoor recreation in America. 1970. Burgess. 285p. illus. $8.25.

Subtitle, *Trends, problems, and opportunities,* explains scope of this volume. One of the few books which interprets the present and future significance of outdoor recreation, describes the responsibilities of the numerous agencies and organizations involved, covers the recent legislation and programs affecting this field, points out the need for interagency cooperation, and identifies current and potential future problems in outdoor recreation. Should be especially meaningful for teachers and students, employees, and administrators in the field, as well as interested laymen.

A1335 **Madow, Pauline,** ed. Recreation in America. 1965. Wilson. 206p. $3.50.

A volume in The Reference Shelf series. Contains reprints of articles and excerpts from books and addresses which give information and comment on the effects of increased leisure in America and on the role of public agencies in providing recreational facilities. Examines the developing profession of recreational management and reviews the various private agencies offering leisure-time opportunities. Useful for background information.

A1336 **Seymour, Whitney N.,** ed. Small urban spaces. 1969. New York. 198p. illus. $6.59.

The subtitle, *The philosophy, design, sociology, and politics of vest-pocket parks and other small urban open spaces,* describes the contents of this book. Written by people involved with the Park Association of New York City, Inc. Of interest to the general reader as well as the specialist in parks and recreation and the landscape architect. Good collateral reading for students in both fields.

A1337* **Smith, Clodus R., Lloyd E. Partain,** and **James R. Champlin.** Rural recreation for profit. 2d ed. 1968. Interstate. 319p. illus. $9.25; paper, $7.25.

A practical guide for rural landowners in planning, developing, operating, and managing rural recreation enterprises. Covers various kinds of recreational facilities, such as vacation farms and ranches, campgrounds, and picnic and sports areas, fishing waters, hunting areas, etc. Should be valuable in vocational/technical agriculture courses, such as agribusiness.

SOCIAL SCIENCES

Agricultural Economics and Agribusiness

A1338 **Beneke, Raymond R.** Managing the farm business. 1955. Wiley. 464p. illus. $7.95.

A discussion of the principles basic to the organization and management of all types of farms, supplemented with physical and economic data. Each chapter centers about a different specific management problem. Some areas covered include obtaining a farm, planning the farm program, managing livestock and poultry enterprises, conservation practices, credit, prices, farm accounting, etc. Prepared primarily for vocational/technical students, but also useful to farm operators.

A1339 **Benjamin, Earl W.,** and others. Marketing poultry products. 5th ed. 1960. Wiley. 327p. illus. $7.95.

Outstanding text and practical reference work designed for students, producers, distributors, and consumers. Covers the entire field, including production, methods of handling, distribution and merchandising, by-products, research, and education. Stresses the fundamental principles underlying the economics of marketing as applied to poultry and eggs.

A1340 **Bishop, Charles E.,** and **William D. Toussaint.** Introduction to agricultural economic analysis. 1958. Wiley. 258p. illus. $7.95.

Provides a theoretical foundation for use in analysis of agricultural economics problems. Intended primarily for second and third year college students to equip them with analytical principles useful in other sciences, such as agronomy, horticulture, animal industry, and poultry. Older book, but may be used as an additional source of information.

A1341 **Bradford, Lawrence A.,** and **Glenn L. Johnson.** Farm management analysis. 1953. Wiley. 438p. illus. $9.95.

An integration of the traditional and the modern approaches to farm management and farm organization in a college textbook for both beginning and advanced students. The first 3 chapters are used to define the subject, the fourth to introduce farm management terms, and the remaining chapters to present and discuss problems of organizing and operating farms. Emphasizes the land-use approach to farm organization and management. Old, but still a much used book.

A1342* Brake, John R., and others. Farm and personal finance. 1968. Interstate. 132p. illus. paper, $2.95.

A discussion of the need for, and sources of, capital, the types and sources of credit, and protection of capital by insurance. Shows such things as how to figure the cost of borrowing money, especially on installment buying, how to estimate monthly payments on a house, etc. A textbook and practical handbook for student and farmer alike—in fact, for just about anyone.

A1343 Breimyer, Harold F. Individual freedom and the economic organization of agriculture. 1965. Illinois. 314p. $6.50.

Author concerns himself with 2 alternative kinds of organization and the policies and problems related to them: the traditional agriculture of the individual farmer versus the modern well-capitalized, large-corporation type of agriculture. Includes discussions of the policies of major national farm groups. A highly readable and comprehensive analysis of an extremely complex subject.

A1344 Case, Harold C. M., and Donald B. Williams. Fifty years of farm management. 1957. Illinois. 386p. illus. $6.95.

A history of farm management, with special reference to research, extension work, and teaching as developed in the U.S. Dept. of Agriculture and the land-grant colleges. Purpose was not to describe the evolution of farm management practices, but rather to show how research procedures have developed and how the science of farm management has found its application in the United States.

A1345* Castle, Emery H., and Manning H. Becker. Farm business management. 1962. Macmillan. 423p. illus. $8.95.

The clarity and simplicity with which the textual material is presented, its orientation toward real problems which managers face, and the completeness of its coverage make this an outstanding introductory textbook on the principles of farm management. Will need the use of supplementary regionally adapted illustrative materials, however. Written primarily on introductory college level.

A1346 Chastain, Elijah Denton, Joseph H. Yeager, and E. L. McGraw. Farm business management. 2d ed. 1968. Auburn Print. 175p. illus. $2.50.

Presents principles, concepts, and problems in farm business management in a brief, but clear style. Covers agriculture as an industry, how to get started, wise use of credit, prices, marketing, farm business analysis, labor and mechanization, and economic development. A textbook for high school or post-high school classes.

A1347 Clark, Colin. Population growth and land use. 1967. St. Martin's. 406p. illus. $14.

Author has brought together material from many diverse fields, including agriculture, to show the causes and problems of population growth and the resulting pressure on land use. He also proposes measures for curing these evils. Many of his views are controversial—making the book interesting to read.

A1348 Clawson, Marion. Man and land in the United States. 1964. Nebraska. 178p. illus. $5.95.

Directed toward a nonprofessional audience, this book dramatizes the history of land use in the United States. Includes chapters on agricultural land use and tenure and on forest landownership and management. Useful for a brief easy-to-read summary.

A1349 ──────, R. Burnell Held, and Charles H. Stoddard. Land for the future. 1960. Johns Hopkins. 570p. illus. $14.

Discusses the history, the present, and the future uses of the major land types in the United States—urban, recreational, agricultural, forestry, grazing, and miscellaneous. The 4 appendixes include many statistical tables and supporting data. A point of departure for continuing study and analysis of land resource issues and policies. Uses a minimum of technical terminology.

A1350 Compton, Henry K. Storehouse and stockyard management. 1970. Business Books. 491p. illus. $35.

Deals with the industrial engineering aspects of store management—the techniques of storage, handling, packing, and transport of goods in day-to-day practice. Uses a minimum of mathematics. Not too technical and therefore useful for the agribusiness student. A British book, and so it should be noted that *stockyard* does not have the American meaning of an enclosure for keeping animals, but rather the British one of a yard in which to store materials.

A1351 Dillon, John L. The analysis of response in crop and livestock production. 1968. Pergamon. 135p. illus. $4.50.

Aim was to give an introductory outline of the analytical principles involved in appraising the efficiency of crop-fertilizer and livestock-feed response. Includes worked examples and exercises for practice in each chapter. Valuable in the vocational/technical collection because it presents the association between the biological and economic aspects of crop and livestock production.

A1352 Dorries, W. L., and J. Roland Hamilton. Economics for modern agriculture. 1965. Exposition. 227p. illus. $5.

A multipurpose textbook in general agricultural economics which applies the principles of economics to the problems of agriculture and related industries. Surveys all aspects of economics for the agricultural student at high school and beginning college levels and stresses the importance of a scientific attitude in the solution of current problems affecting the entire agricultural industry.

A1353 Efferson, John N. Principles of farm management. 1953. McGraw-Hill. 431p. illus. $9.95.

Purpose was to present the basic principles of farm management in simplified form for agriculture students and farmers. Author chose to omit statistical tables and let the instructor select illustrative material based on local situations. Has exercises at end of each chapter sufficiently broad to be used in any region. Intended to be applicable to all areas and all systems of farming and to cover all major problems involved in a farm business enterprise.

A1354 Fowler, Stewart H. The marketing of livestock and meat. 2d ed. 1961. Interstate. 622p. illus. $11; text ed., $8.25.

A systematic survey of the whole field of marketing livestock and meat. Covers such topics as meat production and consumption; the meat packing industry, evaluating livestock and meat; when, where, and how to market livestock; transportation; by-products; government regulations, etc. Intended not only as a textbook for college students, but also as a reference for the stockman, extension specialist, packer buyer, and student at the secondary school level. Includes a glossary of livestock marketing terms.

A1355 Gray, James R. Ranch economics. 1968. Iowa State. 534p. illus. $15.95.

Brings together the peculiarities of the ranching business and formal economic theory. Written primarily for ranchers, students, and businessmen dealing with the range livestock industry. Each chapter is largely a report of research on the subject and is intended to stand alone as a brief monograph. Gives broad coverage of the whole field of ranch economics, land use, and livestock husbandry. Written in semitechnical language.

A1356* Hall, Isaac F., and William P. Mortenson. The farm management handbook. 4th ed. 1963. Interstate. 437p. illus. $8; text ed., $6.

A fairly complete text and handbook for the beginning student in farm management and for the farmer, setting forth the principles of good farm organization, planning, management, and operation. Five appendixes give much practical and useful information on miscellaneous topics ranging from business law to the rate of seeding of farm crops.

A1357 Hamilton, James E., and W. R. Bryant. Profitable farm management. 2d ed. 1963. Prentice-Hall. 394p. illus. $8.44.

Written by vocational agriculture teachers for the beginning farmer and vocational agriculture student. Material organized so that reader can easily find the answers to his specific management problems. In 4 parts entitled: "Planning a start in farming"; "Planning the farm business"; "Improving the home farm"; and "Living in today's world" (covers legal problems, farm organizations, and the farmers' community work). Gives particular attention to management analysis.

A1358 Heady, Earl O., and others. Roots of the farm problem: changing technology, changing capital use, changing labor needs. 1965. Iowa State. 224p. illus. $3.50.

A simplified presentation of research results which help to explain the causes and consequences of farm troubles and policies. Treats overproduction in agriculture as the result of a problem rather than a cause. Also projects future trends in farms and land prices and gives a picture of agriculture in 1980. Provides a readable, nontechnical discussion of a highly complex and highly relevant subject.

A1359 _____, and Harald R. Jensen. Farm management economics. 1954. Prentice-Hall. 645p. illus. $11.25.

Very comprehensive textbook written at an

elementary level for college students. Emphasizes principles, clarified by specific examples from farming operations. Perhaps best suited to the cornbelt area, but contains enough illustrations and data applying to other regions to give it wider appeal. Covers all phases of crop and livestock production, soil management, and engineering applied to machinery and buildings, but omits chapters on records and farm-marketing decisions.

A1360 **Hedges, Trimble R.** Farm management decisions. 1963. Prentice-Hall. 628p. illus. $10.75.

Intended as an introductory text for college students to help them understand and apply economic decision-making principles and procedures in farm management. Also useful for more professionally advanced people. Oriented toward California and the West. Can be used with a manual of exercises or problems. Appendix includes a method for legally describing farm property and a condensed set of tables for calculating the future value of a dollar using varying discount rates and terms of years.

A1361* **Herbst, John.** Farm management: principles, budgets, plans. 2d ed. 1968. Stipes. 294p. illus. paper, $8.20.

A highly useful beginning textbook for teaching the student to use basic farm management tools in decision making. A reasonably complete book on the subject, covering all phases of planning the farm program, budgeting, keeping farm records, and financing. Farm management techniques illustrated with abundant examples from numerous states.

A1362 **Higbee, Edward.** Farms and farmers in an urban age. 1963. Twentieth Century. 183p. illus. paper, $1.45.

Challenges the whole idea of government subsidy for agriculture in the United States, and probably a controversial book. Should be good for general information and background reading. Material described is a clear and interesting account based on the author's own research. Presents facts and cases in an area of major public concern.

A1363* **Hopkins, John A.,** and **Earl O. Heady.** Farm records and accounting. 5th ed. 1962. Iowa State. 377p. illus. $7.50.

A practical text and reference book for college and vocational/technical students and farmers, presenting a system for the use of accounts and records in managing a farm. Divided into 6 major parts: (1) purposes and nature of farm production (and feed) records; (2) farm inventories, depreciation, and net worth statements; (3) farm financial accounts (single-entry); (4) analysis and interpretation of records; (5) income taxes and special accounting problems; (6) budgeting and linear programming.

A1364 **Hunt, Robert L.** Farm management in the South. 1953. Interstate. 581p. illus. $4.50; text ed., $3.50.

Written primarily for vocational agriculture students in the South and Southwest and deals specifically with the problems of farm management in those regions. Covers such topics as characteristics of southern agriculture, selecting the type of farm enterprise, choosing the farm, the farm budget, farm records, financing the farm, farm rental contracts, etc. Includes glossary of terms.

A1365* **Ives, J. Russell.** The livestock and meat economy of the United States. 1966. Amer. Meat Inst. 227p. illus. $6.

Designed chiefly to be used as the basis for a home study course, but also intended as a reference book on the livestock and meat economy. Style is purposely historical and descriptive. Covers material of special interest to those in the meat-packing industry, but not readily available elsewhere. Based on publications of the U.S. Dept. of Agriculture, state experiment stations, sources within the industry, and college textbooks. First chapter provides background of economic concepts, followed by chapters covering production and marketing by stages from farm to table. Last chapter analyzes some important economic aspects of the meat-packing industry.

A1366* **James, Sydney C.** Midwest farm planning manual. 2d ed. 1968. Iowa State. 345p. illus. $6.50.

A practical handbook primarily for use by college students in farm budgeting and programming assignments, but has application for vocational/technical teaching programs and for other farm planners. Contains tables, ready-reckoners, and other information useful in farm business management. Data primarily based on Iowa agriculture, but includes substantial portions from other midwestern states. Emphasis is on results of laboratory and on-the-farm research which may provide useful input-output coefficients as aids in farm planning.

A1367* **Kohls, Richard L.** Marketing of agricultural products. 3d ed. 1967. Macmillan. 462p. illus. $9.95.

A widely adopted, basic textbook on the marketing of food and nonfood agricultural products, useful for college and vocational/technical students. The approach is mixed—partly by function, partly by institution, and partly by commodities. Main focus is on the agricultural industry and selling its products, with little emphasis on the final market for the end products.

A1368 MacGillivray, John H., and Robert A. Stevens. Agricultural labor and its effective use. 1964. Nat. Pr. 107p. illus. paper, $3.

A practical manual on agricultural labor management, especially as it applies to vegetable production. Some of the material is applicable to all farm operations, however. Intended to assist growers in increasing the output of their workers, this easy-to-read, handy primer tells how to hire and keep a good worker anywhere. Contains chapters on training workers and foremen, the effect of fatigue on output, methods to increase worker output, time and motion studies, being a good boss, and farm labor laws and legislation.

A1369 Mellor, John W. The economics of agricultural development. 1966. Cornell. 403p. illus. paper, $3.45.

Deals with the economic development of agriculture in the low-income developing nations of the world. Partially fills the void for a general textbook in agricultural development on the undergraduate level. Author attempts to steer a middle course between an industrial fundamentalism which focuses on developing the nonfarm sector of the economy and an agricultural fundamentalism that focuses only on the need to feed growing numbers of hungry people.

A1370* Moore, John R., and Richard G. Walsh, eds. Market structure of the agricultural industries: some case studies. 2d ed. 1968. Iowa State. 412p. illus. $8.50.

An excellent reference work giving a view of the agricultural industries in relation to their different and evolving market conditions. Each chapter written by 1 or more economists specializing in the area. Covers grocery retailing and such other industries as meat, broiler chickens, fluid milk, ice cream, vegetable processing, apple processing, baking, soybean processing, grain procurement, mixed feed, cotton, farm machinery, and fertilizer. Looks at their actions, measures their impact on the economy, etc.

A1371* Mortenson, William P. Modern marketing of farm products. 2d ed. 1968. Interstate. 277p. illus. $8.50; text ed., $6.

A textbook for vocational agriculture students, written in clear, simple language, telling how our major farm products are marketed. Also discusses transportation and storage, cooperative marketing, marketing margins, marketing agreements and orders, how prices are determined, etc.

A1372* ―――, and Isaac F. Hall. Approved practices in farm management. 3d ed. 1966. Interstate. 260p. illus. $6.25; text ed., $4.75.

This vocational agriculture textbook sets forth the principles and practices of good farm organization, planning, operation and management, and soil conservation. Simply written, but practical and specific, with many helpful illustrations, drawings, and tables. Authors are nationally recognized authorities in the field of farm management.

A1373* Murray, William G. Farm appraisal and valuation. 5th ed. 1969. Iowa State. 534p. illus. $10.50.

A standard textbook in rural appraisal. Presents appraisal principles by following the individual steps in actual appraisal procedure, i.e., first, an inventory of the physical resources of a farm is made, and then this inventory is translated into dollar value. Appraisals by the author of actual farms have been used as illustrations. A well-organized, easy-to-read and -understand text and reference work.

A1374 National Farm Institute. Bargaining power for farmers. 1968. Iowa State. 132p. illus. $3.95.

This book is the published program of the 1968 National Farm Institute at Des Moines, Iowa. Presents a picture of how cooperative bargaining has worked in some areas of the United States. The problems discussed are those relating mainly to overproduction in agriculture.

A1375* Nelson, Aaron G., and William G. Murray. Agricultural finance. 5th ed. 1967. Iowa State. 561p. illus. $9.50.

A standard college textbook and reference work dealing with methods of using capital successfully, including credit. Divided into 2 parts, the first part focuses attention on basic finance principles. The second part is an analysis of different lending institutions and a portrayal of their loan policies and procedures. Could be used as a reference for vocational/technical students, especially those in agribusiness, farm management, and agricultural banking.

A1376 Paarlberg, Donald. American farm policy. 1964. Wiley. 375p. illus. $8.95.

Subtitle, *A case study of centralized decision-*

making, describes the approach of this book. Examines only 1 area of farm policy—farm price supports and production controls. Traces the political and economic origins and consequences of government intervention in the agricultural economy. A stimulating, clearly written book for general background reading.

A1377* Roy, Ewell P. Contract farming U.S.A. 1963. Interstate. 572p. illus. $9.25; text ed., $8.50.

An impartial, thorough analysis of both good and bad facets of contract farming and vertical integration. Broad in scope, covering poultry, cattle, sheep, hogs, dairying, vegetables, etc. Appendixes include a large number of sample contracts. For reference by the student, the farmer, those in agribusiness, and the professional worker.

A1378* ———. Cooperatives today and tomorrow. 2d ed. 1969. Interstate. 612p. illus. $11.25; text ed., $8.50.

An excellent, up-to-date reference and introductory textbook at the post-secondary school level covering all types of cooperatives, both farm and nonfarm. Not restricted to American cooperatives. Discusses such topics as the history, organization, management and taxation of cooperatives, legislation affecting them, cooperatives in the future, and their economic and financial strengths and weaknesses as compared to other types of business organizations.

A1379* ———. Exploring agribusiness. 1967. Interstate. 295p. illus. $9.25; text ed., $6.95.

Appears to be the only textbook in this field so far. Written for students in post-high school institutions with the intention of providing them with a more comprehensive understanding of the nature, scope, importance, and relation to the general economy of agribusiness and to alert them to the occupational opportunities in this field. Has chapters dealing with such things as agribusiness and its relation to farm supplies, to the farmer, the consumer, the government, market competition, the local community, and employment.

A1380 Ruttan, Vernon W. Economic demand for irrigated acreage. 1965. Johns Hopkins. 130p. illus. $5.

Theme is to show that future demand for food and fiber products can best be met by an expansion of irrigated acreage in all the eastern water resource regions and some of the western. Presents some preliminary projections up to 1980. Author concerned with relative profits from public and private investment in irrigation.

A1381 ———, **Arley D. Waldo,** and **James P. Houck,** eds. Agricultural policy in an affluent society. 1969. Norton. 321p. illus. $7.50; paper, $2.50.

Consists of a series of papers presenting prominent positions on economic development, political power, and agricultural policy; the spread between farmer and consumer prices; food marketing policy; bargaining power for farmers and farm workers; rural poverty; agricultural trade, aid, and development policy.

A1382 Schonberg, James S. The grain trade: how it works. 1956. Exposition. 351p. illus. $7.50.

A comprehensive handbook and textbook of the grain trade written by an active member of the Chicago Board of Trade. Author takes up the factors determining the supplies of grain and those creating the demand. He explains the operation of a futures market and the complications of hedges and spreads, describes the operation of the Board of Trade and individual brokerage houses, and discusses world production and trade of principal grains, grain growing, marketing and merchandising, transportation, storage, inspection, exporting, and the mechanics of the market.

A1383 Schultz, Theodore W. Economic crises in world agriculture. 1965. Michigan. 144p. $3.50.

A highly interesting book for background reading by students and others. Deals with the role of agriculture in the economic growth of both the developed and the underdeveloped countries of the world. Author approaches the subject in terms of 2 agricultures—the traditional agriculture of the less developed nations and the modern agriculture of the developed countries.

A1384 Shepherd, Geoffrey S. Agricultural price analysis. 5th ed. 1969. Iowa State. 332p. $7.50.

A well-known, down-to-earth textbook on a difficult, but important, subject. Shows how to explain agricultural price changes in the past, how to predict future changes, and how to estimate effects of new price support programs.

A1385* ———. Farm policy: new directions. 1964. Iowa State. 292p. illus. $6.95.

An easy-to-read summary of the major conventional approaches to solution of the farm problem. Author reviews early concepts of the farm problem in the United States, traces the development of the farm programs, appraises them, and outlines programs of a different kind to replace them. He also discusses the world food problem, and goals and values as determining farm policy.

A1386* _____, and **Gene A. Futrell.** Marketing farm products. 5th ed. 1969. Iowa State. 510p. illus. $10.50.

An up-to-date, standard college textbook on agricultural marketing which should be a valuable source of basic information for technical school students. Section 1 presents the 3 broad agricultural marketing problems and a theory for analyzing them; section 2 discusses overall marketing problems—demand, prices, and costs; and section 3 details specific marketing problems. Written in a concise and readable style.

A1387* **Snodgrass, Milton M.,** and **Luther T. Wallace.** Agriculture, economics, and growth. 2d ed. 1970. Appleton. 436p. illus. $8.50.

A beginning college text in agricultural economics. Covers agricultural history, present agricultural conditions, economic principles, production firms, and agricultural problems. Authors present basic economic principles and demonstrate the importance of making decisions consistent with our developing society.

A1388 **Snowden, Obed,** and **Alvin Donahoo.** Profitable farm marketing. 1960. Prentice-Hall. 403p. illus. $8.44.

A readable, simply written vocational agriculture textbook which should appeal to the technical student concerned with agricultural marketing. Good as a resource book to complement more recent books. Beginning chapters develop understanding of fundamental principles of farm marketing and some factors affecting them. Major portion of book devoted to marketing of specific commodities, with an analysis of movement of each product from farm to consumer. A final chapter covers vertical integration in agriculture.

A1389 **Sorenson, Vernon L.,** ed. Agricultural market analysis. 1964. Michigan State. 344p. illus. $8.50.

A textbook for an advanced undergraduate course in agricultural marketing. Based on papers presented at symposia, and written by many authors. Each chapter a self-contained unit. A compendium of economic principles, facts, viewpoints, and problem settings which should develop analytical insights into the marketing process and decisions.

A1390* **Stone, Archie A.** Careers in agribusiness and industry. 1965. Interstate. 291p. illus. $7; text ed., $5.25.

Indicates job opportunities for vocational/technical school graduates, although the emphasis is on jobs for college graduates. An outstanding vocational guidance book by an authority in the field. Gives an overall view of opportunities in agribusiness, followed by chapters on careers in specific agricultural industries, such as the grain trade and the dairy industry.

A1391 **Talbot, Ross B.,** and **Don F. Hadwiger.** The policy process in American agriculture. 1968. Chandler. 370p. illus. $10.

The politics of food and fiber, relative to production, marketing, and consumption, contained in an easily understood reference for students in introductory courses in agricultural economics and policy and in American government and politics. Intended to provide a structure for analysis rather than a set of answers. Shows how and why farm policy has changed over the years in a well-developed history of the subject.

A1392 **Thomsen, Frederick L.** Agricultural marketing. 1951. McGraw-Hill. 483p. illus. $6.

An older beginning college textbook on farm marketing written in a simple, informative, and very readable style. Divided into parts covering the consumers of farm products and production factors affecting marketing, marketing systems, prices and marketing margins, transportation, market information, grading and other activities, and improvements in marketing.

A1393 _____, and **Richard J. Foote.** Agricultural prices. 2d ed. 1952. McGraw-Hill. 509p. illus. $9.95.

An older, advanced college textbook, probably of most value now as a reference book for various people interested in prices. Was designed to give the student a thorough grounding in price theory and analytical techniques used in studying the movements of commodity prices. The author's style of writing helps to lighten a difficult subject. Divided into 3 parts—price determination and discovery, price analysis and forecasting, and commodity prices.

A1394 **Topel, David G.,** ed. The pork industry: problems and progress. 1968. Iowa State. 236p. illus. $8.

Based on the proceedings of a national conference which discussed the progress and future of the pork industry, with special reference to pork quality characteristics. Of value to producers, college students, and those in technical agriculture programs.

A1395 Voorhis, Horace J. American cooperatives: where they came from, what they do, where they are going. 1961. Harper. 226p. $4.95.

Provides a concise summary of the operations of cooperatives and insight into their role in modern American industry. Although not written as a text, could be used as one in this area where few textbooks are available. Discusses rural electric cooperatives and agricultural cooperatives, as well as other nonfarm oriented cooperatives.

A1396 Wilcox, Walter W., and **Willard W. Cochrane.** Economics of American agriculture. 2d ed. 1960. Prentice-Hall. 538p. illus. $11.50.

Describes each problem area in American agriculture and then introduces modern economic analysis in these areas. Discusses production and marketing activities, the behavior of consumers, the influence of nonfarm agencies and institutions, and the role of government. Special emphasis given to the force of modern technology—its development and adoption and its implications for farm and nonfarm people. An introductory text on the college level.

A1397 Williams, Willard F., and **Thomas T. Stout.** Economics of the livestock-meat industry. 1964. Macmillan. 802p. illus. $14.95.

Offered as a text, a reference, and a report of economic research findings concerning all aspects of the livestock-meat industry. By selection of pertinent materials, could be used as a text for animal science students with limited economics training as well as for agricultural economics students with much more background. Stresses market structure as it relates to operational and pricing efficiency.

A1398 Wilson, Leonard L., ed. Farm and power equipment retailers handbook. 2d ed. 1964. Nat. Farm & Power Equip. Dealers Assn. 516p. illus. $7.50.

A practical handbook of merchandising and management procedures for those in the retail farm and power equipment business. Covers what is important to know in starting such a business—location, insurance, tax aspects, selecting lines of merchandise, management, training employees, advertising, and good public relations.

Rural Sociology

A1399 Bertrand, Alvin L., and **Floyd L. Corty,** eds. Rural land tenure in the United States. 1962. Louisiana State. 313p. illus. $7.50.

Subtitle, *A socio-economic approach to problems, programs, and trends,* describes the contents of this book. Written by 12 specialists in farm economics and rural sociology, it gives a well-balanced evaluation of tenure research and the tenure situation in the early sixties for courses in introductory rural sociology and in land economics.

A1400 Clawson, Marion. Policy directions for U.S. agriculture: long range choices in farming and rural living. 1968. Johns Hopkins. 398p. illus. $10.

A critical reexamination, in nontechnical terms, of overall U.S. agricultural policy, for farm people, farm leaders, public officials, and students. Gives a broad view of what is happening to farming and rural communities, why it is happening, the choices for future agriculture, and some of the problems connected with any choice. Stresses the people involved rather than commodities or property.

A1401 Heaps, Willard A. Wandering workers: the story of American migrant farm workers and their problems. 1968. Crown. 192p. $4.95.

A comprehensive survey of past and present conditions among migrant workers, including taped interviews with typical workers. Includes much useful information on state and federal laws enacted to improve their lot, and on patterns of migration. Excellent supplementary material, written for young adults.

A1402 Kreitlow, Burton W., Edward W. Aiton, and **Andrew P. Torrence.** Leadership for action in rural communities. 2d ed. 1965. Interstate. 346p. illus. $7.25; text ed., $5.75.

Provides information on the principles and practices of leadership in agricultural communities and describes specific programs of action, such as leadership in a homemakers' club and in 4-H clubs, in starting a farmers' cooperative, and a county library, etc. An easily read, valuable book for persons active, or intending to be active, in rural groups of any kind.

A1403* Smith, Thomas L., and **Paul E. Zopf.** Principles of inductive rural sociology. 1970. Davis. 558p. illus. $9.95.

Written for college and university students of rural social organization, but also addressed to people involved with the betterment of life for rural people and to public officials responsible for planning and guiding agricultural policies throughout the world. An up-to-date book based on recent research findings, which would also be useful for vocational/technical students. Has sections covering the rural population—its characteristics, etc.; social ecology, organization, and structure in rural society; and social processes in rural society.

A1404 **Taylor, Miller L.,** and **Arthur Jones, Jr.** Rural life and urbanized society. 1964. Oxford. 493p. illus. $8.50.

A valuable resource book and textbook on social change, especially the process of urbanization of rural society. Draws on a wide range of sources to support its central theme—the marked changes during this century in the relationships between country and town people, especially the urbanization of social organizations.

A1405* **Wright, Dale.** They harvest despair; the migrant farm worker. 1965. Beacon. 158p. illus. $4.95.

Written for the general reader to alert him to the living and working conditions of the migrant worker on the large commercial vegetable and fruit farms. Based on a series of award-winning newspaper articles written by a man who traveled and worked with these people along the Atlantic seaboard during the 1961 harvesting season. A vivid and interesting account which makes plain the human tragedy behind the statistics. Discusses legislative efforts and the work of social service and religious organizations with the migrant laborer and offers helpful suggestions to improve his lot.

B

Periodicals

The term *periodical,* as used in this bibliography, refers to journals, serials, reports, newsletters, and all other continuously issued publications. Periodicals are valuable forms of literature because they not only provide current information frequently but they also contain articles on specific subjects, which normally can be read in a minimal amount of time. Periodical literature is written for readers with different interest levels. In this periodicals section, four terms are used to classify the listed materials according to their complexity. The scientific and technical publications, which include highly specialized journals and reports, association and society publications, proceedings, symposia, and other treatises, are the most advanced. An example of these is the *Soil Science Society of America proceedings.* Popular scientific and technical periodicals, which contain a mixture of news, short reports, and announcements of discoveries as well as longer articles, represent the next level. These semitechnical publications include such titles as *Scientific American* and *Journal of forestry.* House organs (industrial organization publications which are usually free), society and association publications, newsletters, information bulletins, and trade journals (publications dealing with particular industries and their products) constitute the third level, the nontechnical periodicals. These are exemplified by *Agri finance* and *Agricultural education magazine* in this bibliography. The most elementary level is the popular, nontechnical publication—periodicals such as *National wildlife* which are easy-to-read and very general in content.

Selection of titles for this section was based on the overall value of the periodicals' content for vocational agriculture and on the quality of the material included in each. All nondomestic periodicals, most publications representing specific geographic areas, most publications on highly specialized subjects, and almost all serials (transactions, proceedings, etc.) have been excluded in order to provide a basic, comprehensive periodicals list. For information on omitted titles see *Ulrich's international periodicals directory, Irregular serials and annuals: an international directory,* and *The standard periodical directory.* Because of the enormous amount of periodical literature published by the state government agencies, the universities, the private state publication publishers, and the agricultural experiment stations, only one example of each of these has been given. Listed respectively, these are: *South Dakota conservation digest; Land economics; Wallaces farmer,* and *Science in agriculture.* The *Monthly checklist of state publications,* a complete list of available materials published by the states, should be consulted for particular state publications desired.

Purchase of all periodicals should depend on student and faculty demands, the relationship of material to curriculum and to research conducted (if any) in the institution. In most cases, a technical school should only purchase advanced scientific and technical research publications if extra funds are available, if an instructor requests them, or if research is being done in the particular area covered by them. A number of periodicals are free to qualified readers who are usually people actively engaged in the industry or profession to which the publication is devoted. Applications for free subscriptions are available on request.

INDEXING AND ABSTRACTING SERVICES

These titles index or abstract the general periodicals in this section. Most of these services are not annotated because they are too specialized and are often too expensive for the vocational/technical school level. The *Bibliography of agriculture* and the *Biological and agricultural index* are the only titles in this category which merit full description because they are highly useful and general enough to serve any agricultural reader.

B1 Abridged reader's guide to periodical literature. (Abr. R.G.) 1935– . H. W. Wilson Co., 950 University Ave., Bronx, N.Y. 10452. Monthly. $11 a year.

B2 Applied Science and technology index. (A.S. & T. Ind.) 1958– . H. W. Wilson Co., 950 University Ave., Bronx, N.Y. 10452. Monthly (except Aug.). Sold on a service basis.

B3 Art index. (Art Ind.) 1929– . H. W. Wilson Co., 950 University Ave., Bronx, N.Y. 10452. Quarterly. Sold on a service basis.

B4 Bibliography of agriculture. (Bibl. of Agr.) 1942– . CCM Information Corp., 909 3rd Ave., Bronx, N.Y. 10022. Monthly. $85 a year.

An index to the periodicals and other agricultural and allied science literature received in the National Agricultural Library. Each issue is divided into 5 sections: a main entry section; a checklist of new government publications; a list of books recently received by the library; a subject index; and, an author index. This publication is the most comprehensive index to agricultural literature.

B5 Biological abstracts. (Biol. Abstr.) 1926– . Biosciences Information Service of Biological Abstracts, 2100 Arch St., Philadelphia, Pa. 19103. Semimonthly. $640 a year to qualified personnel; $800 to others.

B6 Biological and agricultural index. (Biol. & Agri. Ind.) 1916– . H. W. Wilson Co., 950 University Ave., Bronx, N.Y. 10452. Monthly (except Aug.) $35 a year for individuals; sold on a service basis to libraries.

A subject index to periodicals in the fields of biology, agriculture, and related sciences. Indexes approximately 1,150 periodicals. Publisher attempts to maintain a good subject balance so that no important field will be overlooked and none will be too heavily indexed. The division is fairly equal between the agricultural and biological sciences. All publications indexed are in English although some of them are published in other countries. This is an indispensable index for any agricultural school beyond the high school level.

B7 Business periodicals index. (B.P.I.) 1958– . H. W. Wilson Co., 950 University Ave., Bronx, N.Y. 10452. Monthly (except July). Sold on a service basis.

B8 Chemical abstracts. (Chem. Abstr.) 1907– . American Chemical Society,

1155 16th St. NW, Washington, D.C. 20036. Biweekly. $700 a year to Chemical Society members, colleges, and universities; $1,200 a year to others.

B9 Education index. (Educ. Ind.) 1929– . H. W. Wilson Co., 950 University Ave., Bronx, N.Y. 10452. Monthly (Sept.-June). Sold on a service basis.

B10 Engineering index. (Eng. Ind.) 1884– . Engineering Index, Inc., 345 E. 47th St., New York, N.Y. 10017. Monthly. $350 a year for the monthly issues; $175 for the annual.

B11 Historical abstracts. (Hist. Abstr.) 1955– . American Bibliographical Center-Clio Press, Riviera Campus, 2010 Alameda Padre Serra, Santa Barbara, Calif. 93103. Quarterly. Sold on a service basis.

B12 Index medicus. (Ind. Med.) 1960– . National Library of Medicine, 8600 Rockville Pike, Bethesda, Md. 20014. Monthly. $63 a year.

B13 Mathematical reviews. (Math. R.) 1940– . American Mathematical Society, 321 S. Main St., Providence, R.I. 02904. Monthly. $290 a year.

B14 Metal abstracts. (Met. Abstr.) 1968– . American Society for Metals, Metals Park, Ohio 44073. Monthly. $280 a year; $100 a year to universities, colleges, and public libraries.

B15 Meteorological and geoastrophysical abstracts. (Meteor. & Geoastrophys. Abstr.) 1950– . American Meteorological Society, 45 Beacon St., Boston, Mass. 02108. Monthly. $400 a year; $200 a year to universities and U.S. public libraries.

B16 Psychological abstracts. (Psychol. Abstr.) 1927– . American Psychological Association, Inc., 1200 17th St. NW, Washington, D.C. 20036. Monthly. $40 a year.

B17 Public Affairs Information Service. Bulletin. (P.A.I.S.) 1915– . Public Affairs Information Service, Inc., 11 W. 40th St., New York, N.Y. 10018. Weekly (biweekly in Aug.). $100 a year for the weekly and cumulated bulletins including the annual; $50 a year for the cumulated bulletins and the annual; $25 a year for the annual.

B18 Readers' guide to periodical literature. (R.G.) 1900– . H. W. Wilson Co., 950 University Ave., Bronx, N.Y. 10452. Semimonthly (monthly July-Aug.). $28 a year.

B19 Science citation index. (Sci. Cit. Ind.) 1963– . Institute for Scientific Information, 325 Chestnut St., Philadelphia, Pa., 19106. Quarterly. $1,250 a year.

GENERAL PERIODICALS

B20 Aerial applicator. 1963– . Earth Publications, 8402 Allport Ave., Drawer 2263, Santa Fe Springs, Calif. 90670. 9 issues a year. Free to qualified persons; $4.50 a year to others. Indexed: Bibl. of Agr.

One of the few journals available on aerial applications in the fields of agriculture, fire control, and natural resources. It publishes nontechnical to semitechnical articles on fertilizer application, seed sowing, plant pest control, new equipment, new methods, the pollution problem, and other subjects of vital interest to the industry. Not as technical as the international journal, *Agricultural aviation,* published by the International Agricultural Aviation Centre in the Netherlands.

B21 Agri finance. 1959– . Franchise Publishing Co., Inc., 1722 W. Grand Ave., Chicago, Ill. 60622. Bimonthly. $4 a year. Indexed: Bibl. of Agr.

Intended primarily for officers of farm-lending organizations whose duties include counseling commercial farmers and allied businessmen. Publishes nontechnical articles interpreting agricultural developments as they affect credit available to agricultural businesses and enterprises. Each issue has a special tear-out page listing booklets, brochures, films, and other items free for the asking.

B22 Agricultural chemicals. 1957– . Industry Publications, Inc., Box 31, Caldwell, N.J. 07006. Monthly. $5 a year. Indexed: Bibl. of Agr.; Biol. & Agri. Ind.; Chem. Abstr.; Biol. Abstr.

A trade or news journal containing a mixture of up-to-date articles on the agricultural chemical industry, including news of the commercial fertilizer and plant food industries, short reports, literature reviews, and news of technological interest. Written in nontechnical language and excellent for background information as well as for keeping informed on new chemicals of value to agriculture.

B23* Agricultural education magazine. 1929– . Agricultural Education Magazine, Box 5115, Madison, Wis. 53705. Monthly. $3 a year. Indexed: Educ. Ind.

Most issues of this professional journal cover a particular area of importance to agricultural education. It contains articles on curricula, teaching techniques, resource material, and recent research and developments. Most issues carry reviews on new books of interest to agricultural education.

B24* Agricultural engineering. 1920– . American Society of Agricultural Engineers, 420 Main St., St. Joseph, Mich. 49085. Monthly. $8 a year. Indexed: Bibl. of Agr.; Biol. & Agri. Ind.; Chem. Abstr.; Eng. Ind.

Publishes original papers and condensations of longer papers or research reports dealing with the application of engineering principles and techniques in agriculture. Emphasis is given to the areas of power equipment and machinery, farm structures, and soil and water utilization and management.

B25* Agronomy journal. 1907– . American Society of Agronomy, 677 S. Segoe Rd., Madison, Wis. 53711. Bimonthly. $14 a year. Indexed: Bibl. of Agr.; Biol. & Agri. Ind.; Biol. Abstr.; Chem. Abstr.

A scientific journal containing reports and articles concerned with instruction or research in crop production, soil management, and farmland management. Generally considered one of the most important journals in the field of agronomy. Articles are short, running from 2 to 4 pages, and technical in language and content.

B26 American bee journal. 1861– . American Bee Journal, Hamilton, Ill. 62341. Monthly. $3.50 a year. Indexed: Bibl. of Agr.; Biol. Abstr.; Chem. Abstr.

A journal of up-to-date practical information on beekeeping management, written by and for beekeepers and for classroom use as reference material. The style of writing is generally popular, although technical articles are sometimes included. For those people interested in original research and reviews of present scientific knowledge, the *Journal of apicultural research*, published by the Bee Research Association, England, is recommended.

B27 American beef producer. 1919– . American National Livestock Publishing Co., 1540 Emerson St., Denver, Colo. 80218. Monthly. $3 a year. Indexed: Bibl. of Agr.

This magazine of cattle business management is the official publication of the American National Cattlemen's Association. It contains information about the association, announcements of meetings, and political and commercial news pertaining to the industry. Issues usually contain 3 or 4 short articles on subjects of current interest. For most readers only the recent issues would be of interest.

B28 American dairy review. 1939– . Watt Publishing Co., Mt. Morris, Ill. 61054. Monthly. $10 a year; free to qualified readers (apply for qualified-reader subscription). Indexed: Bibl. of Agr.; Biol. & Agri. Ind.

A trade journal containing up-to-date news articles on the dairy products industry. Not intended for the dairy farmer. Each issue includes specialists' reports on labor relations, state and federal legislation, etc., as it relates to the dairy product industry. Additional information on equipment and supplies described or advertised in this publication can be obtained through its reader service.

B29 American fish farmer and world aquaculture news. 1969– . American Fish Farmer and World Aquaculture News, P.O. Box 1900, Little Rock, Ark. 72203. Monthly. $6 a year.

A new journal in an industry rapidly becoming important. Each issue contains current interest articles, reports on research in progress, short news notes, and other items of interest to the fishery industry. Worldwide in scope, it is valuable for keeping up with applications of new techniques that are resulting from aquacultural work in other areas.

B30 American forests. 1895– . American Forestry Association, 919 17th St. NW, Washington, D.C. 20006. $6 a year. Indexed: Bibl. of Agr.; R.G.

General magazine on conservation, semipopular in style. The subtitle, . . . *The magazine of forests, soil, water, wildlife, and outdoor recreation*, indicates the areas covered. Publication's principle value is to keep the reader aware of our conservation problems and possibilities as well as the relationship of our natural resources to the economic life of our nation.

B31 American fruit grower. 1879– . Meister Publishing Co., Willoughby, Ohio 44094. Monthly. $2 a year. Indexed: Bibl. of Agr.; Biol. Abstr.; Chem. Abstr.

This trade publication, written in a popular

style, contains short articles of current interest, news of various state fruit grower organizations, a calendar of coming events, and other items of interest to the commercial fruit grower. More information on items advertised can be obtained through the publication's reader service.

B32* American horticultural magazine. 1922– . American Horticultural Society, Inc., 2401 Calvert St. NW, Washington, D.C. 20008. Quarterly. $15 a year. Indexed: Bibl. of Agr.; Biol. & Agri. Ind.

Official publication of the society. Concerned with the science and art of growing fruits, vegetables, ornamental plants, and related subjects. Intended for the professional and the serious amateur gardener; the style and language of the articles vary from popular to technical or scientific. At irregular intervals, one of the regular issues is devoted to information on a particular phase of horticulture, and these monographs are also obtainable from the society as separate books. Each publication carries reviews of new books of interest to horticulturists.

B33 American journal of botany. 1914– . Botanical Society of America, Inc., Business Manager, City College of the City University of New York, New York, N.Y. 10031. Monthly (except bimonthly, May-June, Nov.-Dec.). $21.50 a year. Indexed: Bibl. of Agr.; Biol. & Agri. Ind.; Biol. Abstr.; Chem. Abstr.

Scientific journal reporting original research in botanical science and reflecting advances made in the field. Botanical subject areas covered include physiology, morphology, and ecology.

B34 American nurseryman. 1904– . American Nurseryman Publishing Co., 343 S. Dearborn St., Chicago, Ill. 60604. Semimonthly. $6 a year. Indexed: Bibl. of Agr.; Chem. Abstr.

National journal for the nursery industry. Contains a mixture of book reviews, short articles on management practices and new cultural practices, information on new plant materials, a calendar of coming events, and news of interest to the industry.

B35 American vegetable grower. 1953– . Meister Publishing Co., Willoughby, Ohio 44094. Monthly. $12 a year. Indexed: Bibl. of Agr.

National magazine for the commercial vegetable grower and the fresh produce packing industry. Articles must cover the aspects of commercial vegetable production and marketing which include land cultivation, crop raising, control of insect pests and disease, harvesting, packaging, and marketing. Presents current interest items, state news, trends and forecasts, and other general information on management and vegetable culture.

B36 Animal nutrition and health. 1945– . Beeler Publishing Co., 1714 Stockton St., San Francisco, Calif. 94133. Monthly. Free to qualified persons. Indexed: Bibl. of Agr.

Intended for people working in the fields of nutrition, health, or commercial production of animals and poultry. Has no table of contents or index. Each issue contains 2 or 3 technical articles and several short reports as well as a section on new products which is accompanied by a coupon for obtaining free information about them. Available free to libraries.

B37 Appraisal digest. 1950– . New York Society of Real Estate Appraisers, 11 N. Pearl St., Albany, N.Y. 12207. Quarterly. $5 a year.

A generally nontechnical publication written by and for people in the real estate business. Many articles are reprints of speeches or condensed versions of articles printed elsewhere. The scope is broad, covering all phases and types of real estate and not limited to any particular region.

B38 Appraisal journal. 1932– . American Institute of Real Estate Appraisers, 155 E. Superior St., Chicago, Ill. 60611. Quarterly. $10 a year. Indexed: Bibl. of Agr.

An institutional publication containing a wealth of real estate knowledge and analytical articles of both timely and permanent value. Its technical articles range from papers on the valuation of scenic easements to urban renewal parcels and grazing land or timberland bought for speculative investment.

B38a Audubon. 1899– . National Audubon Society, Membership Dept., 950 3rd Ave., New York, N.Y. 10022. Bimonthly. $12 a year. Indexed: Biol. Abstr.; R.G.

A popular nature magazine with outstanding illustrations and authoritative articles on conserving and appreciating wildlife, wilderness, natural resources, and natural beauty. Originally an ornithological publication, now expanded to cover all the biological sciences, with special emphasis on conservation. Includes book reviews. Gives excellent coverage, all in 1 section, of conservation legislation and other news in the fields of conservation, pesticides, ecology, and the environment, as well as Audubon Society activities and viewpoints in these fields.

B39 Avian diseases. 1957– . Avian Diseases, University of Massachusetts, Amherst, Mass. 01002. Quarterly. $11 a year. Indexed: Bibl. of Agr.; Biol. Abstr.; Chem. Abstr.

Publication of the American Association of Avian Pathologists devoted exclusively to publishing original research and case reports on diseases of poultry and game birds. Contributors include world authorities in the field.

B40* Better crops with plant food. 1923– . American Potash Institute, Inc., 1102 16th St. NW, Washington, D.C. 20036. Quarterly. $1 a year. Indexed: Bibl. of Agr.

An up-to-date, informative journal containing papers on fertilizers and their use in field crops, forages, fruit orchards, turfgrass and forestry, written by knowledgable people at state universities and by leaders in the industry. The language and content of the text is nontechnical, sometimes even popular, with coverage generally limited to the United States. Information on AV aids is included in some issues.

B41 Better fruit. Better vegetables. 1906– . Better Fruit Publishing Co., 17150 S.W. Lower Boones Ferry Rd., Lake Oswego, Oreg. 97034. Monthly (except bimonthly for July–Aug. and Sept.–Oct.). $1 a year. Indexed: Biol. Abstr.; Chem. Abstr.

A news magazine published primarily for the western commercial fruit and vegetable grower. Contains short articles of current interest on all aspects of production and marketing. Tends to emphasize fruit production and covers fruit districts of Washington, Oregon, California, Idaho, and Montana.

B42 Big farmer. 1928– . Big Farmer, Inc., 131 Lincoln Hwy., Frankfort, Ill. 60423. 9 issues a year. Free to big income ranchers, $3.50 a year to others. Indexed: Bibl. of Agr.

A farm management magazine for top income farmers and ranchers. Of special value to persons interested in becoming managers or technicians on large farms or ranches. Gives insight into some of the problems commonly encountered. Treats big farming and ranching from the rural sociological, agricultural economic, farm management, farm policy, and the farm change and agricultural adjustment points of view.

B43 Biochemistry. 1962– . American Chemical Society, 1155 16th St. NW, Washington, D.C. 20036. Biweekly. $40 a year. Indexed: Bibl. of Agr.; Chem. Abstr.; Ind. Med.

A basic research journal, technical in language and content, and publishing the results of original research in all areas of fundamental biochemistry. The papers cover enzymes, proteins, carbohydrates, lipids, nucleic acids and their metabolism, genetics, and biosynthesis.

B44* Bioscience. 1951– . American Institute of Biological Science, 3900 Wisconsin Ave. NW, Washington, D.C. 20016. Semimonthly. $18 a year. (Individual subscriptions not acceptable except through membership in the A.I.B.S.). Indexed: Bibl. of Agr.; Biol. Abstr.; Chem. Abstr.

An institutional journal covering a wide range of subjects from ecology, nutrition, veterinary medicine, pesticides, and pollution to agricultural development. The papers, in easy-to-read style and language, are written by authorities in the field. Each issue contains reviews of new books of interest to those in the biosciences.

B45 Bulletin of environmental contamination and toxicology. 1966– . Springer-Verlag New York, Inc., 175 5th Ave., New York, N.Y. 10010. Bimonthly. $20 a year for institutions; special rates available for individual subscriptions for personal use only. Indexed: Bibl. of Agr.; Chem. Abstr.

A technical journal providing rapid publication of short articles reporting important advances in methods, procedures or techniques, and discoveries resulting from current research. Covers air, soil, water, food contamination, and pollution as well as other fields concerned with toxicants in the environment. Of more value to the instructor than the student.

B46 CRC critical reviews in environmental control. 1970– . CRC Critical Review Journals, The Chemical Rubber Co., 18901 Cranwood Pky., Cleveland, Ohio 44128. Quarterly. $56 a year.

Consists of authoritative, logically arranged, comprehensive review articles written by recognized experts in fields relevant to environmental control. These critical evaluations cover new concepts and methods in environmental control extracted from the most significant papers published in current literature.

B47 CRC critical reviews in food technology. 1970– . CRC Critical Review Journals, The Chemical Rubber Co., 18901 Cranwood Pky., Cleveland, Ohio 44128. Quarterly. $56 a year.

Collections of critical reviews by noted experts in food technology. Authors have extracted the most significant papers from current literature and also presented original data and viewpoints which are not available elsewhere. Specific articles are not discussed but rather literature in general on a given subject or process (such as freeze-drying).

B48 Canner/packer. 1895– . Canner/Packer, Vance Publishing Co., 300 W. Adams, Chicago, Ill. 60606. Monthly. $10 a year. Indexed: Bibl. of Agr.; Biol. Abstr.

Published for the canned, glass-packed, frozen, and dry-processed food industries. Each issue carries information on new laws, markets, new equipment, raw products, new packaging and labels, new ready-to-eat products, food prices, and company news.

B49 Cattleman. 1914– . Texas & Southwestern Cattle Raisers Association, Inc., 410 E. Weatherford, Fort Worth, Tex. 76102. Monthly. $4 a year. Indexed: Bibl. of Agr.

A general news magazine on the cattle industry, particularly in the western states. Contains articles on cattle breeds and breeding, cattle production, information on people and events of interest to cattlemen, and political news important to the cattle industry.

B50 Cereal chemistry. 1924– . American Association of Cereal Chemists, 1821 University Ave., St. Paul, Minn. 55104. Bimonthly. $30 a year. Indexed: Bibl. of Agr.; Biol. & Agri. Ind.; Biol. Abstr.; Chem. Abstr.

Scientific periodical publishing papers dealing with analytical procedures, raw materials, processes, products, and fundamental research in the cereal grains, oil seeds, and related industries.

B51* Cereal science today. 1956– . American Association of Cereal Chemists, 1821 University Ave., St. Paul, Minn. 55104. Monthly. $15 a year. Indexed: Bibl. of Agr.; Biol. Abstr.

A society journal with contents ranging from nontechnical to highly technical information on cereal grains, oil seeds, pulses and related materials, their processing and utilization, and their products. Each issue contains association news and events, book reviews, and other items of interest to the industry.

B52 Compost science. 1960– . Rodale Press, Inc., 33 E. Minor St., Emmaus, Pa. 18049. Bimonthly. $6 a year. Indexed: Bibl. of Agr.; Biol. Abstr.; Chem. Abstr.; Eng. Ind.

Magazine of the composting-organic waste recycling industry. Contains technical, scientific, and practical papers which cover the field of large-scale composting and recycling of municipal, agricultural, and industrial organic solid wastes.

B53 Cooperative digest and farm power. 1940– . Southern Trade Publications Co., Box 11393, Greensboro, N.C. 27409. Monthly. $3.50 a year. Indexed: Bibl. of Agr.

A farm business and cooperative electricity trade journal containing short reports and articles on items of current interest to the farmer and others concerned with rural electrification.

B54 Crop science. 1961– . Crop Science Society of America, 677 S. Segoe Rd., Madison, Wis. 53711. Bimonthly. $16 a year. Indexed: Bibl. of Agr.; Biol. & Agri. Ind.; Biol. Abstr.; Chem. Abstr.

A scientific journal concerned with experimentation and research in genetics, physiology, breeding, and culture of major field crops. Too technical in language and content for the beginner.

B55* Crops and soils. 1948– . American Society of Agronomy, 677 S. Segoe Rd., Madison, Wis. 53711. 9 issues a year. $3 a year. Indexed: Bibl. of Agr.; Biol. & Agri. Ind.; Biol. Abstr.

A farm research magazine containing news and views on significant topics in agriculture, reviews of new publications, and nontechnical research reports on crop production and soil management. One of the best periodicals of this type.

B56 Cyanagrams. 1947– . American Cyanamid Co., Agricultural Div., P.O. Box 400, Princeton, N.J. 08540. Quarterly. Free. Indexed: Bibl. of Agr.

A house organ publication covering advances in all areas of agricultural science as well as topics of current interest. Attractive format; many illustrations in color; easily read text.

B57* Dairy herd management. 1965– . Miller Publishing Co., 2501 Wayzata Blvd., Minneapolis, Minn. 55405. Monthly. $10 a year. Indexed: Bibl. of Agr.

The business magazine for top dairy farmers containing practical articles written by knowledgable people both in the dairy field and associated with the universities. Covers feeding, production, equipment, the latest in research, new products, market reports, breeding, herd health, new techniques, etc. Although written primarily for the large-scale dairy farmers, it is equally valuable to the small dairyman and others interested in the business.

B58* Doane's agricultural report. 1963– . Doane Agricultural Service, Inc., 8900 Manchester Rd., St. Louis, Mo. 63144. Weekly. $17.50 a year. Indexed: Bibl. of Agr.

A news service featuring farm real estate activities, government regulations, agricultural law, finance, the futures market, management and new production techniques, in-depth market outlooks, analyses on crops, livestock, and new equipment on the market. Also provides up-to-date information on the economics of all phases of agriculture.

B59 Down to earth. 1945– . Dow Chemical Co., Midland, Mich. 48640. Quarterly. Free to those engaged in agricultural research, extension, and instruction. Indexed: Bibl. of Agr.; Biol. Abstr.; Chem. Abstr.

A review of agricultural chemical progress and results of research done by Dow researchers or carried out in cooperation with other governmental or academic researchers. Every page is perforated so that articles can be torn out and filed for future reference. This is one of the best examples of a house organ published by an industrial organization.

B60* Drovers journal. 1873– . Vance Publishing Co., Food and Agriculture Div., One Gateway Center, 5th at State Ave., Kansas City, Kans. 66101. (Send subscription order to Circulation Dept., Drovers Journal at same address). Weekly. $5 a year.

Business publication (newspaper) of the livestock industry covering all kinds of farm news. Livestock production, livestock marketing, grain and feed trends, livestock futures, market trends for the individual types of livestock such as hogs and sheep, fairs, and livestock shows are some of the features included.

B61 Egg industry. 1968– . Garden State Publishing Co., Garden State Bldg., 4411 Landis Ave., Sea Isle City, N.J. 08243. Monthly. $5 a year.

A national magazine covering commercial egg management from farm to market. Nontechnical in language and content. Regular departments report what's new in prices, government regulations, research, equipment, and on people and events in the egg world.

B62 Electricity on the farm. 1927– . Reuben H. Donnelley Corp., Magazine Publishing Div., 466 Lexington Ave., New York, N.Y. 10017. Monthly. $4.50 a year.

A how-to magazine devoted to practical articles on the use of electric power and electrically powered equipment in various phases of farm operation. Articles are signed but the authority of the author is not always indicated.

B63 Entomological news. 1889– . American Entomological Society, 1900 Race St., Philadelphia, Pa. 19103. Monthly. $7 a year to individuals; $12 to institutions. Indexed: Bibl. of Agr.; Biol. Abstr.; Chem. Abstr.

This journal, which gives more attention to the biological than the economic aspects of entomology, publishes articles of wide interest to the general entomologist and the nonspecialist on insect life and related terrestrial arthropods. Each issue contains bibliographical references to current literature. Of most value to those who have had some basic courses in entomology.

B64 Farm chemicals. 1894– . Meister Publishing Co., 37841 Euclid Ave., Willoughby, Ohio 44094. Monthly. $10 a year. Indexed: Bibl. of Agr.; Biol. & Agri. Ind.; Chem. Abstr.

A national business magazine for the fertilizer and pesticide industries serving primarily those persons responsible for management, marketing, and production but valuable to anyone concerned with farm chemicals. The language and style of presentation is nontechnical. Each issue has a listing of free information concerning fertilizer and pesticide problems and a book review section.

B65* Farm journal. 1877– . Farm Journal, Inc., Washington Sq., Philadelphia, Pa. Monthly. $2 a year. Indexed: Abr. R.G.; R.G.

A popular magazine published in separate editions for each section of the nation. Its stated purpose is to be of service to families who own, rent, or operate farmland and to those doing business with farmers. It contains a broad collection of articles on farm equipment, cultural practices, and production techniques. These are accompanied by news notes and editorials on farm events and developments.

B66* Farm quarterly. 1946– . F. & W. Publishing Co., 22 E. 12th St., Cincinnati, Ohio 45210. Bimonthly. $2 a year. Indexed: Bibl. of Agr.; Biol. & Agri. Ind.

A nontechnical journal giving a well-rounded, up-to-date view of agriculture and the allied industries. Categories covered include management, livestock, buildings and equipment, soils and crops, book reviews, governmental laws and regulations, and other news items.

B67 Farm supplier. 1927– . Watt Publishing Co., Mt. Morris, Ill. 61054. Monthly. $10 a year. Indexed: Bibl. of Agr.

Trade magazine featuring information on

selling, merchandising, service, and management for farm supply, feed, and fertilizer dealers. Excellent source for new ideas on farm supply selling and for information on new products and farm facts and forecasts.

B68*   Farmer's digest. 1937– . Farmer's Digest, P.O. Box 363, Brookfield, Wis. 53005. Monthly (except bimonthly in June-July and Aug.-Sept.). $3.60 a year.

Digest of best articles from more than 100 farm magazines with source of original article indicated. Includes practical articles on all phases of farming. Very good for maintaining an up-to-date, overall picture of farming.

B69   Feed industry. 1925– . Communications Marketing, Inc., 5100 Edina Industrial Blvd., Edina, Minn. 55435. Monthly. $3 a year. Indexed: Bibl. of Agr.; Biol. Abstr.

A magazine for companies and individuals engaged in the manufacture of feed in the major poultry and livestock producing regions. Covers the formulation, manufacturing, and marketing of feed.

B70*   Feed management. 1950– . Garden State Publishing Co., Garden State Bldg., Sea Isle, N.J. 08243. Monthly. $3 a year. Indexed: Bibl. of Agr.; Biol. Abstr.

National magazine published in eastern, southern, central, and western editions which covers feed formulation, production, and marketing. Also covers political news and farm trends that affect the feed industry, new research, new products, and the trade notes of workers and companies.

B71*   Feedstuffs. 1929– . Miller Publishing Co., 2501 Wayzata Blvd., Minneapolis, Minn. (Send subscription order to Circulation Manager, Feedstuffs, Box 1289, Minneapolis, Minn. 55440). Weekly. $10 a year. Indexed: Bibl. of Agr.; Biol. & Agri. Ind.; Biol. Abstr.; Chem. Abstr.

The subtitle for this periodical is *The weekly newspaper for agribusiness*. Research, foods, grain and feeds, animal nutrition, dairying, commodity futures prices, agribusiness stock summaries, and new technological methods in use are among the areas covered. Quite valuable for an up-to-date, bird's-eye view of agricultural industries.

B72   Fertilizer solutions. 1959– . National Fertilizer Solutions Association, 910 Lehmann Bldg., Peoria, Ill. 61602. Bimonthly. Free to qualified persons. Indexed: Bibl. of Agr.

Official journal of the association containing articles and picture stories written by people in the industry and those associated with academic institutions. Interest level of content ranges from nontechnical to semitechnical. Covers specific uses of liquid fertilizers, general business news, pesticides, pollution, legal facts, news from Washington, and agronomy tips. Editorial content slanted toward liquid fertilizer manufacturers and dealers.

B73   Food engineering. 1928– . Food Engineering, Chilton Co., Chestnut & 56th Sts., Philadelphia, Pa. 19139. Monthly. $25 a year; free to qualified individuals in the food engineering field. Indexed: Bibl. of Agr.; A.S. & T. Ind.; Biol. Abstr.; Eng. Ind.

Semitechnical journal primarily for individuals in administrative, plant operating, engineering, or technical management functions in companies manufacturing foods or beverages.

B74*   Food technology. 1947– . Institute of Food Technologists, Suite 2120, 221 N. LaSalle St., Chicago, Ill. 60601. Monthly. $20 a year. Indexed: Bibl. of Agr.; Biol. & Agri. Ind.; A.S. & T. Ind.; Biol. Abstr.; Chem. Abstr.; Psychol. Abstr.

Gives comprehensive coverage of the food industry field from the basic food sciences to engineering and technological problems and procedures, new food products, and current literature. Style and content range from nontechnical to scientific.

B75   Forest history. 1957– . Forest History Society, Inc., 733 River St., Santa Cruz, Calif. 95060. Quarterly. $7.50 a year. Indexed: Bibl. of Agr.

Disseminates history of the North American forests and related forest activities as well as current news of interest to those in the forestry field. Written in simple, easy-to-read style. The literature section has signed reviews of new books.

B76*   Forest products journal. 1951– . Forest Products Research Society, 2801 Marshall Ct., Madison, Wis. 53705. Monthly. $25 a year. Indexed: Bibl. of Agr.; Biol. & Agri. Ind.; Biol. Abstr.; Chem. Abstr.; Eng. Ind.

A society journal with articles ranging from nontechnical to scientific in content. Covers timber production, fiber processes, glues, treatments and coatings, wood drying, wood engineering, production management, and marketing. Each issue carries society news, information on new products, lists of technical literature available, and business briefs.

B77 Fruit varieties and horticultural digest. 1946– . L. D. Tukey, Business Manager, Fruit Varieties and Horticultural Digest, Tyson Hall, University Park, Pa. 16802. Quarterly. Available to individuals only through membership in the American Pomological Society; $5 a year to libraries and institutions. Indexed: Bibl. of Agr.; Biol. Abstr.

Magazine written by horticulturists for horticulturists about the science and practice of fruit growing. Reports on all types of fruits grown in the United States but gives greater coverage to common fruits such as apples, peaches, and pears. Thus, also contains much to interest the student and fruit gardener.

B78* Garden journal. 1951– . Garden Journal, The New York Botanical Garden, Bronx, N.Y. 10458. Bimonthly. $5 a year. Indexed: Bibl. of Agr.; Biol. & Agri. Ind.

Official publication of the New York Botanical Garden. Articles on botanical and horticultural developments, discoveries, concepts, history, and instruction written in a nontechnical but scientifically accurate manner. Articles, intended for the layman, student, and professional, are composed by botanists, horticulturists, gardeners, and conservationists.

B79* Gleanings in bee culture. 1873– . A. I. Root Co., 623 W. Liberty St., Medina, Ohio 44256. Monthly. $3.50 a year. Indexed: Bibl. of Agr.; Biol. & Agri. Ind.; Biol. Abstr.

Trade journal, in nontechnical language, covering bee culture and research in the field and in the laboratory. A monthly honey report, news and events, and other items of current interest to beekeepers are featured in each issue.

B80 The golf superintendent. 1933– . The Golf Superintendent, 3158 Des Plaines Ave., Des Plaines, Ill. 60018. 10 issues a year. $5 a year. Indexed: Bibl. of Agr.

Official publication of the Golf Course Superintendents Association of America which is devoted to turfgrass science and golf course management. Carries practical, technical, and scientific articles on all areas of turf maintenance written by leading agronomists, scientists, and golf course superintendents.

B81 Grain and feed journals. 1898– . Grain and Feed Journals, 1115 Board of Trade Bldg., 141 W. Jackson Blvd., Chicago, Ill. 60604. Semimonthly. $6 a year. Indexed: Bibl. of Agr.

Trade journal devoted to business management in the grain and feed industry. In addition to articles on production, marketing, and management, each issue features business briefs, tips for feeders, nutrition news, and new equipment. Of value to the agricultural producer as well as the agribusinessman.

B82* Hoard's dairyman. 1870– . W. D. Hoard & Sons Co., 28 Milwaukee Ave. W., Ft. Atkinson, Wis. 53538. Semimonthly. $1 a year. Indexed: Bibl. of Agr.; Biol. & Agri. Ind.

A popular type of magazine covering all phases of the dairy industry such as breeding, feeds and feeding, disease, equipment, production, and marketing. Many of the articles are written by people in the field or by those associated with agricultural colleges. Valuable for an up-to-date, overall view of the dairy industry.

B83 Hog farm management. 1964– . Miller Publishing Co., 2501 Wayzata Blvd., Minneapolis, Minn. 55440. Monthly. $10 a year.

National journal emphasizing better swine management for large volume, specialized pork producers. Carries practical articles on labor, feeding, breeding, health, marketing, in addition to current swine research reviews and industry news items.

B84 Home and garden supply merchandiser. 1950– . Miller Publishing Co., 2501 Wayzata Blvd., Minneapolis, Minn. 55440. Monthly. $10 a year.

A business magazine devoted to management and merchandising. Intended for retail garden stores, garden supply wholesalers, and people in the manufacturing, processing, marketing, and selling of garden and lawn supplies. In addition to articles, each issue contains information on free and inexpensive literature, equipment displays, the need for competition, and other news and ideas of value to the student as well as those in the business.

B85* Home garden and flower grower. 1914– . Home Garden, Portland Pl., Boulder, Colo. 80302. Monthly. $7 a year.

A popular magazine covering all phases of gardening. Includes information on specific plants, fertilizers, greenhouses, planting, lawn seeding, landscaping and construction, garden guidelines, new books, and other subjects of interest to the layman and the horticulturist.

B86* Horticulture. 1923– . Massachusetts Horticultural Society, Horticultural Hall, 300 Massachusetts Ave., Boston, Mass. 02115. Monthly. $6 a year. Indexed: Biol. & Agri. Ind.; Biol. Abstr.

A society publication slanted toward flower gardening although it does have material on subjects such as fruit trees and food plants. The articles are practical descriptions of plants or results of experience with them and are not limited to a particular geographic area of the United States.

B87*   Implement and tractor. 1876– . Intertec Publishing Corp., 1014 Wyandatte St., Kansas City, Mo. 64105. Semimonthly (except Jan. and Mar. with 3 issues each). $3 a year. Indexed: Bibl. of Agri.

A trade journal for the farm and industrial equipment industry. Each issue contains several short articles, industry news briefs, and descriptions of new equipment designed. The new literature section is an excellent source for free and inexpensive materials available from commercial industries. This periodical gives the reader a comprehensive view of the entire industry.

B88   Iowa veterinarian. 1930– . Iowa Veterinary Medical Association, 826 Fleming Bldg., Des Moines, Iowa 50309. Bimonthly. $3 a year. Indexed: Bibl. of Agr.

The practical articles in this journal, which is published for the practicing veterinarian, are useful in any geographical area, but other sections such as "Association news" are limited to Iowa. This is a good example of a state veterinary association publication.

B89   Journal of agricultural and food chemistry. 1953– . American Chemical Society, 1155 16th St. NW, Washington, D.C. 20036. Bimonthly. $20 a year. Indexed: Biol. & Agri. Ind.; A.S. & T. Ind.; Biol. Abstr.; Chem. Abstr.

A basic research journal placing chief emphasis on the chemical aspects of plant nutrients and regulators, pesticides, food processing, nutrition, flavors, and compounds isolated from food materials. Serves agronomists, entomologists, nutritionists, advanced students, and others interested in the broad fields of agricultural and food chemistry.

B90*   Journal of animal science. 1942– . American Society of Animal Science, Business Manager, Claude Cruse, 425 Illinois Bldg., 113 Neil St., Champaign, Ill. 61820. Monthly. $24 a year. Indexed: Bibl. of Agr.; Biol. & Agri. Ind.; Biol. Abstr.; Chem. Abstr.

Primarily devoted to the solution of livestock production problems. At intervals publishes review papers on subjects of general interest to those in the animal science field. These articles and papers are technical in style and content. A "News and notes" section in each issue covers society announcements and other items of professional interest.

B91   Journal of the American Society for Horticultural Science. 1969– . American Society for Horticultural Science, Business and Circulation Manager, 615 Elm St., St. Joseph, Mich. 49085. Bimonthly. $18 a year. Indexed: Bibl. of Agr.; Biol. & Agri. Ind.; Biol. Abstr.; Chem. Abstr.; Sci. Cit. Ind.

A society publication primarily concerned with reports of results from original research on horticultural plants and their products. Original papers presenting new approaches to teaching and extension work in horticulture included.

B92   Journal of the American Society of Farm Managers and Rural Appraisers. 1937– . American Society of Farm Managers and Rural Appraisers, P.O. Box 295, DeKalb, Ill. 60115. Semiannual. $5 a year. Indexed: Bibl. of Agr.

This journal, nontechnical in style and content, serves as a medium for the expression of individual opinion concerning farm management and rural appraisal theory and practice.

B93   Journal of the American Veterinary Medical Association. 1877– . American Veterinary Medical Association, 600 S. Michigan Ave., Chicago, Ill. 60605. Semimonthly. $30 a year. Indexed: Bibl. of Agr.; Biol. & Agri. Ind.; Biol. Abstr.; Chem. Abstr.; Ind. Med.

A research journal with emphasis on practical rather than theoretical research. It is divided into these sections: articles on small animals; articles on large animals; news from Washington; news within the industry; book reviews; and association announcements. The association's other publication, *American journal of veterinary research,* stresses theoretical research.

B94   Journal of the Association of Official Analytical Chemists. 1915– . Association of Official Analytical Chemists, Box 540, Benjamin Franklin Station, Washington, D.C. 20044. Bimonthly. $20 a year. Indexed: Biol. Abstr.; Chem. Abstr.

A society publication with special emphasis on new or improved methods of analyzing fertilizers, foods, feeds, drugs, cosmetics, and other products related to agriculture and public health. Articles range from semitechnical to highly technical in content. Each issue includes reviews of significant books in the field.

B95*   Journal of dairy science. 1917– . American Dairy Science Association, 113 N.

Neil St., Champaign, Ill. 61820. Monthly. $20 a year. Indexed: Bibl. of Agr.; Biol. & Agri. Ind.; Biol. Abstr.; Chem. Abstr.

A scientific journal containing results of research in all phases of the dairy science field including herd management, feeds and feeding, genetics, and breeding. Each issue offers news of association affairs, book reviews, and short technical notes on results of experiments.

B96* Journal of economic entomology. 1908– . Entomological Society of America, 714 E. Pratt St., Baltimore, Md. 21202. Bimonthly. $23 a year to nonmembers. Indexed: Bibl. of Agr.; Biol. & Agri. Ind.; Biol. Abstr.; Chem. Abstr.; Psychol. Abstr.

The leading journal in the field of economic entomology. Primarily a research journal, it is concerned with the principles of insect damage, population, and distribution. Various methods of natural and applied control, including problems of insecticide toxicity, resistance, and residue, are covered.

B97* Journal of food science. 1936– . Institute of Food Technologists, Suite 2120, 221 N. LaSalle St., Chicago, Ill. 60601. Bimonthly. $20 a year. Indexed: Bibl. of Agr.; Biol. & Agri. Ind.; Biol. Abstr.; Chem. Abstr.; Psychol. Abstr.

An official publication of the institute, containing abstracts of important papers as well as original research articles that deal with food processing and technology. Intended for nutritionists, food technologists, and others interested in the broad field of food science.

B98* Journal of forestry. 1917– . Society of American Foresters, 1010 16th St. NW, Washington, D.C. 20036. Monthly. $18 a year. Indexed: Bibl. of Agr.; Biol. & Agri. Ind.; Biol. Abstr.; Chem. Abstr.; Eng. Ind.; P.A.I.S.

A semitechnical publication with easy-to-read articles devoted to the advancement of forestry. Covers the science of managing and operating forest lands, including forest products, recreation, natural resources, and environment. Each issue carries forestry news, the national outlook, book reviews, and a listing of current forestry literature available from the U.S. Dept. of Agriculture and the state experiment stations.

B99 Journal of milk and food technology. 1937– . International Association of Milk, Food, and Environmental Sanitarians, Inc., H. L. Thomasson, Executive Secretary, Box 437, Shelbyville, Ind. 46176. Monthly. $10 a year; $8 to libraries. Indexed: Bibl. of Agr.; Biol. & Agri. Ind.; Biol. Abstr.; Chem. Abstr.

A technical journal devoted to food protection and environmental health. Publishes results of research and experimentation on food contamination during processing, good manufacturing practices, food storage and shipment, and other subjects of interest to the food industry. Also gives current information on sanitary standards for food processing, food products, food containers, and new processing equipment.

B100 Journal of nutrition. 1928– . American Institute of Nutrition, 9650 Rockville Pike, Bethesda, Md. 20014. Monthly. $30 a year. Indexed: Bibl. of Agr.; Biol. & Agri. Ind.; Biol. Abstr.; Chem. Abstr.; Ind. Med.

The major research journal in the nutrition field. Publishes papers reporting results of original research and other scientific investigations in human and animal nutrition.

B101 Journal of range management. 1948– . American Society of Range Management, Journal of Range Management, Managing Editor, 2120 S. Birch St., Denver, Colo. 80222. Bimonthly. $15 a year. Indexed: Bibl. of Agr.; Biol. & Agri. Ind.; Biol. Abstr.; Chem. Abstr.

Primarily concerned with the presentation of ideas and facts related to the study, management, and use of range ecosystem resources and with the encouragement of public appreciation for the social and economic benefits to be derived from these resources. The articles are nontechnical to semitechnical in style and content.

B102 Journal of soil and water conservation. 1946– . Soil Conservation Society of America, 7515 N.E. Ankeny Rd., Ankeny, Iowa 50021. Bimonthly. $10 a year. Indexed: Bibl. of Agr.; Biol. & Agri. Ind.; Biol. Abstr.; Chem. Abstr.

A semitechnical journal giving attention to the science and art of efficient land use. Contains articles written by authorities in the field and by researchers associated with academic institutions. Also has short notes on conservation news, society news items, and reviews of significant books on natural resources, outdoor recreation, water resources, forestry, fish, and wildlife.

B103 Journal of wildlife management. 1937– . The Wildlife Society, Suite S-176, 3900 Wisconsin Ave. NW, Washington, D.C. 20016. Quarterly. $20 a year. Indexed: Bibl. of Agr.; Biol. & Agri. Ind.; Biol. Abstr.; Chem. Abstr.

Official journal of the society, publishing concise, scientifically accurate, original papers of research, interpretations, or ideas on all phases of wildlife management. Each issue presents society news and reviews of new books important in the field.

B104* Kiplinger agricultural letter. 1929– . The Kiplinger Washington Editors, 1729 H St. NW, Washington, D.C. 20006. Biweekly. $24 a year.

A newsletter publication which makes the latest farm news available to the reader. Provides details on current trends in farming, farm business, prices, products, methods, and farm legislation. Written in letter format with main subjects underlined for easy scanning.

B105 Land economics. 1925– . Journals Dept., University of Wisconsin Press, Box 1379, Madison, Wis. 53701. Quarterly. $10 a year. Indexed: Bibl. of Agr.; B.P.I.; P.A.I.S.

A source of scholarly studies in land economics. Its scope includes air and water use, housing, and land reform. It reflects the growing emphasis on land use planning and views land economics as economics in the broadest sense. A good example of a university press publication.

B106* Landscape architecture. 1910– . Circulation Manager, Landscape Architecture, Schuster Bldg., 1500 Bardstown Rd., Louisville, Ky. 40205. Quarterly. $8.50 a year. Indexed: Bibl. of Agr.; Art Ind.; P.A.I.S.

A publication of the American Society of Landscape Architects devoted to the planning, techniques, and art of land use and landscape development. Most articles, nontechnical in language and content, give practical, how-to information and results of experience.

B107 Landscape industry. 1955– . Brantwood Publications, 850 Elm Grove Rd., Elm Grove, Wis. 53122. Bimonthly. $10 a year.

A trade magazine for the landscape/nursery industry. Nontechnical in language and content, it gives special emphasis to environmental improvement. In addition to articles, it contains reports on industry trends and events, new products, and new equipment.

B108 Meat management. 1934– . Harcourt Brace Jovanovich Publications, 757 3rd Ave., New York, N.Y. 10017. (Circulation Office: 1 E. 1st St., Duluth, Minn. 55802). Monthly. $5 a year. Indexed: Bibl. of Agr.

The business magazine for management, marketing, and production executives in the meat industry. In addition to articles of current interest, it reports on livestock trends and political news of importance to the industry. More information on new products or equipment advertised can be obtained through the reader service department.

B109* Modern veterinary practice. 1958– . American Veterinary Publications, Inc., 114 N. West St., Wheaton, Ill. 60187. Monthly (except twice in July). $15 a year. Indexed: Bibl. of Agr.

A practical journal with principal articles written by veterinarians and members of the publication's staff. Contains items of interest from recent literature, reviews news books, reports new developments in veterinary medicine, and lists opportunities in the field. An important feature is the "Animal health technologist" section published for technologists and assistants in veterinary practice.

B110 National hog farmer. 1956– . Webb Publishing Co., 1999 Sheperd Rd., St. Paul, Minn. 55116. Monthly. $4 a year.

Trade magazine with practical articles on breeding, feeding, housing, marketing, and other phases of hog farm production and management. Serves the pork industry as well as the hog farmer.

B111 National provisioner. 1891– . National Provisioner, Inc., 15 W. Huron St., Chicago, Ill. 60610. Weekly. $6 a year. Indexed: Bibl. of Agr.

Essentially a news journal for the meat processing, packing, and rendering industries. Covers all aspects of packing, wholesaling, sanitation, processing, transportation, and meat industry news from across the nation. Contains longer articles on research, new laws, technological practices, and other subjects of current value to the industry.

B112 National wildlife. 1962– . National Wildlife Federation, Inc., 1412 16th St. NW, Washington, D.C. 20036. Bimonthly. $5 a year.

A national magazine, popular in style, for the general reader, with short, authoritative articles, photo-essays, news items, and other features concerned with the wise use of our natural resources. Natural resource management, natural history, and conservation of soil, water, forests, plants, and wildlife are within its scope.

B113 National wool grower. 1911– . National Wool Growers Association, 600 Crandall Bldg., Salt Lake City, Utah 84101. Monthly. $5 a year. Indexed: Bibl. of Agr.

Designed to inform and educate those working in the sheep industry. Features practical, how-to articles on all phases of the industry and reports on individual production procedures and practices. Regular departments offer industry news, state association news, Washington news, and a calendar of coming events.

B114 Northern logger and timber processer. 1952– . Northeastern Loggers' Association, Inc., Old Forge, N.Y. 13420. Monthly. $4 a year. Indexed: Bibl. of Agr.

General publication on the lumber industry. Of interest to foresters, loggers, and those in the wood production industries. Subjects covered include logging safety and engineering, forest management, sawmilling methods, and wood utilization. Geographical coverage is broader than the title implies.

B115* Nursery business. 1956– . Nursery Business, 850 Elm Grove Rd., Elm Grove, Wis. 53122. Bimonthly. $5 a year. Indexed: Bibl. of Agr.

A magazine designed to help the nurseryman improve his business skills and production techniques. Features practical articles on successful nursery operations, production and marketing, advertising, personnel management, and new developments in the business. Each issue includes book reviews and lists of free materials available.

B116* Organic gardening and farming. 1954– . Rodale Press, Inc., 33 E. Minor St., Emmaus, Pa. 18049. Monthly. $5.85 a year. Indexed: Bibl. of Agr.; R.G.

A popular-type magazine dealing with gardening by natural methods. Offers how-to information on turning weeds, manure, garbage, and other organic wastes into rich compost and on controlling garden pests and diseases without using pesticide sprays and chemicals.

B117 Packer. .1901– . Vance Publishing Corp., 300 W. Adams St., Chicago, Ill. 60606. Weekly. $12 a year.

National business newspaper of the fresh fruit and vegetable industry covering commercial production and marketing. Emphasizes merchandising and packaging. Intended for growers, shippers, receivers, distributors, and retailers of fruits and vegetables.

B118 Park maintenance. 1948– . Madisen Publishing Div., P.O. Box 409, Appleton, Wis. 54911. Monthly. $4 a year.

A journal of practical information on park administration, including structures, equipment, supplies, and grounds maintenance. Intended for persons responsible for parks, golf courses, and other outdoor recreation areas. Other features include information on people, new books, free literature, and industry news.

B119 Parks and recreation. 1966– . National Recreation and Park Association, 1700 Pennsylvania Ave. NW, Washington, D.C. 20006. Monthly. $7.50 a year. Indexed: P.A.I.S.

A popular publication concerned with the technical advances in conservation, park, and recreation fields and with the effective utilization of natural resources.

B120 Pest control. 1933– . Harvest Publishing Co., 9800 Detroit Ave., Cleveland, Ohio 44102. Monthly. $5 a year. Indexed: Bibl. of Agr.; Biol. & Agri. Ind.; Biol. Abstr.; Chem. Abstr.

National magazine intended primarily for members of the pest control industry. Emphasizes pest control in relation to public health rather than to agricultural production. It covers all aspects of the industry from restaurant roach control and rat problems in the slums to bats and prairie dogs.

B121 Phytopathology. 1911– . American Phytopathological Society, 1821 University Ave., St. Paul, Minn. 55104. Monthly. $40 a year. Indexed: Bibl. of Agr.; Biol. & Agri. Ind.; Biol. Abstr.; Chem. Abstr.

Scientific journal presenting original research papers on some phases of plant disease and review papers developing a new concept, hypothesis, or theory of general importance to the study of plant disease.

B122* Plants and gardens. 1945– . Brooklyn Botanic Garden, Brooklyn, N.Y. 11225. Quarterly. $3 a year. Indexed: Bibl. of Agr.; Biol. & Agri. Ind.; Biol. Abstr.

The first 3 issues of the year are small handbooks, each covering a particular kind of plant group, such as ferns, conifers, and herbs, or a particular field of concentration, such as miniature gardens. Over a period of several years these handbooks will cover the entire field. The last issue of the year is a digest of current, worthwhile horticultural literature consisting of reprints or condensations of articles from other publications. Contributors are authorities in their fields, and these include horticulture, entomology, botany, agronomy, biology, and the nursery business.

B123* Poultry digest. 1939– . Garden State Publishing Co., Garden State Bldg., Sea Isle City, N.J. 08243. Monthly. $3 a year. Indexed: Biol. & Agri. Ind.

A digest of condensed versions of practical articles appearing in other journals. (Sources of original articles are given.) Covers all phases of value or interest to poultry managers and servicemen, including flock health, feeds, nutrition, building construction, equipment, financing, and breeding.

B124* Poultry meat. 1964– . Watt Publishing Co., Mt. Morris, Ill. 61054. Monthly. $10 a year. Indexed: Bibl. of Agr.

National magazine for the broiler business presenting informative and practical ideas on all phases of production, processing, and marketing. *Poultry meat,* if consulted with *Poultry digest* and *Poultry tribune,* gives a fairly complete coverage of the entire poultry industry.

B125 Poultry science. 1908– . Poultry Science Association, C. B. Ryan, Secretary-Treasurer, Texas A. & M. University, College Station, Tex. 77843. Bimonthly. $30 a year. Indexed: Bibl. of Agr.; Biol. & Agri. Ind.; Biol. Abstr.; Chem. Abstr.

A research journal devoted to advancing the scientific study of poultry. Publishes research findings on all phases of poultry science including management, health and disease, physiology, marketing, teaching, and technology.

B126* Poultry tribune. 1895– . Watt Publishing Co., Mt. Morris, Ill. 61054. Monthly. $10 a year. Indexed: Bibl. of Agr.; Biol. Abstr.

Designed to serve the egg industry with informative, practical comments and ideas on management, production, and marketing. Regular features are news on products and equipment, ongoing research, and marketing trends.

B127 Quarterly review of biology. 1926– . State University of New York at Stony Brook, Stony Brook, L.I., N.Y. 11790. Quarterly. $10 a year. Indexed: Bibl. of Agr.; Biol. & Agri. Ind.; Biol. Abstr.; Chem. Abstr.; Ind. Med.; Psychol. Abstr.

Each issue contains 2 or 3 articles reviewing current knowledge on subjects in the biological sciences. Primarily valuable for the critical reviews of new books written by subject specialists. Each review gives a brief indication of the character, the content, and the value of the book. Botanical sciences, zoological sciences, environmental sciences, and earth sciences are some of the areas covered by reviews.

B128* Rural sociology. 1936– . Rural Sociological Society, Howard M. Saver, Secretary-Treasurer, South Dakota State University, Brookings, S.D. 57006. Quarterly. $12 a year. Indexed: Bibl. of Agr.; Biol. & Agri. Ind.; Hist. Abstr.; P.A.I.S.; Psychol. Abstr.; Sci. Cit. Ind.

Designed to promote the development of rural sociology. Contains reports, brief articles and commentary, and book reviews on research, teaching, and extension work in the field. Reports on such topics as community development, sociological concepts, migration, farm tenancy, and community attitudes.

B129 Science in agriculture. 1944– . Pennsylvania Agricultural Experiment Station, Room 106, Armsby Bldg., University Park, Pa. 16802. Quarterly. Free. Indexed: Bibl. of Agr.

Serves as a medium for reporting accomplishments and programs of research, education, and extension work in the Pennsylvania State University, College of Agriculture. A good example of experiment station publications printed by most state universities.

B130 Science news. 1921– . Science Service, Inc., 231 W. Center St., Marion, Ohio 43302. Weekly. $7.50 a year. Indexed: Bibl. of Agr.; Abr. R.G.; Eng. Ind.

Essentially a news magazine covering all areas of science including aerospace, earth sciences, environmental sciences, medical sciences, and physical sciences. The annotated lists of new books and new films are valuable features.

B131* Scientific American. 1859– . Scientific American, Inc., 415 Madison Ave., New York, N.Y. 10017. Monthly. $10 a year. Indexed: Bibl. of Agr.; A.S. & T. Ind.; Abr. R.G.; Biol. Abstr.; Chem. Abstr.; Ind. Med.; Math. R.; Met. Abstr.; R.G.; Meteor. & Geoastrophys. Abstr.; Psychol. Abstr.

A generally circulated publication covering biological, behavioral, and physical sciences. Contributors are scientists, who describe their own and closely related work in a form easily understood by the nonspecialist. The outstanding illustrations accompanying the articles are valuable as teaching aids. Includes a bibliography for further reading on subjects covered by articles in each issue.

B132 Seed world. 1915– . Seed World, 434 S. Wabash Ave., Chicago, Ill. 60605. Semimonthly (except monthly in June, July, Aug. and Dec.). $5 a year. Indexed: Bibl. of Agr.; Biol. Abstr.; Chem. Abstr.

A news magazine for the seed industry. Contains information on new developments, new equipment and products, trade news, and other subjects of current interest to seedsmen. An added feature is the section containing bulletin board suggestions for retail seed stores and garden centers.

B133 Seedsmen's digest. 1950– . Skarien & Associates, 1910 W. Olmos, San Antonio, Tex. 78201. Monthly. $4 a year.

Intended for commercial seed firms and related services. Gives short reports on seedsmen associations, new seed varieties, seed processing methods and equipment, crops, planting methods, fertilizers, agricultural trade, and allied industry news.

B134 Soil Science Society of America proceedings. 1936– . Soil Science Society of America, 677 S. Segoe Rd., Madison, Wis. 53711. Bimonthly. $16 a year. Indexed: Bibl. of Agr.; Biol. & Agri. Ind.; Biol. Abstr.; Chem. Abstr.

A scientific journal containing papers and notes on original research concerned with the revision of existing concepts or the development of new concepts and techniques in various phases of soil science. Some issues contain short, critical essays or reviews on timely subjects.

B135 South Dakota conservation digest. 1934– . South Dakota Dept. of Game, Fish, & Parks, Pierre, S.D. 57501. Bimonthly. Free. Indexed: Biol. Abstr.

Dedicated to the wise use of the state's natural resources and to the betterment of fishing, hunting, camping, and other forms of outdoor recreation. Although written primarily for South Dakota, much of the material would be equally valuable in other states, particularly those in the surrounding area. Similar publications are available for most all of the states.

B136* Successful farming. 1902– . Meredith Corp., 1716 Locust St., Des Moines, Iowa 50303. Monthly. $2 a year. Indexed: Biol. Abstr.

A farm management magazine published primarily for the farmer and his family, which covers all phases of farming. Special issues are entirely devoted to one subject, and these give in-depth treatment to decision-making in each subject area. Some issues contain how-to articles accompanied by explanatory illustrations.

B137 Trees magazine. 1937– . Scanlon Publishing Co., 7621 Lewis Rd., Olmsted Falls, Ohio 44138. Bimonthly. $3 a year.

Devoted to the aesthetic and functional use of ornamental trees, to the pathological, entomological, and physical disorders of trees, and to the technical and business administration of trees and parks. Discussions, comments, notes, and news items contributed by people engaged in either the scientific advancement or the commercial development of the field.

B138 Trends in parks and recreation. 1964– . National Conference on State Parks, 1700 Pennsylvania Ave., NW, Washington, D.C. 20006. Quarterly. $10 initial subscription; $3.50 a year renewal.

Concerned with studies, concepts, philosophies, and projections related to the many aspects of parks and recreation. In addition to original articles, this nontechnical journal carries reprints of speeches and reprints from other publications.

B139 Turf-grass times. 1965– . Turf-Grass Publications, Inc., P.O. Box 51088, 218 19th Ave. N, Jacksonville Beach, Fla. 32250. 8 issues a year. $5 a year.

Articles, comments, notes, and news items on irrigation, seeding, sodding, fertilizers, weed control, pests and diseases, equipment, and management of turfgrass. Contributors are persons in the industry or those associated with academic institutions.

B140 Vegetable crop management. 1965– . Vegetable Crop Management, P.O. Box 67, Minneapolis, Minn. 55440. Monthly. $10 a year.

National magazine of practical articles on commercial vegetable production and management. Coverage includes all kinds of vegetables and some specialty crops which are not actually vegetable. Reports on economic trends, marketing developments, new equipment available, and other news of importance to the industry.

B141* Veterinary medicine/small animal clinician. 1905– . Veterinary Medicine Publishing Co., 144 N. Nettleton, Bonner Springs, Kans. 66012. Monthly. $12 a year. Indexed: Bibl. of Agr.; Biol. & Agri. Ind.; Biol. Abstr.; Chem. Abstr.

A practical journal with articles, many having full-color illustrations, written by and for practicing veterinarians and covering both medical and management aspects. Medical abstracts, case studies, and industry news are some of its other features. Interestingly, in this journal, covering both large and small animals, every effort is made to relate advertising to editorial content.

B142 Wallaces farmer. 1874– . Wallace-Homestead Co., 1912 Grand Ave., Des Moines, Iowa 50305. Semimonthly. $3 a year.

A general agricultural news journal for farmers of Iowa. Serves to keep farmers advised of new methods and procedures, new products, marketing trends, legislative news concerning agriculture, and helpful hints about agricultural production and marketing. Similar, privately published journals are available for many of the states.

B143 Weed science. 1951– . Weed Science Society of America, F. W. Slife, Dept. of Agronomy, University of Illinois, Urbana, Ill. 61801. Bimonthly. $20 a year. Indexed: Bibl. of Agr.; Biol. & Agri. Ind.; Biol. Abstr.; Chem. Abstr.

Scientific journal publishing original manuscripts about weeds, weed control research, effect of herbicides on plants, and other related topics. Contributors include biological and chemical scientists and engineers.

B144 Weeds, trees, and turf. 1962– . Harvest Publishing Co., 9800 Detroit Ave., Cleveland, Ohio 44102. Monthly. $10 a year. Indexed: Bibl. of Agr.

Intended for persons engaged in the vegetation maintenance and control industry and related fields. Contains practical information on weed control, irrigation, turf maintenance, commercial sod growing, and tree protection and care.

B145 Wood and wood products. 1923– . Vance Publishing Co., 300 W. Adams St., Chicago, Ill. 60606. Monthly (except 2 issues in Oct.). $5 a year. Indexed: Bibl. of Agr.; Eng. Ind.

Practical magazine devoted to wood and allied products management and operations. Gives up-to-date information on wood industry trends, technological procedures, production ideas, patents, new products, and recent literature published.

B146 World wood. 1960– . World Wood, Circulation Dept., 500 Howard St., San Francisco, Calif. 94105. Monthly (except 2 issues in July). $10 a year. Indexed: Bibl. of Agr.

Practical journal presenting a worldwide view of forestry and the forest products industry. Areas covered include forestry, logging, logging machinery and equipment, and the lumber and timber industry. Valuable for information on new technology being used in countries outside of the United States.

C

Pamphlets and Other Inexpensive Materials

These materials are especially useful in small collections, in library vertical file holdings, and for educators who desire instructional materials for teaching in the classroom. Libraries with limited funds for purchasing books and other materials may keep their collections up-to-date and reasonably complete by obtaining and organizing pamphlets, announcements, reports, and other inexpensive materials for ready reference and for circulation. Most libraries maintain vertical files of pamphlets, maps, pictures, etc., usually arranged by subject and weeded periodically to remove outdated items. These files may be particularly valuable for housing materials of local interest, irregular items, such as maps, catalogs, and photographs not to be cataloged, and clippings from newspapers and magazines. Because companies and agencies provide these materials inexpensively, teachers can not only order copies of an item for every student but also an enormous amount of diversified matter for general use from them. In this section, only the sources which are most commonly consulted for these media are annotated. Others are mentioned in C10, Publications from Commercial, Industrial, and Organizational Sources and in C11, Publications Lists of State Experiment Stations, Cooperative Extension Services, and State Boards of Agriculture. The directories and guides in section F should be consulted for lists of specific organizations, associations, firms, institutions of learning, and individuals who provide this type of material.

C1* Bimonthly list of publications and motion pictures. 1909– . Office of Information, U.S. Dept. of Agriculture, Washington, D.C. 20250. Bimonthly. Free.

Annotated list of Dept. of Agriculture publications grouped by series. Includes blank for ordering of free materials. Those listed *For Sale Only* must be purchased from the U.S. Government Printing Office. Titles included on this list are those received during the 2 months prior to the date of the list's publication. Also can serve as a checklist for vertical file holdings.

C2 Catalogue of FAO publications, 1945–1968. 1969. Issued by the Food & Agriculture Organization of the United Nations. Available from Distribution and Sales Section, Food and Agriculture Organization of the United Nations, Via delle Terme de Caracalla, 00100 Rome, Italy. 117p. Free. Supplement. 1969– . Quarterly. Free.

This catalog lists all priced publications issued by FAO since 1945 and includes titles in preparation on the date that the catalog was issued. Updated by quarterly supplements. Listed by broad subjects and subdivided by types of publication such as manuals, maps, papers, and series. Annotations included for some titles. Several publications which are worldwide in scope are particularly useful to those interested in agriculture in other countries. Others, which are concerned only with subject matter, make no reference to geographical area and may be of immense value for supplemental reading in general agriculture, economics and statistics, fisheries, forestry, nutrition, or agricultural legislation.

C3 Educators' grade guide to free teaching aids. 1955– . Educators Progress Service, Inc., Randolph, Wis. 53956. Annual. $8 (approx.)

See section D, no.7.

C4 Free and inexpensive learning materials. 1941– . Div. of Surveys and Field Services, George Peabody College for Teachers, Nashville, Tenn. 37203. Latest ed. $3.

Includes pamphlets, maps, pictures, posters, charts, and other educational aids grouped by subject. Material is screened before being listed and is factual, free from exaggeration and propaganda, and has a minimal amount of advertising. Each annotation gives basic information on the item and the full name and address of its distributor. Although this publication is primarily intended for elementary and secondary schools, many of the titles listed are on the college or adult level. In addition, this publication serves as a source of great value for information on private industries, institutes, foundations, etc., which issue free and inexpensive educational materials.

C5* List of available publications of the United States Department of Agriculture. 1929– . Office of Information, Publications Div., U.S. Dept. of Agriculture. Available from the Superintendent of Documents, U.S. Government Printing Office, Washington, D.C. 20402. Latest ed. 45¢ (also called List no.11)

A catalog of U.S. Dept. of Agriculture publications that were available when the list was prepared which gives general information on how to obtain these materials. Contains a separate, annotated listing of periodicals. Arranged by broad subject and indexed by subject and title. Valuable for quick reference to publications obtainable on a given subject. While not as up-to-date as the *Monthly catalog of United States government publications,* it is considerably easier to use, and it is revised every few years. Especially helpful for organizing a basic subject file of pamphlets, brochures, and other inexpensive materials.

C6 List of publications. Canada Dept. of Agriculture. 1912– . Available from Information Div., Canada Dept. of Agriculture, Ottawa, Ontario, Canada. Latest ed. Free.

List of available publications prepared by the Canada Dept. of Agriculture, arranged under broad subject headings, or in some instances, by type of publication such as research reports. No annotations. Contains general instructions for obtaining free titles desired and for ordering purchasable items. Similar lists are published by most, if not all, of the provinces. Useful for obtaining supplemental reference materials for subject files, especially in those states having similar climate, etc.

C7* Monthly catalog of United States government publications. 1895– . Issued by and available from the Superintendent of Documents, U.S. Government Printing Office, Washington, D.C. 20402. Monthly. $7 a year (including index).

A bibliography of current publications issued by all branches of the government which is arranged by department and bureau and which lists full title, date of publication, paging, price, etc., for each entry. Includes books, films, and other AV media as well as pamphlets and ephemeral material. Valuable to the classroom teacher and a must for the library. Each issue contains an index, and in Dec., a cumulative index is published. Index entries are under subject, title, key words, or issuing body. The Feb. issue contains a list of U.S. government periodicals, serials, and subscription publications. General instructions for ordering are printed in each issue. By writing to the bureaus or departments, librarians can request that their library be placed on the mailing lists of these agencies to receive all published materials free of charge.

C8* Monthly checklist of state publications. 1910– . Comp. by Exchange and Gift Div., Processing Dept., U.S. Library of Congress. Available from the Superintendent of

Documents, U.S. Government Printing Office, Washington, D.C. 20402. Monthly. $8 a year.

A bibliography of state publications issued during the last 5 years. Although limited to those publications received by the Library of Congress, it is a fairly complete listing. Monographs are added as they are received each month and are arranged alphabetically by state and issuing agency. Periodicals are listed semi-annually in the June and Dec. issues. Includes publications prepared by state agencies, state agricultural experiment stations, and state institutions of higher learning. Of greatest value to the librarian who collects this type of material from several states. The normal time lag from publication until inclusion in *Monthly checklist* is about 6 months, but because this may extend for 1 to 2 years, other sources should also be consulted for current materials. State publications listed consist of all types of material from books and periodicals to pamphlets, brochures, maps, and brief progress reports.

C9 Over 2,000 free publications; yours for the asking. 1968. Comp. and ed. by Frederic J. O'Hara. New American Library, 1301 Ave. of the Americas, New York, N.Y. 10019. 352p. paper, 95¢

An up-to-date list of free or inexpensive U.S. government publications, organized by department, bureau or other division. Has a subject index and a list of depository libraries in the United States at which by law these publications must be retained and made available to residents of the area for borrowing. Gives price, format, and date as well as an annotation for each entry.

C10 Publications from commercial, industrial, and organizational sources.

An enormous amount of material can be obtained from these companies at little or no cost. Since this material is prepared from a public-relations point of view, it may be somewhat slanted in presentation, and the reliability of individual items must be determined by the educator. These sources can be invaluable because copies can often be obtained in large quantities for classroom use. Some examples of these companies are International Harvester, American Forest Products Industries, Inc., and American Potash Institute, Inc., and hundreds of others can be located in directories, bibliographies, advertisements, and in periodicals.

C11 Publications lists of state experiment stations, cooperative extension services, and state boards of agriculture.

Most of these publications eventually appear in the *Monthly checklist of state publications;* however, most of the state agricultural agencies do issue a list of their publications which are currently available. These lists are easier to use and in most instances, are more up-to-date. Instructions are usually included for obtaining titles. Entered items are those considered to be of general interest to citizens of the state, but they are often of value to people in surrounding states or regions. These publications coupled with those of the USDA are the best sources for agricultural information. Addresses of the state agricultural colleges, state departments of agriculture, and the experiment stations are available from the Office of Information, U.S. Dept. of Agriculture, on request. Many of these agencies maintain mailing lists on which libraries may be placed by request.

C12* Selected United States government publications. 1928– . Available from Superintendent of Documents, U.S. Government Printing Office, Washington, D.C. 20402. Biweekly. Free.

An annotated list of all types of materials for sale by the Superintendent of Documents that is comprised of publications considered to be of general interest. The agricultural publications form only a small part of the list, but it is free and can serve as a handy checklist against vertical file holdings. A detachable form is appended to the list to facilitate ordering of publications desired. Not only includes free and inexpensive pamphlets and other ephemeral materials but also lists higher priced items, such as books, etc.

D

Filmstrips, Transparencies, and Other Audiovisual Aids

A wealth of audiovisual material is available for the agricultural sciences in all forms ranging from slides and transparencies to 8, 16, and 35mm educational films. Because these aids are so numerous and often transitory in nature, this section does not attempt to list particular titles. Instead, it includes only easily obtainable sources for audiovisual aids known to have adequate listings for agriculture. Examples only are given for certain types of sources including educational and audiovisual periodicals, state college and university catalogs and lists, industrial and supply house catalogs, and state and U.S. government audiovisual publications. Many educational and AV magazines have sections on new films and other similar aids but few provide agriculturally related materials in every issue. The periodical guides, such as *The standard periodical directory* and *Ulrich's international periodicals directory,* provide a comprehensive listing of these magazines. Almost all of the state university and the U.S. and state government sources for audiovisual media are indexed in the *Monthly checklist of state publications* and in the *Monthly catalog of U.S. government publications.* The U.S. government AV catalogs are especially good, and particular attention should be given to the examples of them in D2, D16, and D17. Industrial and supply house catalogs, valuable sources of information on purchasing audiovisual equipment as well as buying AV aids, are numerous but often difficult to locate. The directories and other guides to institutions and firms in section F may be helpful to anyone seeking company or association names and addresses.

D1 The booklist. 1905– . American Library Association, 50 E. Huron St., Chicago, Ill. 60611. Bimonthly. $10 per year; single copy 65¢

An example of an educational magazine containing reviews and announcements of AV materials as well as of other types of learning aids. AV review sections include 16mm films, 8mm film loops, filmstrips, and recordings. Authoritatively annotated because each entry is reviewed by the *Booklist* editorial and reviewing staff with the assistance of librarians and school personnel who are familiar with such media and the methods for properly evaluating them. Entries indicate title, producer, distributor, date, lengthy descriptions of content, and intended audience. The number of agriculturally related entries is minimal but those included are of high quality.

D2 Color filmstrips and slide sets of the United States Department of Agriculture. (USDA. Miscellaneous publication no. 1107). 1970. Superintendent of Documents, U.S. Government Printing Office, Washington, D.C. 20402. 13p. Free.

A good example of the AV catalogs published by the USDA. Exclusively agricultural coverage. Includes filmstrips, slide sets, charts, etc., on agricultural economics, beautification, civil defense, conservation, cooperatives, farming, food, forestry, and many other subjects. Entries provide such pertinent information as the title, release date, number of frames, price, and a summary of contents. The filmstrips and slide sets are available for purchase only, but most are very low in price.

D3 Combined 16mm film catalog. 1970. U.S. Atomic Energy Commission, Washington, D.C. 20545. 79p. Free.

In this 1970 revision, the popular-level and the professional-level film catalogs of the Atomic Energy Commission are joined. The films available in this catalog provide information on general agriculture, waste disposal, plant growth, etc. Entries give such details as year produced, time length, color, producer, source, subject content, and honors awarded for the film. Materials are available on free loan when requested for educational purposes or for other nonprofit public uses.

D4 EFLA bulletin. 1945– . Educational Film Library Association, Inc., 250 W. 57th St., New York, N.Y. 10019. Monthly. By membership.

A good example of an AV periodical which contains current materials arranged by type of media. Includes motion pictures, filmstrips, and phonodiscs. Agricultural AV aids are included but not distinguished by subject.

D5 Educational motion pictures: descriptive catalog of 16mm motion pictures containing a subject and grade-level index. 1970. Audio-Visual Center, Indiana University, Bloomington, Ind. 47401. 1120p. $5; free with the purchase of $25 worth of materials.

This state university catalog, which contains nearly 8,800 films for all levels ranging from preschool to general adult, is updated frequently by supplements. It is in 2 parts: (1) a subject and grade-level index, which groups materials by broad subjects, and (2) a list of the films available from Indiana University arranged alphabetically by title. In part 2, each entry gives the film length, sound, color, price, subject headings, level, producer, and a detailed description of the film. A very useful AV source for those interested in vocational agriculture because materials listed are quite numerous, are of good quality, and relate to many contemporary problems in the field.

D6 Educational motion pictures for school and community. Audio-Visual Services, Bureau of School Services, College of Education, University of Kentucky, Lexington, Ky. 40506. Annual. Free.

Exemplifies catalogs prepared by most all state university AV service departments. Arranged alphabetically by title and indexed by subject. Coverage in agriculture and related fields is good. Lists all films and other AV aids available for rental, and gives for each the price, a summary, the interest level, television clearance, and other standard information. Updated by supplements during the year.

D7* Educators' grade guide to free teaching aids. 1955– . Educators Progress Service, Inc., Randolph, Wis. 53956. Annual. $8 (approx.)

An annotated catalog of free materials including maps, pamphlets, bulletins, exhibits, books, and charts. Arranged by broad subject and indexed by title, subject, and sources for obtaining teaching aids. Interest level is not limited to grade school, and most of the entries, which do not specify particular grade limitations, may be of interest to any group. Agricultural media are found in the general science and the general science-nature study sections. Cost of the guide is minimal in relation to availability of listed materials free of charge.

D8* Educators' guide to free films. Educators Progress Service, Inc., Randolph, Wis. 53956. Annual. $10 (approx.)

Complete, current list of all free films available in the United States and Canada. (Other types of AV media are covered in separate but related works, e.g. *Educators' guide to free filmstrips*.) An extensive section on agriculture includes not only films from film foundations and motion picture agencies, but also those available from industrial firms, government departments, and research services. Entries give standard bibliographic information as well as comprehensive annotations that more than adequately describe content. Most of the films serve no one age group because interest level is largely determined by the purpose for which each picture is used. The eminent quality of the included media is evident in the high percentage of films cleared for television. The relatively high cost of the educators' guides is

D9*   Educators' guide to free filmstrips. 1949– . Educators Progress Service, Inc., Randolph, Wis. 53956. Annual. $8 (approx.)

justifiable because the media are attainable free of charge.

A companion catalog to *Educators' guide to free films,* this annual guide lists all free filmstrips and many slides and transparencies currently available in the United States and Canada. Has an excellent comprehensive section on agriculture which includes high quality media produced and distributed by various commercial and private agencies. Entries have informative annotations, standard bibliographic descriptions, and often indicate interest level, availability of accompanying scripts or phonodiscs, and approval for television. Well worth its price because materials listed are free.

D10*   Educators' guide to free science materials. 1960– . Educators Progress Service, Randolph, Wis. 53956. Annual $8 (approx.)

Also arranged by broad subject, this list, another of the educators' guides, includes most all free science materials available in the United States and Canada. In format and entry information given, this listing is similar to its companion catalogs. Agricultural materials are found under biology, conservation, and general science-nature study. Especially valuable because it excludes all nonscientific media.

D11   Elementary, secondary, college educational materials catalog. Learning Arts, P.O. Box 917, Wichita, Kans. 67201. Annual. Free.

A catalog of phonograph records, sound filmstrips, tapes, transparencies, flat pictures, and books which lists media by type and then by broad subject and advisable grade level. Agricultural materials are scattered throughout the science sections. Many of the media may be used interchangeably for high school and college, and these are especially good for vocational school and 2-year college instructional programs. Lengthy descriptions of all entries and suggestions for audience age-level are most helpful to the selector. An excellent example of a major producer-distributor industrial catalog.

D12   Encyclopaedia Britannica Films, Inc. catalog. Encyclopaedia Britannica Educational Corp., 425 N. Michigan Ave., Chicago, Ill. 60611. Annual. Price varies.

This catalog, prepared by the Educational Corp., has films in many agriculturally related areas such as the earth, biological, and environmental sciences. Format used and information given in the entries are similar to that usually found in AV catalogs. Coded notations indicate level of interest or difficulty. Media indexed are: filmstrips; 8mm movie loops, 16mm movie reels, multimedia, and printed matter.

D13*   Farm film guide. 1965. Business Screen Magazines, Inc., 7064 Sheridan Rd., Chicago, Ill. 60626, or the Farm Film Foundation, Inc., 1425 H St. NW, Washington, D.C. 20005. 49p. $1.25.

Primarily a film guide for agribusiness, this catalog is a comprehensive list of motion pictures in such areas as field and garden crops, orchards, farm home and family life, agribusiness, and farm management. Visual aids describe the businesses and services available to the farmer and rancher and thus offer good introductory educational material for students interested in the field. Though coverage is limited to a specialized area, the low price of the guide merits its purchase. Media may be used for various age levels depending on the intended purpose.

D14   Film evaluation guide, 1946–65. 1965. Educational Film Library Association, Inc., 17 W. 60th St., New York, N.Y. 10023. 528p. $30; $25 to EFLA members. Supplement 1, 1965–67. 1968. 163p. $12; $10 to EFLA members. Supplement 2, 1967–71. 1972. 133p. $12; $10 to EFLA members.

A valuable, highly reputable source for 16mm films because media are evaluated by an Educational Film Library Association preview committee before inclusion. The guide is a mere compilation of the *EFLA evaluations* (cards issued monthly) and thus, it is neither a comprehensive nor a truly selective list. Entries are included just as they are received and kept current by supplements. In addition to standard bibliographic information, entries give a detailed content synopsis, Dewey decimal number, the evaluator(s), uses, age level (elementary to adult), rating, and any comments made during the evaluation. Extensive coverage in agriculture in all areas including general agriculture, land and soil, field crops, forestry, gardening, animal husbandry, landscape planning, and conservation.

D15   Film news: the newsmagazine of films, filmstrips, recordings, educational TV. 1939– . Film News Co., 250 W. Fifty-seventh St., New York, N.Y. 10010. Bimonthly. $5.

An AV periodical which reports news on recent developments in the AV field, meetings, society proceedings, and, especially, new media available. Provides descriptive, critical summaries of films, filmstrips, and phonodiscs, and though agricultural aids are not treated separately, they are included in various issues.

D16* Films of the U.S. Department of Agriculture. (Agriculture handbook no.14). 1968. Superintendent of Documents, U.S. Government Printing Office, Washington, D.C. 20402. 81p. 40¢

Exemplifies USDA AV aid sources indexed in *Monthly catalog of U.S. government publications*. Includes sections on state film lending libraries, films available, and how to handle, borrow, and buy materials. Arranged by title, entries are also indexed by subject. Titles appear in boldface black print followed by information such as color or black and white, release date, approval for TV, length in time, and descriptions of the content, awards, and other details.

D17* Forest service films, available on loan for educational purposes to schools, civic groups, churches, television. 1969. Forest Service, U.S. Dept. of Agriculture, Washington, D.C. 20250. 34p. Free.

Another example of the AV catalogs prepared by the U.S. government and listed in *Monthly catalog*. Entries are arranged by broad subject area. Title, length of time, release date, interest level, and subject content are given for each film. An excellent source for educational AV materials in forestry because it includes all films available from the regional film libraries of the Forest Service free of charge. These regional collections are listed separately in this publication.

D18 Free and inexpensive learning materials. 1941– . Div. of Surveys and Field Services, George Peabody College of Teachers, Nashville, Tenn. 37203. Latest ed. $3.

See section C, no.4.

D19 Index to 8mm motion cartridges. 1969. R. R. Bowker, 1180 Ave. of the Americas, New York, N.Y. 10036. 402p. $19.50; NICEM, University of Southern California, University Park, Los Angeles, Calif. 90007. paper, $8.50.

An index prepared by the National Information Center for Educational Media which excludes all media except instructional 8mm motion cartridges. Covers several agricultural areas such as soil management and conservation, farm work and products, and general agriculture. Each entry gives standard details about the cartridge but interest level is not specified. Purchase of the catalog should depend on equipment available and demand for this particular type of media.

D20 Index to overhead transparencies. 1969. R. R. Bowker, 1180 Ave. of the Americas, New York, N.Y. 10036. 552p. $22.58; NICEM, University of Southern California, University Park, Los Angeles, Calif. 90007. paper, $8.50.

Another index prepared by the National Information Center for Educational Media, this catalog lists over 18,000 items. The listed transparencies incorporate many titles that have never before been combined in a single work. Entries are arranged by title, and the format is basically the same as in the other NICEM indexes. The NEA Association for Educational Communications and Technology standards serve as a foundation for the composition of each annotation. Agricultural coverage is adequate, and a directory of producers and directors is provided.

D21 Index to 16mm educational films. 2d ed. 1969. R. R. Bowker, 1180 Ave. of the Americas, New York, N.Y. 10036. 1111p. $39.50; NICEM, University of Southern California, University Park, Los Angeles, Calif. 90007. paper, $8.50.

Excluding all AV media except 16mm educational films, this index offers excellent coverage in agriculture in such areas as general agriculture, animal husbandry, animal farm products, plant farm products, farm work, plant science, and soil management. Entries provide a content summary, the title, time length, producer-distributor, year of production, and other particulars. No interest level is indicated for films. Even though this is one of the most comprehensive listings of all 16mm films compiled, several of the materials listed in the index may also be found in other sources, and because of its high cost, the guide should be carefully reviewed before purchase. (Probably available at nearby college or university library.)

D22 Index to 35mm educational filmstrips. 2d ed. 1970. R. R. Bowker, 1180 Ave. of the Americas, New York, N.Y. 10036. 872p. $34; NICEM, University of Southern California, University Park, Los Angeles, Calif. 90007. paper, $12.

This index covers only 35mm educational filmstrips. Arrangement and entry content are almost identical to that in *Index to 16mm educational films* except for additional details given such as number of frames and inclusion of filmstrip captions. Subject sections on agriculture are soil conservation, farm work, general agriculture, agriculture-crops, and agriculture-livestock. Due to its expensiveness, this guide should also be personally reviewed and evaluated by the librarian, teacher, or administrator prior to purchasing.

D23* Industrial and agricultural education. (v.9 of *Educational media index*). 1964. McGraw-Hill Book Co., 330 W. 42nd St., New York, N.Y. 10036. 235p. $5.

One of the most comprehensive indexes to the available educational media specifically adapted to agricultural vocational education. Includes kinescopes, charts, maps, models, mock-ups, phonodiscs, and other less common AV materials as well as films, filmstrips, and slides. In addition to a description of the technical characteristics for each medium, a brief, informative annotation and a suggested age-interest level are given. A list of sources and their addresses is provided for teachers, librarians, or administrators wishing to rent or purchase media. Supersedes the *Educational film guide, 1936–62*, and the *Filmstrip guide, 1948–62*, published by H. W. Wilson Co.

D24 Learning directory 1970–1971. 1970. Westinghouse Learning Corp., 100 Park Ave., New York, N.Y. 10017. 7v. $75. Supplement. 1972. 914p. $19.50.

Includes 200,000 items representing all types of educational materials ranging from books to films and filmstrips. Items are arranged by subject topics in columns. Information given for each entry includes topic, level, medium, title, color, sound, source, price, date, size, and reference notes, which refer to the reader's guidance and source index sections in volume 1. The directory's greatest value is in enabling coordination of all types of media on a particular subject. However, the source is somewhat handicapped by its lack of a detailed subject or title index and its lack of comprehensiveness. Because of its high cost, this guide should be carefully studied and evaluated before purchasing. (Copy is probably available at any nearby college or university library.)

D25 Library of Congress catalog: motion pictures and filmstrips. 1953– . Card Div., U.S. Library of Congress, Bldg. No.159, Navy Yard Annex, Washington, D.C. 20541. Quarterly, with annual and 5-year cumulations. $8 a year.

Often disregarded because of the high cost of its cumulative volumes, this catalog is an annual, inexpensive, and comprehensive guide to educational motion pictures, filmstrips, and theatrical films. Listed media have been cataloged, and cards are available for them from the library. Bibliographic details provided by film producers and distributors include authors and titles of works on which films are based, if any, availability of accompanying teacher's guides, series titles, credits, suggested subject headings, LC card numbers, and the standard details, such as title, running time, and color or black and white. The more current issues of the catalog offer an index of producers and distributors and their addresses. Many sources, industrial firms, for example, which are not routinely covered in other guides, are included on this list. Agricultural films and filmstrips are easily accessible using the detailed subject index.

D26 Mountain plains educational media council film catalog. 1962/63– . Supplement available for 1970. Obtainable from Colorado, Nevada, Wyoming, and Utah universities. Biennial. No price available.

This film catalog has somewhat limited agricultural coverage but it is nevertheless valuable for its separate section on vocational/technical agriculture. Media described are 16mm films on all levels from elementary to adult. Title, producer, source, date, length, summary, level, LC number, and other data are given for each film. Kept up-to-date by supplements.

D27* Nasco agricultural sciences technical catalog. (catalog no.106). [n.d.] Nasco, Ft. Atkinson, Wis. 53538. 300p. Free.

One of the most up-to-date sources for instructional materials as well as equipment in the agricultural sciences. Nasco, originally founded by a vocational agriculture instructor, provides agricultural educational aids which are reviewed carefully by noted experts in this field. In addition to listing agricultural supplies, the catalog also includes slides, filmstrips, film loops, overhead transparencies, records, and video tape recording systems. Each entry contains a helpful description of content and the order price. A most valuable catalog because it is one of the few producer-distributor supply guides whose coverage is exclusively agricultural.

D28* National audio/tape catalog. 1967. National Education Association, 1201 16th St. NW, Washington, D.C. 20036. 114p. $3.

Short annotations, running times, series titles, interest levels (college, secondary, elementary), and broadcast restrictions are among the items of information given for each entry listed in this source. Tape recordings include programs for lectures, discussions, interviews, etc. Overall, a comprehensive listing of magnetic tapes available from the National Tape Repository. There are agricultural tapes in the sections on animals, natural history, water, timber, safety, botany, biology, natural resources, and oceanography. One of the few sources for audio materials on agriculture.

D29* Science materials catalog. CCM School Materials, Inc., 2124 W. 82d Pl., Chicago, Ill. 60620. Annual. Free.

An excellent example of a producer-distributor supply catalog which includes agricultural films, filmstrips, film loops, charts, graphs, maps, kits, models, pictures, mock-ups, slides, transparencies, live or preserved biological specimens, and AV equipment. Levels of interest range from elementary to college and are largely determined by purpose for use of materials. A brief description of each entry is given, which includes content, type of media, price, age level, and title. One of the most complete catalogs for all types of scientific AV materials, this source should be available to all teachers in the physical and life sciences.

D30 Selected films and filmstrips on food and nutrition. 1961. Food & Nutrition Council of Greater New York, Inc., Bureau of Publications, Teachers College, Columbia University, New York, N.Y. 10027. $1.25. Supplement. 1964. Available from Mr. O. F. Pye, Nutrition Dept., Teachers College, Columbia University, New York, N.Y. 10027.

An inexpensive, selective guide to films and filmstrips on food and nutrition updated by supplements. Arranged by type of media and then listed under broad subject headings alphabetically by title. Subject headings include general nutrition, career guidance, and gardening. Entries not only offer details on sound, color, date, and distributor but also give intended audience, content, and appraisal comments. Selections are highly creditable because they are reviewed by an evaluation committee prior to inclusion.

D31* State films on agriculture. 1962. Office of Information, U.S. Dept. of Agriculture, Washington, D.C. 20250. 86p. Free.

A useful guide to agricultural 16mm films of state colleges, universities, and state agencies in the states and Puerto Rico. Some of the more broadly covered subject areas include agricultural engineering, beef cattle, dairying, pests and diseases of plants, forestry, nutrition, corn, and agricultural economics. Although some films are loaned beyond the state or region where produced, many are available only within that defined area. Thus, the arrangement by state is especially serviceable for acquisition of media. Various films are loaned for preview and others are available only for purchase. Entries are comprised of brief annotations, dates of production, titles, running times, TV clearance, terms of sale or rental, and other standard information. In all, an excellent source for inexpensive educational media.

D32 Turtox biological supplies catalog. General Biological, Inc., 8200 S. Hoyne Ave., Chicago, Ill. 60620. Biennial. Free.

A supply catalog of guaranteed, excellent materials with a detailed subject index which provides direct access to specific items. Comprised of 10 sections: living materials; zoological materials; botanical materials; demonstration materials; skeletal preparations; microscopic slides; models; visual aids; projection slides; and, apparatus. Materials are for educational and professional use. Gives short, helpful descriptions for each entry. A very good purchasing guide for materials in agriculture.

D33 U.S. government films. 1969. National Audio-Visual Center, National Archives and Records Service, General Services Administration, Washington, D.C. 20408. 165p. No price available.

A comprehensive listing of available films and filmstrips, slides and photographs produced by government agencies and departments and distributed by the National Audio-Visual Center. For vocational agriculture, media on farm management, forestry, livestock, meat, and other related subjects are available. Films entitled, *Poultry processing inspection,* and *Reconditioning a mower* are specific examples of the technical educational materials listed. An accompanying list gives sources of U.S. government films and filmstrips as well as information on types of media available, where they are distributed, prices, and terms of purchase, format, scarcity of media, and television clearance.

D34 Ward's catalog: biology and the earth sciences. Ward's Natural Science Establishment, Inc., P.O. Box 1712, Rochester, N.Y. 14603. Biennial. Free.

A comprehensive supply catalog listing living materials, dissection materials, jar mounts, microscope slides, visual aids, models, and other supplies which are available to teachers, librarians, and administrators. This beautifully illustrated catalog not only offers brief, helpful descriptions of each entry but also provides pictured examples of them. Materials are applicable for most grade levels, and are of very high quality.

E

Government Documents

Federal and state government publications are excellent sources of authoritative information on any subject. The Department of Agriculture covers all phases of the agricultural sciences in the myriad of printed matter which it issues. Much of the statistical and indexing material on agriculture is available nowhere else, nor is much of the detailed, practical, how-to-do-it type of information. These materials are usually inexpensive and written on a level easily understood by the general reader. Thus, the agricultural government publications are some of the best sources for information on the vocational/technical school level.

The documents on agriculture are listed in the *Monthly checklist of state publications* and the *Monthly catalog of U.S. government publications,* two basic general guides to materials published by governmental agencies. Both of these bibliographies are fully annotated in section C, nos.7 and 8. The *Monthly checklist* indexes state publications by subject and issuing body, and the *Monthly catalog* indexes its materials by subject, issuing agency, individual author, and title. Neither bibliography is easy to use, but with practice the searcher will be able to locate desired titles in a short time. Both contain annual lists of frequently published periodicals, and the *Monthly catalog* provides additional sections on preview publications, new classification numbers, and an annual list of depository libraries. These two sources are exceedingly useful, inexpensive reference and selection tools.

The sources for obtaining agricultural government documents are given in the *Monthly catalog,* the *Monthly checklist,* and publications lists prepared by state and federal agencies separately. Most state titles are available free or for a small price from the issuing body. Federal materials, usually sold by the U.S. Government Printing Office, are often free to qualified people if requested from the agency who prepares them. Librarians, administrators, teachers, and others involved with agriculture may be placed on mailing lists of departments or institutions to receive publications regularly. Any documents, even those no longer being printed, may also be borrowed from the depository libraries, which are listed in the September issue of the *Monthly catalog.* In addition, librarians should not overlook the fact that some publications may be obtained free from their local county agricultural agent or from their congressman.

The documents in this section have been selected according to certain criteria. Materials excluded are: all state publications (which in most cases are only of local interest); monographs having a later copyright date than 1970 (with the exception of the important 1971 book by Sylvia Mechanic); series known to be no longer in print; and, titles considered to be of little value to vocational/technical agricultural schools.

The section's format also requires some explanation. Titles are arranged in three subsections: guides to government publications; general monographs; and, periodicals and other serials. Only the most useful items for vocational/technical agricultural education are annotated. The others have possible value, but they are not essential for the basic collection. Most entries provide ordering information including the address of the distributor, the Superintendent of Documents classification number, and the depository library item number. When this information is not given it is because the authors were unable to find it. When a publication can be obtained only from the issuing body, it is usually free.

The first subsection—on catalogs, indexes, and other guides to government publications—covers mostly general works that include agricultural subjects. This subsection is given to help librarians and other users with background information in a difficult field.

CATALOGS, INDEXES, AND OTHER GUIDES

E1 Andriot, John L. Guide to popular U.S. government publications. 1960. Documents Index, Box 195, McLean, Va. 22101. 125p. $7.50.

Contains 2 sections: one, arranged by departments and agencies of the U.S. government, covers popular periodicals and serials, history or current activities of the agency, and lists of publications; the other, arranged by subject, covers about 2,000 popular publications in print in 1960. Annotated when necessary. Gives issuing body, Superintendent of Documents classification number, and price. A forerunner of the author's other work, *Guide to U.S. government serials and periodicals*. Has a very useful introduction on how to order government publications.

E2 ———. Guide to U.S. government serials and periodicals. 1962– . Documents Index, Box 195, McLean, Va. 22101. Annual. $60 a year.

A comprehensive, annotated list of the serial publications of the federal government, its agencies, and branch offices. Arranged aphabetically by issuing body. Indexed by agency, subject, and title. For each entry, Andriot gives the following information when it is available: starting date; Superintendent of Documents classification number; depository item number; frequency; price; place available; Library of Congress classification number; Library of Congress card number; and a brief annotation. Has been issued biennially in a 3-volume set through 1969. Proposes to be an annual publication thereafter.

E3 Boyd, Anne M., and **Rae E. Rips.** United States government publications. 3d ed. 1949. H. W. Wilson Co., 950 University Ave., Bronx, N.Y. 10452. 627p. $7.50.

One of the best-known standard guides to U.S. government publications. Although dated, it is still useful for an understanding of retrospective works and, because of its arrangement, for understanding the relationships of the departments and agencies of the federal government and their publications. Arranged according to the organization of the government as follows: Congress and the laws; the courts; the executive branch; the ten departments; and the independent agencies. Each division of the government is described, and an annotated list of selected publications that is prepared by the division is given. Includes index.

E4* Jackson, Ellen. Subject guide to major United States government publications. 1968. American Library Association, 50 E. Huron St., Chicago, Ill. 60611. 175p. $5.50.

An annotated list of U.S. government publications, which the author has found to be of permanent value during her many years of work with them. Dates of titles included range from the earliest periods to early 1968. Arrangement is by subject. The author gives more emphasis to the scholarly type of publication than does Leidy in his work, *A popular guide to government publications.* Not indexed. A very helpful concluding chapter by Dr. W. A. Katz describes guides, catalogs, and indexes pertaining to government publications for the librarian.

E5 Leidy, W. Philip. A popular guide to government publications. 3d ed. 1968. Columbia University Press, 440 W. 110 St., New York, N.Y. 10025. 365p. $12.

A well-recommended list of popular U.S. government publications which mainly covers those published between 1961 and 1966 but which also includes older works that are still of great value. Contains about 3,000 titles published by the Government Printing Office, other governmental agencies, and the Pan American Union. Arranged by broad subjects with a detailed subject index. Entries include issuing body, price, Superintendent of Documents classification number, and brief descriptive annotations.

E6* Mechanic, Sylvia. Annotated list of selected United States government publications available to depository libraries. 1971. H. W. Wilson Co., 950 University Ave., Bronx, N.Y. 10452. 407p. $16.50.

A list of about 500 U.S. government publications, primarily comprised of serials, which the author considers basic for all depository libraries. Each entry consists of a reproduction of the Library of Congress catalog card, an annotation, and, in some instances, a list of 4 or 5 recent titles in the series. Arrangement is by depository item number with the Superintendent of Documents classification number also included. An extremely useful guide for both depository and nondepository libraries as a selection tool for ordering government publications and for borrowing from depository libraries by interlibrary loan. Includes a series or title index and several helpful appendixes, such as an explanation of the Superintendent of Documents classification system.

E7 O'Hara, Frederic J. Over 2,000 free publications: yours for the asking. 1968. New American Library, Inc., 1301 Ave. of the Americas, New York, N.Y. 10019. 352p. paper, 95¢

See section C, no.9.

E8* Schmeckebier, Laurence F., and R. B. Eastin. Government publications and their use. 2d ed. 1969. Brookings Institution, 1775 Massachusetts Ave. NW, Washington, D.C. 20036. 502p. $8.95.

One of the more up-to-date, standard reference works for studying U.S. government publications. Intended as a guide to the acquisition and use of government documents. Arrangement is by form of publication (i.e., catalogs and indexes, bibliographies, maps, periodicals, etc.) and by type (i.e., presidential papers, congressional publications, etc.). The appendix contains a listing by state of depository libraries. Has an excellent index.

E9* U.S. Dept. of Agriculture. A guide to understanding the United States Department of Agriculture. rev. ed. 1968. Prepared by the Personnel Office, U.S. Dept. of Agriculture. Available from the Superintendent of Documents, U.S. Government Printing Office, Washington, D.C. 20402. unpaged. illus. Dep. item no.10, S.D. no.A49.8/968. 40¢

A booklet for employees, students, and general readers which explains the organizational structure and programs of the department. In his government documents section that is published monthly in *Wilson bulletin,* Dr. F. J. O'Hara recommends this guide as one of the titles on agriculture that all libraries should have.

E10* ———, Office of Information. List of available publications of the United States Department of Agriculture. 1929– . Office of Information, Publications Div., U.S. Dept. of Agriculture. Available from the Superintendent of Documents, U.S. Government Printing Office, Washington, D.C. 20402. Latest ed. Dep. item no.91, S.D. no.A21/9/8:(nos.). 45¢ (also called List no.11)

See section C, no.5.

E11* U.S. Government Printing Office. Price lists of government publications. 1898– . Superintendent of Documents, U.S. Government Printing Office, Washington, D.C. 20402. Irregular. Dep. item no.554, S.D. no.GP3.9:(nos.). Free.

A series of sales catalogs containing in-print U.S. government publications. Each catalog covers a separate subject field. Entries give prices, Superintendent of Documents classification numbers, and, occasionally, annotations. For the most part, they list the more popular publications. Frequently revised. Useful for ready reference, for students writing papers, and for the specialist as a point of departure. About 47 are now in print, the latest being No.87. Those of special interest for agriculture schools include the following: PL 10, *Laws, rules, and* regulations; PL 21, *Fish and wildlife;* PL 31, *Education;* PL 33A, *Occupations;* PL 35, *National parks;* PL 36, *Government periodicals and subscriptions services;* PL 38, *Animal industry;* PL 41, *Insects;* PL 42, *Irrigation, drainage, and water power;* PL 43, *Forestry;* PL 44, *Plants;* PL 46, *Soils and fertilizers;* PL 48, *Weather, astronomy, and meteorology;* PL 53, *Maps;* PL 68, *Farm management;* PL 70, *Census publications;* PL 86, *Consumer information;* and, PL 87, *States and territories of the United States and their resources.*

E12* U.S. Superintendent of Documents. Selected United States government publications. 1928– . Superintendent of Documents, U.S. Government Printing Office, Washington, D.C. 20402. Biweekly. Dep. item no.556, S.D. no.GP3.17:(nos.). Free.

See section C, no.12.

E13 Wynkoop, Sally. Government reference books: a biennial guide to U.S. government publications. 1970– . Libraries Unlimited, Inc., Box 263, Littleton, Colo. 80120. Biennial. $7.50.

A guide to reference works which have been published by the U.S. Government Printing Office. Divided into 3 sections: the social sciences; science and engineering; and the humanities. It is arranged by broad subjects within each section and has a title-subject index. Each entry gives complete bibliographical information, price, Library of Congress card number, Superintendent of Documents classification number, and very informative annotations. Appendixes include information on ordering and a directory of issuing agencies.

MONOGRAPHS

E14 Administration of vocational education, rules and regulations; regulations for the administration of vocational education programs under the provisions of the federal vocational education acts. rev. ed. 1967. Prepared by the U.S. Office of Education. Available from the Superintendent of Documents, U.S. Government Printing Office, Washington, D.C. 20402. 102p. Dept. item no.460-A-80, S.D. no.FS5.280: 80017-A. 45¢

Contains federal laws and regulations up to 1966 which relate to the administration of vocational education programs in the states and territories. Appendixes include the texts of the Vocational Education Act of 1963, the Smith-Hughes Act of 1917, the Vocational Education Act of 1946, the Appalachian Regional Development Act of 1965, acts concerning Puerto Rico, the Virgin Islands, Guam, and American Samoa, and the Fitzgerald Act of 1937.

E15 Agricultural equipment technology: a suggested 2-year post high school curriculum. 1970. Prepared by the U.S. Office of Education. Available from the Superintendent of Documents, U.S. Government Printing Office, Washington, D.C. 20402. 112p. illus. Dep. item no.406-A-81, S.D. no.FS5.281:81015. $1.25.

Features course outlines; text and reference book lists; farmland requirements; laboratory equipment, costs, and organization; library services for faculty and students; technical education procedures; and a directory of scientific, trade, and technical societies. One of the numbers in the Office of Education's series, OE-81,000-81,999, this guide is intended for teachers and administrators concerned with planning programs on this level. Also a useful guide for students who should benefit from its bibliographies, job descriptions, and outlines.

E16 Century of service: the first 100 years of the United States Department of Agriculture. 1963. Written by Gladys L. Baker, and others. Available from the Superintendent of Documents, U.S. Government Printing Office, Washington, D.C. 20402. 560p. Dep. item no.10, S.D. no.A1.2:Ag8/22. $2.75.

A history of the U.S. Dept. of Agriculture, written to commemorate its centennial. Describes the department's organizational development and its response to changing national, international, scientific, and economic conditions from 1862, the date of its establishment, to 1962. A concluding chapter provides an outlook for the new century.

E17 Climatic atlas of the United States. 1968. Prepared by the U.S. Environmental Science Services Administration. Available from the Superintendent of Documents, U.S. Government Printing Office, Washington, D.C. 20402. 80p. illus. Dep. item no.208-B-1, S.D. no.C52.2: C61/2. paper, $4.25.

As stated in the preface, the purpose of this atlas is to depict the climate of the United States in terms of the distribution and variation of constituent climatic elements. It presents in uniform format a series of 271 climatic maps and 15 tables, which show the national distribution of mean, normal, and/or extreme values of temperature, precipitation, wind, barometric pressure, relative humidity, dew point, sunshine, sky cover, solar radiation, and evaporation. Individual sheets and sets from this atlas may be purchased separately.

E18a A directory of information resources in the United States: federal government. 1967. Prepared by the National Referral Center for Science and Technology, U.S. Library of Congress. Available from the Superintendent of Documents, U.S. Government Printing Office, Washington, D.C. 20402. 411p. Dep. item no. 787-A, S.D. no.LC1.31: D62/4. paper, $2.75.

E18b A directory of information resources in the United States: general toxicology. 1969. Prepared by the National Referral Center for Science and Technology, U.S. Library of Congress. Available from the Superintendent of Documents, U.S. Government Printing Office, Washington, D.C. 20402. 293p. Dep. item no. 787-A, S.D. no.LC1.31:(nos.). paper, $3.

E18c A directory of information resources in the United States: physical sciences, biological sciences, engineering. 1965. Prepared by the National Referral Center for Science and Technology, U.S. Library of Congress. Available from the Superintendent of Documents, U.S. Government Printing Office, Washington, D.C. 20402. 352p. Dep. item no.787-A, S.D. no. LC1.31: D62. paper, $2.25.

E18d A directory of information resources in the United States: social sciences. 1965. Prepared by the National Referral Center for Science and Technology, U.S. Library of Congress. Available from the Superintendent of Documents, U.S. Government Printing Office, Washington, D.C. 20402. 218p. Dep. item no. 787-A, S.D. no.LC1.31: D62/2. paper, $1.50.

E18e A directory of information resources in the United States: water. 1966. Prepared by the National Referral Center for Science and Technology, U.S. Library of Congress. Available from the Superintendent of Documents, U.S. Government Printing Office, Washington, D.C. 20402. 248p. Dep. item no.787-A, S.D. no.LC1.31: D62/3. paper, $1.50.

Each volume lists in alphabetical order organizations from which information concerning the subject matter of the volume can be obtained. Includes libraries and other similar bibliographic services, centralized information centers capable of providing processed and evaluated data, professional societies and other organizations through which contact can be made with individual specialists, industrial firms willing to extend their information services beyond the organization, government agencies or offices able to provide assistance in specific areas of interest, and many other resources which make scientific and technical information available in some form. For each source, the volumes give the address and telephone number, areas of interest, publications, information services, and holdings, if any. Extensive subject indexes for all of the volumes are included. See also section F, nos.16-20.

E19 Directory of schools offering technical education programs under Title III of the George Barder Act and the Vocational Education Act of 1963. 1968. Prepared and distributed by the Office of Education, U.S. Dept. of Health, Education, and Welfare, Washington, D.C. 20202. 108p. Dep. item no.461 (rev. 1965), S.D. no.FS5.2: D62/2. Free from issuing body.

Arranged by states. Contains names of schools and their locations, types of programs conducted, and designation of the level of the program (secondary, post-secondary, or supplementary). Especially useful because it lists schools offering programs or courses in agriculture and related fields, including agricultural business, fisheries, food technology, forestry, horticulture, pulp and paper technology, welding, and wood utilization.

E20 Farm crop production technology, field and forage crop, and fruit and vine production options: a suggested 2-year post high school curriculum. 1970. Prepared by the U.S. Office of Education. Available from the Superintendent of Documents, U.S. Government Printing Office, Washington, D.C. 20402. 179p. illus. Dep. item no.406-A-81, S.D. no.FS5.280: 81016. $1.50.

An illustrated guide to a suggested curriculum for farm crop production technology. It provides laboratory equipment, costs, and layout; course outlines; suggested texts and supplementary readings; a section on library use, and several other aids to help teachers and administrators instigate educational programs in this area. A number in the Office of Education's series, OE-81,000-81,999.

E21 Farm labor fact book. 1959. Prepared by the U.S. Dept. of Labor. Available from the Superintendent of Documents, U.S. Government Printing Office, Washington, D.C. 20402. 240p. Dep. item no.745, S.D. no.L1.2: F22. $1.

A somewhat dated guide to the farm labor of the United States which describes the different groups comprising the farm work force. These include migrant workers, foreign workers, hired workers, farm operators, and unpaid members of their families. Discusses the various problems of each group, their interrelationships, and the whole economic situation.

E22 **Food processing technology: a suggested 2-year post high school curriculum.** 1967. Prepared by the U.S. Office of Education. Available from the Superintendent of Documents, U.S. Government Printing Office, Washington, D.C. 20402. 97p. illus. Dep. item no. 460-A-82, S.D. no.FS5.282: 82016. paper, 50¢.

Number OE-82,016 of the Office of Education's series entitled vocational education (OE-82,000-82,999), this curriculum guide contains a wealth of program ideas and teaching aids for food processing technological education. Excludes dairy products, dietary, and artificial foods and emphasizes canning methods. The bibliography and list of associations are very good although some of the materials are no longer in print.

E23 **Forest technology: a suggested 2-year post high school curriculum.** 1968. Prepared by the U.S. Office of Education. Available from the Superintendent of Documents, U.S. Government Printing Office, Washington, D.C. 20402. 142p. illus. Dep. item no.460-A-80, S.D. no.FS5.280: 80054. $1.25.

For school administrators, supervisors, counselors, teachers, and advisory committees responsible for planning and instigating forestry technological course work. Gives course outlines, text and reference lists, technical education methods, laboratory costs, equipment and arrangement, a discussion of library, student, and faculty services, and forest land requirements. An important number in the Office of Education's vocational education series.

E24 **From sea to shining sea, a report on the American environment: our natural heritage.** 1968. Prepared by the U.S. President's Council on Recreation and Natural Beauty. Available from the Superintendent of Documents, U.S. Government Printing Office, Washington, D.C. 20402. 304p. Dep. item no.851-J, S.D. no.Pr36.8:R24/Sel. paper, $2.50.

This report describes the progress in environmental improvement programs since the 1965 White House Conference on Natural Beauty, presents proposals and recommendations to stimulate federal, state, local, and private action to further progress, and also contains a guide for action by local officials and citizen groups. These proposals and recommendations in essence represent a statement of long-term, comprehensive goals for the nation. Includes a subject index and lists of books and pamphlets, films, state and federal agencies, and private organizations which can help in promoting these programs.

E25 **Grain, feed, seed, and farm supply technology: a suggested 2-year post high school curriculum.** 1968. Prepared by the U.S. Office of Education. Available from the Superintendent of Documents, U.S. Government Printing Office, Washington, D.C. 20402. 185p. illus. Dep. item no.460-A-81, S.D. no.FS5.281: 81014. $1.50.

A curricular program for technical school training in farm supply, this Office of Education guide is written to aid administrators and faculty members in planning courses for students aspiring to be skilled technicians. Among other features, it includes course outlines, text and reference book lists, and a directory of societies associated with supplying grain, feed, seed, and farm equipment. This is number 81,014 of the Office of Education's series entitled Vocational education (OE81,000-81,999).

E26 **Ornamental horticulture technology: a suggested 2-year post high school curriculum.** 1970. Prepared by the U.S. Office of Education. Available from the Superintendent of Documents, U.S. Government Printing Office, Washington, D.C. 20402. 206p. illus. Dep. item no.406 A-81, S.D. no.HE5.281:81017. paper, $1.75.

This guide is not only useful for planning ornamental horticulture technological programs but also for analyzing and evaluating those being used in schools at the present time. Suggests curriculum for 5 areas: floriculture, landscape development, turfgrass management, nursery operation, and arboriculture. The bibliographies are comprehensive, and the list of societies interested in technological aspects of ornamental horticulture is a useful inclusion.

E27 **Parks for America, a survey of park and related resources in the fifty states, and a preliminary plan.** 1964. Prepared by the Bureau of Outdoor Recreation, U.S. Dept. of the Interior. Available from the Superintendent of Documents, U.S. Government Printing Office, Washington, D.C. 20402. 485p. illus. Dep. item no.648, S.D. no.I29.2:P21/11. paper, $5.25.

An attractively illustrated survey of the parks and recreational areas for each state. Indicates where these areas are located on maps. Topics of interest discussed for each state are general facts, existing public areas, the need for parks and related facilities, and recommendations for the future. Tables are also given that compare the acreage, significant features, type of use, and activities of the existing areas and the potential recreational areas.

E28   People of rural America. 1968. Written by Dale E. Hathaway, J. Allan Beegle, and Keith Bryant. Prepared by the U.S. Bureau of the Census in cooperation with the Social Science Research Council. Available from the Superintendent of Documents, U.S. Government Printing Office, Washington, D.C. 20402. 289p. illus. Dep. item no.154-A(rev.), S.D. no.C3.30: R88. $3.50.

Describes the rural population as reported in the 1960 census of population and evaluates the residence categories used. Develops the hypothesis that proximity to large cities plays a crucial role in determining the characteristics of rural areas. Contains many graphs, charts, statistical tables, and maps that are quite useful for further interpretation of the text.

E29*   Style manual. rev. ed. 1967. Prepared by the U.S. Government Printing Office. Available from the Superintendent of Documents, U.S. Government Printing Office, Washington, D.C. 20402. 512p. Dep. item no.548, S.D. no.GP1.23/4:St9/967. $3.

An indispensable reference for students, typists, printers, authors, and anyone else desiring information on writing style and proper usage of the English language. Gives rules concerning capitalization, spelling, punctuation, abbreviations, compound words, plant and insect names, numerals, italics, signs and symbols, tabular work, and foreign languages. There is also an abridged edition (1967) which contains almost all of the material listed above. A supplement to the *Style manual,* entitled *Word division* (1962), contains rules for the proper division and spelling of more than 20,000 commonly used terms.

E30   USDA summary of registered agricultural pesticide chemical uses. 3d ed. 1968–69. Prepared by the Agricultural Research Service, Pesticides Regulation Div., U.S. Dept. of Agriculture. Available from the Superintendent of Documents, U.S. Government Printing Office, Washington, D.C. 20402. 3v. Dep. item no.44, S.D. no.A77.302:P43/4. v.1 $3.50; v.2 $3.50; v.3 $6.95.

A 3-volume work giving the chemical uses of agricultural pesticides in a well-organized format. For every pesticide, the chemical name, principal formulations, type of pesticide, uses on different crops, tolerance, dosage, and limitations that should be known to enable proper utilization are noted. Each volume covers different subjects. Volume 1 is concerned with herbicides, defoliants, desiccants, and plant regulators; volume 2, with fungicides and nematicides; and volume 3, with insecticides, repellents, and acaricides. Amendments to update this third edition are issued as replacement pages.

E31   Vocational Education Act of 1963. 1965. Prepared by the U.S. Office of Education. Available from the Superintendent of Documents, U.S. Government Printing Office, Washington, D.C. 20402. 29p. illus. Dep. item no. 460-A-80, S.D. no.FS5.280:80034. 15¢

Based on an article in *School life,* this booklet describes the benefits of the Vocational Education Act of 1963. The original publication of the act itself is no longer in print.

E32   Vocational education: the bridge between man and his work. 1968. Prepared by the U.S. Office of Education. Available from the Superintendent of Documents, U.S. Government Printing Office, Washington, D.C. 20402. 220p. Dep. item no.460-A-80, S.D. no.FS5.280:80052. paper, $2.25.

A report by the U.S. Advisory Council on Vocational Education. Divided into 3 sections: first, it reviews the situation of vocational education since the Vocational Education Act of 1963; then it evaluates and assesses the situation; and finally, makes legislative and administrative recommendations for the future.

PERIODICALS AND OTHER SERIALS

E33   Agricultural conservation program: handbook. Prepared and distributed by Agricultural Stabilization and Conservation Service, U.S. Dept. of Agriculture, Washington, D.C. 20250. Annual. Dep. item no. 103-A- (nos.1–50; one for each state), S.D. no.A82.37:(nos.). Free.

A separate handbook is issued for each state.

E34 **Agricultural economics reports.** 1961– .
Prepared by the Economic Research Service, U.S. Dept. of Agriculture. Available from the Superintendent of Documents, U.S. Government Printing Office, Washington, D.C. 20402. Irregular. Dep. item no.42-C, S.D. no. A93.28:(nos.). $2 a year.

Supplies current information on agricultural economics including research results, agricultural production, marketing, and utilization of products. Reports range from semipopular to semitechnical in approach. Intended for agricultural workers on the professional and technical levels.

E35 **Agricultural economics research.** 1949– .
Prepared by the Economic Research Service, U.S. Dept. of Agriculture. Available from the Superintendent of Documents, U.S. Government Printing Office, Washington, D.C. 20402. Quarterly. Dep. item no.18, S.D. no.A93.26:(nos.). paper, $1.50 a year.

Written in fairly technical terms. Describes the complex techniques, developments, and results of agricultural economic and statistical research. Usually free only to professionals in the field, but it is inexpensive for any library or individual reader to purchase.

E36 **Agricultural finance review.** 1938– . Prepared and distributed by the Economic Research Service, U.S. Dept. of Agriculture, Washington, D.C. 20250. Annual. Dep. item no. 18-A, S.D. no.A93.9/10:(nos.). Free.

Gives a report of the news, research progress, and recent data collected in the field of agricultural finance. Covers farm credit, farm insurance, and farm taxation. Beneficial to educators, economists, government officials, and many agricultural organizations. Annual issues are supplemented during the year.

E37* **Agricultural marketing.** 1956– . Prepared by the Consumer and Marketing Service, U.S. Dept. of Agriculture. Available from the Superintendent of Documents, U.S. Government Printing Office, Washington, D.C. 20402. Monthly. Dep. item no.24-K, S.D. no. A88.26/3:(nos.). $2 a year.

Free to libraries and officials interested in consumer and marketing news. Activities of the Consumer and Marketing Service, including consumer protection, consumer food programs, marketing regulatory plans, and marketing services programs on the federal, state, city, and commercial level, are discussed in these issues.

E38 **Agricultural outlook digest.** 1947– . Prepared and distributed by the Economic Research Service, U.S. Dept. of Agriculture, Washington, D.C. 20250. Irregular. S.D. no. A93.10:(nos.). Price varies.

Single sheets which summarize the latest developments in farm production, income, farm employment, demand, exports, prices paid and received by farmers, and the outlook for farm products. Issued to update product or general outlook reports and to provide a condensed version of the USDA's current reports.

E39 **Agricultural prices, (year), annual summary Prl-3.** 1959– . Prepared and distributed by the Crop Reporting Board, U.S. Dept. of Agriculture, Washington, D.C. 20250. Annual. Dep. item no.18-C, S.D. no.A92.16/2:(nos.). Free.

E40* **Agricultural research.** 1953– . Prepared by the Agricultural Research Service, U.S. Dept. of Agriculture. Available from the Superintendent of Documents, U.S. Government Printing Office, Washington, D.C. 20402. Bimonthly. Dep. item no.25-A, S.D. no.A77.12:(nos.). $1.50 a year.

A popular periodical written for the general public and for those who must deal directly with marketers, farmers, and processors of agricultural products. Discusses research done in crops, livestock management, soils, poultry, fruits, vegetables, and other related areas. Available free to agricultural extension workers, vocational agriculture teachers, federal and state workers, trade associations, the press and radio, farm organizations, and advisory committees.

E41 **Agricultural science review.** 1963– . Prepared by the Cooperative State Research Service. Available from the Superintendent of Documents, U.S. Government Printing Office, Washington, D.C. 20402. Quarterly. Dep. item no.40-A-1, S.D. no.A94.11:(nos.). $1.25 a year.

Acquaints the reader with the current status of agricultural research. Includes reports on research in progress, results that have been published, and general trends. This material is presented in authoritative review articles which should not only keep the general agricultural worker informed of new developments in the field but also should improve the regulation of research.

E42* **Agricultural situation.** 1921– . Prepared by the Statistical Reporting Service, U.S. Dept. of Agriculture. Available from the Superintendent of Documents, U.S. Government Printing Office, Washington, D.C. 20402.

Monthly. Dep. item no.19, S.D. no.A92.23:(nos.). $1 a year.

A nontechnical periodical on current agricultural developments written from the economic point of view. Consists of brief reports which summarize and analyze agricultural marketing and economics. Free to libraries and various agricultural agencies and workers.

E43* **Agricultural statistics.** 1936– . Prepared by the U.S. Dept. of Agriculture. Available from the Superintendent of Documents, U.S. Government Printing Office, Washington, D.C. 20402. Annual. Dep. item no.1, S.D. no. A1.47:(date). Price varies.

Because this is one of the most valuable statistical reference works available, it belongs in every agricultural library. It provides statistics on all phases of the agricultural and allied fields, and it is well indexed for quick access to specific information. Data are compiled from actual counts, estimates by the USDA, and census statistics. This series is especially useful for comparative studies because it has tables of data representing several years for most subjects, and it has been published annually since 1936. It was part of the Yearbook of Agriculture prior to 1936.

E44 **Agriculture decisions.** 1942– . Prepared by the U.S. Dept. of Agriculture. Available from the Superintendent of Documents, U.S. Government Printing Office, Washington, D.C. 20402. Monthly. Dep. item no.2, S.D. no.A1.58/a:(nos.). $6.50 a year.

A monthly publication describing the judicial decisions made by the secretary of agriculture excluding those of a general nature and those which must be published in the *Federal register.* Recent court action concerning the department's regulatory laws under which the secretary's decisions are made is also reported. This publication is annually indexed, and it offers several special features such as a table of statutes and orders, and lists of decisions described.

E45* **Agriculture handbook.** 1950– . Prepared by the U.S. Dept. of Agriculture. Available from the Superintendent of Documents, U.S. Government Printing Office, Washington, D.C. 20402. Irregular. Dep. item no.3, S.D. no.A1.76:(nos.). Price varies.

Designed for the student and the professional or technical worker. Presents information on various home economic and agricultural subjects in quick-reference format. The handbooks are written in several forms: manuals; glossaries; guidebooks; specification descriptions; or lists. These are very inexpensive sources for comprehensive coverage of specific topics. Some titles, such as *Selected weeds of the United States,* no.366, are so useful as texts and guides that they are also published as separate monographs. The following are a few of the important volumes of this series: Little, Elbert L., *Checklist of native and naturalized trees of the United States,* no.41; Musil, Albina F., *Identification of crop and weed seeds,* no.219; Thornberry, Halbert H., *Index of plant virus diseases: plant pests of importance to North American agriculture,* no.307; U.S. Agricultural Research Service, *Crops Research Div., Index of plant diseases in the United States,* no.165; U.S. Bureau of Plant Industry, Soils, and Agricultural Engineering, *Soil survey manual,* no.18; U.S. Dept. of Agriculture, Economic Research Service, *Handbook of agricultural charts,* [nos. vary]; U.S. Div. of Timber Management Research, *Silvics of forest trees of the United States,* no.271; U.S. Forest Products Laboratory, Madison, Wis., *Wood: colors and kinds,* no.101; U.S. Soil Conservation Service, *A manual on conservation of soil and water,* no.61; Watt, Bernice K., *Composition of foods, raw, processed,* no.8.

E46* **Agriculture information bulletin.** 1949– . Prepared by the U.S. Dept. of Agriculture. Available from the Superintendent of Documents, U.S. Government Printing Office, Washington, D.C. 20402. Irregular. Dep. item no.4, S.D. no.A1.75:(nos.). Free.

Intended for those interested in a more specialized view of agriculture than that provided by the U.S. Dept. of Agriculture's *Farmers' bulletins, Home and garden bulletins,* and *Leaflets.* Though more technical in approach, the style is popular, and the material is understandable and useful for people in urban as well as rural situations.

E47* **Bimonthly list of publications and motion pictures.** 1909– . Prepared by the U.S. Dept. of Agriculture. Available from the Office of Information, U.S. Dept. of Agriculture, Washington, D.C. 20250. Bimonthly. Dep. item no.92, S.D. no.A21.6/5. Free.

See section C, no.1.

E48 **C & MS.** 1954– . Prepared and distributed by the Consumer and Marketing Service, U.S. Dept. of Agriculture, Washington, D.C. 20250. Irregular. Dep. item no.19-A, S.D. no.A88.40/2:(nos.). Free.

E49 **CCC price support program data.** 1968– . Prepared and distributed by the

Periodicals and Serials

Commodity Credit Corp., Agricultural Stabilization and Conservation Service, Office of the Deputy Administrator, Commodity Operations, U.S. Dept. of Agriculture, Washington, D.C. 20250. Irregular. S.D. no.A82.210/3:(date). Free.

E50 **CDC veterinary public health notes.** Prepared and distributed by the U.S. Center for Disease Control, 1600 Clifton Rd. NE, Atlanta, Ga. 30333. Monthly. S.D. no.HE-20.2312:(date). Free.

E51 **CR crops research.** (series). Prepared and distributed by the Agricultural Research Service, Crops Research Div., U.S. Dept. of Agriculture, Washington, D.C. 20250. Irregular. Dep. item no.26, S.D. no.A77.534:(nos.). Free.

E52 **Catalog of federal assistance programs** (title varies). 1965– . 1971 catalog prepared and distributed by the U.S. Management and Budget Office, Executive Office of the President, Washington, D.C. 20506. (former issues prepared by various offices). Biennial. Dep. item no.853-A-1, S.D. no.PrEx2.20:(year). 1971 ed. $7.25.
See section F, no.8.

E53 **Census of agriculture.** 1840– . Prepared by the U.S. Bureau of the Census. Available from the Superintendent of Documents, U.S. Government Printing Office, Washington, D.C. 20402. Every 5 years in the years ending with 4 and 9. Dep. item nos.152 and 153, S.D. no.C3.31/4:(nos.). Price varies per volume.

E54 **Census of commercial fisheries.** Prepared by the U.S. Bureau of the Census. Available from the Superintendent of Documents, U.S. Government Printing Office, Washington, D.C. 20402. Dep. item no.133-C, S.D. no.C3.245:(nos.). paper, 30¢

E55 **Census of irrigation.** Prepared by the U.S. Bureau of the Census under the supervision of the chief of the Agriculture Div. Available from the Superintendent of Documents, U.S. Government Printing Office, Washington, D.C. 20402. Decennially in the years ending with 9. Dep. item no.152, S.D. no. C3.31/12:(nos.). $2.50.

E56 **Census of population.** 1780– . Prepared and distributed by the Bureau of the Census, U.S. Dept. of Commerce, Suitland, Md. 20233. Decennially. Dep. item nos.154 and 159, S.D. no.C3.223/3:(nos.). Price varies.

E57 **Clearinghouse announcements in science and technology: agriculture and food.** Prepared and distributed by the National Technical Information Service, U.S. Dept. of Commerce, Springfield, Va. 22151. S.D. no.C51.7/5:(nos./date). paper, $5 a year.

E58 **Climates of the states.** 1959– . Prepared by the Environmental Data Service, Environmental Science Services Administration, U.S. Dept. of Commerce. Available from the Superintendent of Documents, U.S. Government Printing Office, Washington, D.C. 20402. Irregular. Dep. item no.273-B, S.D. no.C30.71/3:(nos.). Price varies from 15¢ to 30¢ among states.
Consists of separate publications, issued for each state and territory, which average 15 to 20 pages in length. Infrequently revised. Each issue gives a detailed description of the climate of the state, tables of freeze data, tables of the norms, means, and extremes of the weather, and temperature, precipitation, and storm maps.

E59 **Climatological data, national summary.** 1950– . Prepared by the Environmental Data Service, Environmental Science Services Administration, U.S. Dept. of Commerce. Available from the Superintendent of Documents, U.S. Government Printing Office, Washington, D.C. 20402. Monthly. S.D. no.C20.51:(nos.). $2.50 a year.
Summarizes the weather for the states, the Pacific area, and the West Indies. Specifies the degree days, general conditions, severe storm data, upper air statistics, river stages, and flood conditions. Each section represents a region or state, and the statistical information for it is collected at the weather station in that area. An annual volume, a condensation of the monthly publications, is also issued as part of this series.

E60 **Commercial fisheries review.** 1939– . Prepared by the Fish and Wildlife Service, U.S. Bureau of Commercial Fisheries. Available from the Superintendent of Documents, U.S. Government Printing Office, Washington, D.C. 20402. Monthly. Dep. item no. 609-A, S.D. no.149.10:(nos.). paper, $7 a year.

E61 **Conservation bulletin.** Prepared and distributed by the U.S. Dept. of the Interior, Washington, D.C. 20240. Irregular. Dep. item no.601, S.D. no.I1.72:(nos.). Free.

E62 **Conservation research reports.** 1965– . Prepared and distributed by the Forest Service, U.S. Dept. of Agriculture, Washington, D.C. 20250. Irregular. illus. Dep. item no.25-D, S.D. no.A77.23:(nos.). 20¢ (approx.)

An irregularly published series of reports which deals with research in conservation. Includes reports on research projects that are not as complete as those in the *Technical bulletin* issued by the department.

E63 Conservation yearbook. 1965– . Prepared by the U.S. Dept. of the Interior. Available from the Superintendent of Documents, U.S. Government Printing Office, Washington, D.C. 20402. Annual. Dep. item no.601-A, S.D. no.I1.95:(nos.). paper, $1.

E64 Costs and returns (FCR 1–). 1962– . Prepared and distributed by the Economic Research Service, U.S. Dept. of Agriculture, Washington, D.C. 20250. Annual. S.D. no. A93.9/11:(nos.). Free.

E65 County and city data book. 1949– . Prepared by the U.S. Bureau of the Census. Available from the Superintendent of Documents, U.S. Government Printing Office, Washington, D.C. 20402. Irregular. Dep. item no.151, S.D. no.C3.134/2:(nos.). $5.

E66 Crop and livestock reports. Prepared and distributed by the Crop Reporting Board, U.S. Dept. of Agriculture, Washington, D.C. 20250. Monthly. S.D. no.A92:(nos.). Price varies.

E67 Crop production (CrPr2-2). Prepared and distributed by the Crop Reporting Board, U.S. Dept. of Agriculture, Washington, D.C. 20250. Monthly. Dep. item no.20-B, S.D. no. A92.24:(date). Free.

E68* Extension service review. 1930– . Prepared by the Cooperative Extension Service, U.S. Dept. of Agriculture and cooperating state land-grant colleges and universities. Available from the Superintendent of Documents, U.S. Government Printing Office, Washington, D.C. 20402. Monthly. Dep. item no.60, S.D. no.A43.7:(nos.). paper, $1.50 a year.

An outstanding source of ideas and practical information intended for extension workers whose main duty is to encourage better farming, improvement of home and community, and to provide young people with the opportunity for learning away from school. Especially concerned with the important agricultural extension programs including 4-H club work, home demonstrations, cooperation in communities, and conservation.

E69* The farm index. 1962– . Prepared by the Economic Research Service, U.S. Dept. of Agriculture. Available from the Superintendent of Documents, U.S. Government Printing Office, Washington, D.C. 20402. Monthly. Dep. item no.42-F, S.D. no.A93.33: (nos.). $2 a year.

Presents the research results of the Economic Research Service's investigations in nontechnical language. Arranged by broad subject heading, such as rural life, the consumer, the foreign market, farming, and marketing. Free distribution is limited to economists, the press, extension workers, and other groups directly involved with farmers and farming businesses.

E70 Farm labor. Prepared by the Statistical Reporting Service, U.S. Dept. of Agriculture. Available from the Superintendent of Documents, U.S. Government Printing Office, Washington, D.C. 20402. S.D. no.A93.28:(nos.). paper, 20¢ (approx.)

E71 Farm labor (La 1). Prepared and distributed by the Crop Reporting Board, U.S. Dept. of Agriculture, Washington, D.C. 20250. Monthly. S.D. no.A92.12:(date). Free.

E72 Farm real estate market developments. 1942– . Prepared and distributed by the Economic Research Service. U.S. Dept. of Agriculture, Washington, D.C. 20250. Semiannual. Dep. item no.42-A, S.D. no.A93.9/4:(nos.). Free.

E73* Farmers' bulletin. 1889– . Prepared by the U.S. Dept. of Agriculture. Available from the Superintendent of Documents, U.S. Government Printing Office, Washington, D.C. 20402. Irregular. Dep. item no.9, S.D. no. A1.9:(nos.). Price varies.

One of the best-known series issued by the U.S. government. Gives information in simple language on the practical methods of applying scientific research to farming as well as on how to solve the everyday problems encountered in the home and garden. Intended for farmers and ranchers but also of interest to the general public.

E74 Fertilizer trends. 1956– . Prepared and distributed by the U.S. National Fertilizer Development Center, Muscle Shoals, Ala. 35660. Biennial. S.D. no.Y3.T25:(nos.). Free.

E75 Fire control notes. 1936– . Prepared and distributed by the Superintendent of Documents, U.S. Government Printing Office, Washington, D.C. 20402. Quarterly. Dep. item no.82, S.D. no.A13.32:(v. & no.). paper, 75¢ a year; single copy 20¢

Periodicals and Serials

E76 Fishery leaflet. 1941– . Prepared and distributed by the Fish and Wildlife Service, Bureau of Commercial Fisheries, U.S. Dept. of the Interior, Washington, D.C. 20240. Irregular. Dep. item no.611-D, S.D. no.I49.28:(nos.). Free.

E77 Forest pest leaflet. 1955– . Prepared and distributed by the Forest Service, U.S. Dept. of Agriculture, Washington, D.C. 20250. Irregular. Dep. item no.82-A, S.D. no.A13.52:(nos.). Free.

E78* Home and garden bulletin. 1950– . Prepared by the U.S. Dept. of Agriculture. Available from the Superintendent of Documents, U.S. Government Printing Office, Washington, D.C. 20402. Irregular. Dep. item no.11, S.D. no.A1.77:(nos.). Price varies.

Offers details on home building, vegetable, fruit, and flower growing, common gardening problems, how to control pests, and many other related topics of interest. A popular series written in nontechnical language. Equally useful for the rural and the urban dweller.

E79 Home economics research reports. 1957– . Prepared by the U.S. Dept. of Agriculture. Available from the Superintendent of Documents, U.S. Government Printing Office, Washington, D.C. 20402. Irregular. Dep. item no.11-B, S.D. no.A1.87:(nos.). Price varies.

Written for professional and technical workers in the home economics and allied fields. Most of these reports are on a technical and semitechnical level, but some are popular in approach. They present all of the important aspects of home economics research in a condensed, easy-to-read form.

E80 Laws relating to agriculture, May 29, 1884– (date). Prepared by the U.S. Congress, House, Document Room. Available from the Superintendent of Documents, U.S. Government Printing Office, Washington, D.C. 20402. Irregular. Dep. item no.998, S.D. no. Y1.2:(nos.). paper, $3.50.

E81 Laws relating to forestry, game conservation, flood control, and related subjects, March 1, 1911– (date). Prepared by the U.S. Congress, House, Document Room. Available from the Superintendent of Documents, U.S. Government Printing Office, Washington, D.C. 20402. Irregular. Dep. item no.998, S.D. no. Y1.2:(nos.). paper, $3.

E82 Laws relating to vocational education and agricultural extension work, May 8, 1914– (date). Prepared by the U.S. Congress, House, Document Room. Available from the Superintendent of Documents, U.S. Government Printing Office, Washington, D.C. 20402. Irregular. Dep. item no.998, S.D. no.Y1.2:(nos.). paper, $4.

E83* Leaflet. 1927– . Prepared by the U.S. Dept. of Agriculture. Available from the Superintendent of Documents, U.S. Government Printing Office, Washington, D.C. 20402. Irregular. Dep. item no.12, S.D. no.A1.35:(nos.). Price varies.

Presents practical information on farming in a style that is easily understandable by the general reader. Although similar to the *Farmers' bulletin* in content, the leaflets are limited to specific directions and methods, and they range from only 1 to 8 pages in length. Some numbers are available free from members of Congress and state and county extension officers.

E84 Library lists. 1942– . Prepared and distributed by the National Agricultural Library, U.S. Dept. of Agriculture, Washington, D.C. 20250. Irregular. Dep. item no.95-A, S.D. no.A17.17:(nos.). Free.

One of special interest to librarians is the subseries, Library list no.1, Selected list of American agricultural books in print and current agricultural periodicals, S.D. no.A17.17:1/date. Another similar publication which should be noted is: Significant books about U.S. agriculture, 1860–1960, S.D. no.A17.18/2:Ag8(1860–1960).

E85 List of published soil surveys. Prepared and distributed by the Soil Conservation Service, U.S. Dept. of Agriculture, Washington, D.C. 20250. Annual. Dep. item no.102-A, S.D. no.A57.38:(list/year). Free.

E86 Manual of meat inspection procedures of the Department of Agriculture. Prepared by the Consumer and Marketing Service, Meat Inspection Div., U.S. Dept. of Agriculture, Washington, D.C. 20250. Available from the Superintendent of Documents, U.S. Government Printing Office, Washington, D.C. 20402. Irregular. Dep. item no.24-P, S.D. no.A88.6/4:M46(nos.). paper, $9.25.

E87 Market news reports. Prepared and distributed by the Consumer and Marketing Service, U.S. Dept. of Agriculture, Washington, D.C. 20250. Dep. item no.19-A.

E88 Marketing bulletin. 1959– . Prepared by the U.S. Dept. of Agriculture. Available from the Superintendent of Documents, U.S. Government Printing Office, Washington, D.C. 20402. Irregular. Dep. item no.13-G, S.D. no. A1.95:(nos.). Price varies.

Intended for the general reader, this series on marketing is written in popular style. Contains all kinds of agricultural marketing information. Recent titles, such as *USDA yield grades for beef* and *How to buy poultry by USDA grades,* have been concerned with grades of meat and poultry.

E89 Marketing research reports. 1952– . Prepared by the U.S. Dept. of Agriculture. Available from the Superintendent of Documents, U.S. Government Printing Office, Washington, D.C. 20402. Irregular. Dep. item no. 13-B, S.D. no.A1.82:(nos.). Price varies.

Acquaints the professional and technical worker with the recent marketing research results. Semitechnical to semipopular in approach. Concerned with the major aspects of marketing such as transportation, processing, and distributing. Free to qualified libraries and persons associated with marketing farm products.

E90* Miscellaneous publication. 1927– . Prepared by the U.S. Dept. of Agriculture. Available from the Superintendent of Documents, U.S. Government Printing Office, Washington, D.C. 20402. Irregular. Dep. item no. 13-A, S.D. no.A1.38:(nos.). Price varies.

These publications are issued in different forms, including special reports, indexes, bibliographies, directories, lists, or catalogs that are not adaptable to the other department series. Thus, size and format vary according to the type of material presented. Technical to nontechnical in nature. This is one of the best-known agricultural monographic series. Some of the more important volumes are: Hitchcock's *Manual of the grasses of the United States,* no.200; *Fact book of U.S. agriculture,* no.1063; *Profiles, careers in the U.S. Department of Agriculture,* no.1071; *Woody-plant seed manual,* no.654; and *Native woody plants of the United States, their erosion control and wildlife values,* no.303.

E91 National union catalog. 1956– . Prepared and distributed by the U.S. Library of Congress, Card Div., Bldg. 159, Navy Yard Annex, Washington, D.C. 20541. Monthly, cumulative quarterly, annual, and quinquennial. S.D. no.LC30.8:(date). paper (monthly), $600 a year.

E92 New serial titles. 1961– . Prepared and distributed by the U.S. Library of Congress, Card Div., Bldg. 159, Navy Yard Annex, Washington, D.C. 20541. Monthly. S.D. no. LC1.23/5:(date & nos.). paper, $25 a year.

E93 News for farmers cooperatives. 1934– . Prepared and distributed by the Farm Cooperative Service, U.S. Dept. of Agriculture, Washington, D.C. Monthly. Dep. item no.66, S.D. no.A89.8:(nos.). paper, $3 a year.

Considers the current developments in the major cooperatives as well as the practical applications for cooperative policies and principles. A well-illustrated magazine intended for cooperative members and students desiring information on the cooperative movement.

E94* Occupational outlook handbook. 1949– . Prepared by the U.S. Bureau of Labor Statistics. Available from the Superintendent of Documents, U.S. Government Printing Office, Washington, D.C. 20402. Biennial. illus. Dep. item no.768, S.D. no.L2.3:(nos.). single copy $4.25.

A biennially revised guide to almost all of the main occupations, trades, skills, and professions in the United States. Covers the duties, training, education, chances for advancement, work location, benefits, compensation, and demand for each occupation considered. An excellent vocational guidance reference for the student and the counselor.

E95 Occupational outlook quarterly. 1957– . Prepared by the U.S. Bureau of Labor Statistics. Available from the Superintendent of Documents, U.S. Government Printing Office, Washington, D.C. 20402. Quarterly. Dep. item no.770-A, S.D. no.L2.70/4: (v. & no.). paper, $1.50 a year; single copy 45¢

E96* Our public lands. 1951– . Prepared by the Bureau of Land Management, U.S. Dept. of the Interior. Available from the Superintendent of Documents, U.S. Government Printing Office, Washington, D.C. 20402. Quarterly. Dep. item no.633-A, S.D. no.I53.12: (nos.). $1 a year.

This periodical is the official organ of the U.S. Dept. of the Interior, Bureau of Land Management. It provides information on the development, utilization, and conservation of the resources on the public lands of America. Articles are concerned with such specific subjects as forest conservation, exploitation of minerals, land use, and wildlife and livestock range management. The style is nontechnical.

E97 Outdoor recreation action. 1966– . Prepared by the Bureau of Outdoor Recreation, U.S. Dept. of the Interior. Available from the Superintendent of Documents, U.S. Government Printing Office, Washington, D.C. 20402. Quarterly. Dep. item no.657-B-5, S.D. no.I66.17:(nos.). paper, $2 a year; single copy 55¢

E98 Outdoor recreation research. 1966– .
Prepared by the Bureau of Outdoor Recreation, U.S. Dept. of the Interior. Available from the Superintendent of Documents, U.S. Government Printing Office, Washington, D.C. 20402. Dep. item no.657-B-8, S.D. no.I66.18:(nos.). paper, $1.25.

E99 PA [Program aid]. 1946– . Prepared and distributed by the U.S. Dept. of Agriculture, Washington, D.C. 20250. Irregular. Dep. item no.14-A, S.D. no.A1.68:(nos.). Price varies.

The *PA* series gives recent details about farm programs for teachers, extension agents, students, and those immediately affected by government plans and programs. Of special value to teachers as supplementary reading material for classes because the program aids are available free in large quantities for interested groups. Because these series numbers are published so sporadically and because they contain information that is often ephemeral in nature, they have limited usefulness for library collections.

E100 Pesticides monitoring journal. 1967– .
Prepared by the U.S. Federal Committee on Pest Control. Available from the Superintendent of Documents, U.S. Government Printing Office, Washington, D.C. 20402. Quarterly. Dep. item no.1061-C, S.D. no.Y3.F31/18:9 (v. & no.). paper, $1.75 a year; single copy 50¢

E101 Plant disease reporter. 1917– . Prepared by the Agricultural Research Service, Crops Research Div., U.S. Dept. of Agriculture. Available from the Superintendent of Documents, U.S. Government Printing Office, Washington, D.C. 20402. Semimonthly. Dep. item no.100-B, S.D. no.A77.511:(nos.). $5 a year.

An illustrated periodical concerned with recent advances and discoveries in plant pathology. Includes research papers written by specialists and news on plant disease outbreaks and epidemics.

E102 Poultry inspector's handbook. Prepared and distributed by the Consumer and Marketing Service, Poultry Div., U.S. Dept. of Agriculture, Washington, D.C. 20250. Irregular. Dep. item no.24-P, S.D. no.A88.6/4:(nos.). Free.

E103* Production research reports. 1956– .
Prepared by the U.S. Dept. of Agriculture. Available from the Superintendent of Documents, U.S. Government Printing Office, Washington, D.C. 20402. Irregular. Dep. item no.13-C, S.D. no.A1.84:(nos.). Price varies.

Deals with the research done on agricultural production and emphasizes the utilization of the derived results for improving this phase of farming. These reports are in semitechnical and semipopular style. Distributed free to professional and technical workers.

E104 Rural development newsletter. Prepared and distributed by the Office of Information, U.S. Dept. of Agriculture, Washington, D.C. 20250. Irregular. S.D. no.A21.25/3:(nos.). Free.

E105 SCS national engineering handbook.
Prepared and distributed by the Soil Conservation Service, U.S. Dept. of Agriculture, Washington, D.C. 20250. Irregular. Dep. item no.120-A, S.D. no.A57.6/2:En3/(nos.). Price varies.

E106 Sanitation handbook of consumer protection programs. Prepared by the Consumer and Marketing Service, U.S. Dept. of Agriculture. Available from the Superintendent of Documents, U.S. Government Printing Office, Washington, D.C. 20402. Irregular. Dep. item no.24-P, S.D. no.A88.6/4:(nos.). paper, $1.50.

E107 Service and regulatory announcements, SRA-C & MS—(series). 1915– . Prepared and distributed by the Consumer and Marketing Service, U.S. Dept. of Agriculture, Washington, D.C. 20250. Irregular. Dep. item no.111, S.D. no.A88.6/3:(nos.). Free.

E108 Situation and outlook reports. Prepared by the Economic Research Service, U.S. Dept. of Agriculture. Available from the Superintendent of Documents, U.S. Government Printing Office, Washington, D.C. 20402. paper, price varies.

E109* Soil conservation. 1935– . Prepared by the Soil Conservation Service, U.S. Dept. of Agriculture. Available from the Superintendent of Documents, U.S. Government Printing Office, Washington, D.C. 20402. Monthly. Dep. item no.122, S.D. no.A57.9:(nos.). paper, $2.50 a year.

An attractively illustrated magazine whose articles are, for the most part, popular in style. Deals with soil and water conservation, land use, news of the Soil Conservation Service, watershed programs, and activities of the local soil conservation districts. Articles have appeal for the general reader as well as for the student and teacher in agriculture.

E110 Soil survey investigations reports. 1928– . Prepared and distributed by the Soil Conservation Service, U.S. Dept. of Agriculture, Washington, D.C. 20250. Irregular. Dep. item no.121-A, S.D. no.A57.52:(nos.). Free.

A USDA series concerned with the properties of the soil in certain counties or regions. Gives soil maps, detailed descriptions of the soil, land productivity, management suggestions, and discussions on various soil problems. Reports soil investigations and laboratory data.

E111 Soil survey reports. 1928– . Prepared and distributed by the Soil Conservation Service, U.S. Dept. of Agriculture, Washington, D.C. 20250. Irregular. S.D. no.A57.38:(nos.). Price not listed because of a highly limited distribution.

Soil survey reports may be obtained for certain counties or regions within the states. Detailed soil maps, information on climate, technical descriptions of soil morphology, productivity estimates, principles of soil classification, and other facets of agriculture that are related to soils are features of this series. An excellent reference for information on local soil characteristics.

E112* Statistical abstract of the United States. 1878– . Prepared by the U.S. Bureau of the Census. Available from the Superintendent of Documents, U.S. Government Printing Office, Washington, D.C. 20402. Annual. Dep. item no.150, S.D. no.C3.134:(nos.). $5.75.

E113* Statistical bulletins. 1923– . Prepared and distributed by the Economic Research Service, U.S. Dept. of Agriculture, Washington, D.C. 20250. Irregular. Dep. item no.15, S.D. no.A1.34:(nos.). Price varies.

For the most part, these irregularly published bulletins contain statistical tables on agricultural products. Each issue covers one or more commodities that are similar in nature and includes statistics on production, movement from farms and reshipment, marketing receipts, prices, exports and imports, and foreign production for each product. Of interest to anyone involved with agricultural marketing and production.

E114 Statistical summary. 1940– . Prepared and distributed by the Statistical Reporting Service, U.S. Dept. of Agriculture, Washington, D.C. 20250. Monthly. S.D. no.A92.13:(nos.). Free.

Summarizes statistical data estimated for or collected on the prices, sales, stocks, and production of agricultural materials. Reports on all kinds of products such as fibers, grains, vegetables, nuts, fruit, seeds, livestock, milk, and dairy products. Only a single sheet in length but it gives an adequate monthly review of the economic situation.

E115 Summaries of federal milk marketing orders. Prepared and distributed by the Consumer and Marketing Service, Milk Marketing Orders Div., U.S. Dept. of Agriculture, Washington, D.C. 20250. Irregular. S.D. no. A88.2:M59/29/(nos.). Free.

E116* Technical bulletin. 1927– . Prepared by the Agricultural Research Service, U.S. Dept. of Agriculture. Available from the Superintendent of Documents, U.S. Government Printing Office, Washington, D.C. 20402. Irregular. Dep. item no.16, S.D. no.A1.36:(nos.). Price varies.

Written in technical style for the scientist and student in agriculture, this series presents research results on specific crops, industries, and localities. The papers in the bulletins are reports on original research done by the Dept. of Agriculture and other cooperating departments and agencies. Free distribution of the *Technical bulletin* is limited because only a small number of copies are printed for each edition. However, these are inexpensive to buy, and certain numbers, such as 1082, entitled, *Major uses of land in the United States,* are especially useful as individual books in agricultural collections.

E117 Technical equipment report. 1958– . Prepared and distributed by the Forest Service, U.S. Dept. of Agriculture, Washington, D.C. 20250. Irregular. S.D. no.A13.49/3:(nos.). Free.

E118 Tree planters' notes. 1950– . Prepared by the Forest Service, U.S. Dept. of Agriculture. Available from the Superintendent of Documents, U.S. Government Printing Office, Washington, D.C. 20402. Quarterly, with annual summary. Dep. item no.86-D, S.D. no. A13.51:(v. & no.). paper, $1 a year; single copy 30¢

E119 United States standards for grades of (commodity). Prepared by the Consumer and Marketing Service, U.S. Dept. of Agriculture. Available from the Superintendent of

Documents, U.S. Government Printing Office, Washington, D.C. 20402. Irregular. S.D. no. A88.6/2:[CT]. paper, price varies.

E120 **Utilization research report.** 1957– . Prepared by the U.S. Dept. of Agriculture. Available from the Superintendent of Documents, U.S. Government Printing Office, Washington, D.C. 20402. Irregular. illus. Dep. item no.16-A, S.D. no.A1.88:(nos.). Price varies.

An infrequently published series dealing with the research done on farm product utilization. Ranges from semipopular to semitechnical in style. The 1961 report title, *Processing of poultry by-products and their utilization in feeds,* exemplifies the specific subject matter of the items within this series.

E121* **Vocational education, agricultural education** (OE 81,000-81,999). Prepared by the U.S. Office of Education. Available from the Superintendent of Documents, U.S. Government Printing Office, Washington, D.C. 20402. Irregular. Dep. item no.460-A-81, S.D. no.FS5.281: (nos.). Price varies.

An extremely useful series of publications dealing specifically with vocational agricultural education. Indispensable as teaching and learning aids for teachers, counselors, administrators, and students. Some recent titles, which indicate the type of material included, are: *Objectives for vocational and technical education in agriculture; Instruction in farm mechanics; Farm business analysis,* and the 2-year post high school curriculum guides cited elsewhere.

E122 **Vocational education, miscellaneous** (OE 80,000-80,999). Prepared and distributed by the Office of Education, U.S. Dept. of Health, Education, and Welfare, Washington, D.C. 20202. Available from the Superintendent of Documents, U.S. Government Printing Office, Washington, D.C. 20402. Irregular. Dep. item no.460-A-80, S.D. no.FS5.280: (nos.). Price varies.

E123 **Water spectrum.** 1969– . Prepared by the Corps of Engineers, U.S. Dept. of the Army. Available from the Superintendent of Documents, U.S. Government Printing Office, Washington, D.C. 20402. Quarterly. Dep. item no.342-B, S.D. no.D103.48: (v. & no.). paper, $2.50 a year; single copy 65¢.

E124 **Weekly weather and crop bulletin, national summary.** 1924– . Prepared by the Environmental Data Service, Environmental Science Services Administration, U.S. Dept. of Commerce. Available from the Superintendent of Documents, U.S. Government Printing Office, Washington, D.C. 20402. Weekly. S.D. no.C30.11: (v. & no.). paper, $5 a year; single copy 10¢

E125 **Working reference of livestock regulatory establishments, stations, and officials.** 1954– . Prepared by the Consumer and Marketing Service, U.S. Dept. of Agriculture. Available from the Superintendent of Documents, U.S. Government Printing Office, Washington, D.C. 20402. Monthly. Dep. item no.32-A, S.D. no.A88.16/20: (date). paper, $9 a year; single copy $1.50.

E126* **Yearbook of agriculture.** 1894– . Prepared by the U.S. Dept. of Agriculture. Available from the Superintendent of Documents, U.S. Government Printing Office, Washington, D.C. 20402. Annual. Dep. item no.17, S.D. no.A1.10:(year). Price varies.

From 1894 to 1936, this yearbook contained a summary report of the secretary of agriculture, a review of miscellaneous developments in the agricultural sciences, and an appendix of agricultural data, which was published as a separate series entitled *Agricultural statistics* after 1936. The secretary of agriculture report was also omitted after 1937. The *Yearbook* then assumed a new format, which it has retained up to the present time, with each volume devoted to a single subject. It now includes articles on agricultural problems, new developments in the field, and research findings. The articles are written by specialists; they vary from technical to popular in approach, and the volumes offer excellent bibliographies for further reading. Free copies may be obtained from congressional representatives and senators because each is allotted by law 400-550 copies for distribution to his district or state. Individual titles are: 1936 *Better plants and animals I;* 1937 *Better plants and animals II;* 1938 *Soils and men;* 1939 *Food and life;* 1940 *Farmers in a changing world;* 1941 *Climate and man;* 1942 *Keeping livestock healthy;* 1943-47 *Science in farming;* 1948 *Grass;* 1949 *Trees;* 1950-51 *Crops in peace and war;* 1952 *Insects;* 1953 *Plant diseases;* 1954 *Marketing;* 1955 *Water;* 1956 *Animal diseases;* 1957 *Soil;* 1958 *Land;* 1959 *Food;* 1960 *Power to produce;* 1961 *Seeds;* 1962 *After a hundred years;* 1963 *A place to live;* 1964 *Farmer's world;* 1965 *Consumers all;* 1966 *Protecting our food;* 1967 *Outdoors USA;* 1968 specialized types of sources may be consulted: 1970 *Contours of change.*

F

Guides to Societies, Organizations, and Industries Concerned with Agriculture and Related Fields

This section reviews the most useful guides to institutions, organizations, and firms which may provide free and inexpensive educational materials for vocational/technical agriculture students, teachers, or librarians. It also reviews directories of equipment and supply distributors, individuals of importance to agriculture, and sources of miscellaneous information. In addition to the section titles listed, other more specialized types of sources may be consulted:

1. Directories published by most of the state and federal government agencies which are indexed in *Monthly checklist of state publications* and *Monthly catalog of U.S. government publications*

2. Yearbooks and handbooks of periodicals (represented by only a few randomly chosen examples in this section)

3. Individual association and society guides such as, the American Society of Agricultural Engineers' *Agricultural engineers yearbook*, which gives the official ASAE standards, recommendations, and data as well as other pertinent information on agricultural engineering supplies, equipment, test summaries, and soil and water management

4. Source guides of the National Academy of Sciences, National Research Council, and other general scientific organizations. (The publications of the Research Council on Agriculture, listed in its catalog and issued as part of an annual numbered series consisting of books, monographs, and reports, are excellent references for vocational/technical agricultural education.)

F1 ACS laboratory guide to instruments, equipment, and chemicals. 1955– . American Chemical Society Publications, 1155 Sixteenth St. NW, Washington, D.C. 20036.

Annual. illus. Included with annual subscription to *Analytical chemistry* for $5 a year.

A valuable publication for anyone interested in obtaining additional information on chemicals, equipment, instruments, etc. Contains lists of advertised products, supply houses, instruments and equipment, chemicals, trade-names, new books, and companies. The company directory is the most useful feature because full addresses are provided for the reader to request materials and services directly from their supplier. A good reference for the vocational/technical school whose curriculum stresses the chemical aspects of agriculture.

F2 Agri-business buyers reference. 1945– . Food & Feed Grain Industries Publishing Co., 1117 Board of Trade Bldg., Chicago, Ill. 60604. Annual. illus. Included with annual subscription to *Grain and feed journals, consolidated* for $6 a year.

An annual directory of better manufacturers and suppliers which was published earlier as the *Grain trade buyers' guide and management reference*. Contains grain elevators, feed mill machinery, equipment and supplies, advertisers and their complete addresses, and short articles on some of the current problems in the agri-business world and how to solve them. A good source of manufacturers and suppliers who may print pamphlets and other materials useful for the student, teacher, and administrator.

F3 Agriculture teachers directory and handbook. 1953– . Agriculture Teachers Directory, Faulkner Publications, 3042 Overlook Dr., Montgomery, Ala. 36109. Annual. illus. $7.50.

Provides a wealth of information on teachers, staff members, and associations related to vocational/technical agriculture. An especially helpful guide to the Livestock Record Associations, American Dairy Association state managers, texts and reference books, free films and filmstrips available, and advertisers. About 11,000 teachers and their staff members are listed as well as the representatives of the Vocational Agricultural Teachers' Association. The section on the agriculture "Technologies," those associations and institutions which provide information on agriculture and their addresses, is perhaps the most outstanding asset of this annual.

F4 The almanac of the canning, freezing, preserving industries. 1916– . E. E. Judge, 79 Bond St., Westminster, Md. 21157. Annual. illus. $10.

Important as a guide to information sources in food science as well as a monographic text and reference work for the student and teacher. Gives associations in the food industry, their members, boards of arbitration, and food and drug inspectors. Has a buyers' guide which includes the addresses, officers, convention dates, and phone numbers of the food industry associations and companies.

F5 American fruit grower buyer's directory. 1943– . Meister Publishing Co., Willoughby, Ohio 44094. Annual. illus. Included with annual subscription to *American fruit grower* for $2 a year.

A well-known, frequently used reference for sources of material on fruit growing. This annual issue of *American fruit grower* lists organic chemicals; spraying equipment; nursery stock; tractors; irrigation, pruning, harvesting, and fruit processing equipment; and the addresses of the producers and suppliers of these items. There is an accompanying classified subject index which indicates the nearby dealers in each fruit-producing state.

F6 American vegetable grower buyer's directory. 1955– . Meister Publishing Co., Willoughby, Ohio 44094. Annual. illus. Included with annual subscription to *American vegetable grower* for $2 a year.

Sections of this buyers' guide to products and machinery are arranged according to the type of equipment or process used in commercial vegetable production. The entries are adequately indexed, and addresses are supplied for the 2,000 companies included. Of the few specialized sources on this subject, this is one of the most inexpensive and useful.

F7 Canner/packer yearbook. Vance Publisher Corp., 300 W. Adams St., Chicago, Ill. 60606. Annual. illus. $2 or included with annual subscription to *Canner/packer* for $10 a year; free to qualified readers.

Not only provides details on about 850 manufacturers of equipment and supplies used in food production but also gives statistical information relevant to food processing for the convenience of the food industry worker. The manufacturers are arranged alphabetically by firm names, followed by their addresses. The equipment, supplies, and services are presented in subject categories for easy reference.

F8 Catalog of federal assistance programs; a description of the federal government's domestic programs to assist the American people in furthering their social and economic progress. 1967. U.S. Office of Economic Opportunity, Executive Office of the President, Washington, D.C. 20506. 701p. Free.

An especially helpful description of national governmental programs that have been adopted to promote social and economic improvement in America's society. With each program, the title, nature and purpose, people eligible for assistance from it, officials to contact for details, authorization, legislation, administering agency, and printed information available are given. A very comprehensive work. See also section E, no.52.

F9 Chem sources. 1958– . Directories Publishing Co., Flemington, N.J. 08822. Annual. $35.

For the most part, this work is comprised of a list of about 40,000 chemicals. It is primarily valuable in this section, however, because of its guide to about 600 producers and suppliers of organic and inorganic chemicals, their branch offices, and addresses. These companies are arranged alphabetically by name and chemical produced or supplied.

F10 City and county directories. R. L. Polk & Co., Polk Bldg., 431 Howard St., Detroit, Mich. 48231 (and other similar publishers). Annual. Paging varies. Price varies.

One of the best sources for local companies and associations which might offer literature or services to the student and teacher seeking additional help with educational projects. Usually revised every year, these references typically provide classified business directories, alphabetical lists of businesses and private citizens, street guides, lists of householders and other occupants of city or county buildings, and numerical telephone directories. Most of these guides are expensive, and they should be closely examined and evaluated before purchasing. (Available at the Chamber of Commerce in most cities)

F11 Commercial fertilizer yearbook. 1912– . Walter W. Brown Publishing Co., Inc., 75 3rd St. NW, Atlanta, Ga. 30308. Annual. illus. $10.

A valuable guide to the fertilizer industry associations and their addresses. Other information given for each company includes officials, telephone numbers, rail and other shipping data, major plant equipment, and fertilizer production capacity. Has good articles on specific items of interest, a roster of the state fertilizer control officials, grade regulations, and other helpful features.

F12 Conservation directory. 1969. National Wildlife Federation, 1412 16th St. NW, Washington, D.C. 20036. 146p. $1.50.

A valuable source of agencies, organizations, and individual officials interested in conservation and management of natural resources. Sections are on the U.S. government, international, regional, and interstate organizations and commissions, government agencies, and citizen groups in the states and Canadian provinces. Also lists the major colleges and universities that offer programs in conservation.

F13 County agents directory. 1915– . C. L. Mast Jr. & Associates, 2041 Vardon La., Flossmoor, Ill. 60422. Annual. illus. $5.50.

Specifically designed to serve agricultural extension workers, this reference is useful to anyone seeking information on the county agents, home demonstration agents, 4-H club representatives, state extension specialists, land-grant colleges and experiment stations, the officials of the U.S. Dept. of Agriculture, the National Association of State Depts. of Agriculture, and those of National Register Associations. Includes articles on the county agents and other phases of agricultural extension work.

F14 Dairy industries catalog. 1927– . Magazines for Industry, Inc., 777 Third Ave., New York, N.Y. 10017. Annual. illus. $10.

An annual directory of dairy industry supply and equipment manufacturers, dairy product manufacturers associations, agricultural institutions offering training in dairy product processing, officials in the U.S. and Canada who enforce the dairy and food laws, and the federal and state standards. Dairy production statistics are also supplied in this catalog.

F15 Directory of communicators in agriculture. 1969. Agricultural Relations Council, 18 S. Michigan Ave., Chicago, Ill. 60603. Unpaged. $5.

A who's who approach to the field of agricultural communication, which gives the names and addresses of the leading communicators. Arranged according to organization with the names of members and their positions held. The groups described are: the Agricultural Relations Council, American Agricultural Editors Association, American Association of Agricultural College Editors, Canadian Farm Writers Federation, National Association of Farm Broadcasters, and the Newspaper Farm Editors of America.

F16* A directory of information resources in the U.S.: federal government. 1967. Prepared by the National Referral Center for Science and Technology. U.S. Library of Congress. Available from the Superintendent of Documents, U.S. Government Printing Office, Washington, D.C. 20402. 411p. $2.75.

For each association, library, museum, agency, and information center resource included in this directory, gives areas of interest, holdings, publications, and organization. An excellent guide to the federal government activities developed to meet demands for information in science and technology. See also section E, no.18.

F17* A directory of information resources in the United States: general toxicology. 1969. Prepared by the National Referral Center for Science and Technology, U.S. Library of Congress. Available from the Superintendent of Documents, U.S. Government Printing Office, Washington, D.C. 20402. 293p. $3.

With the exception of individuals and general libraries, this directory lists associations, agencies, institutions, and other sources related to toxicology which provide information on the subject. Each entry consists of the company's complete address, a description of their toxicology interests, their holdings, publications, and specific information services. The appendixes contain highly useful data including poison control centers, professional organizations having substantial interest in toxicology, and some periodicals which emphasize this subject. See also section E, no.18.

F18* A directory of information resources in the United States: physical sciences, biological sciences, engineering. 1965. Prepared by the National Referral Center for Science and Technology, U.S. Library of Congress. Available from the Superintendent of Documents, U.S. Government Printing Office, Washington, D.C. 20402. 352p. $2.25.

This directory, concerned with universities, associations, professional societies, industrial firms, and other sources which offer information on the physical, biological and engineering sciences, is a fairly comprehensive list of sources. Details such as organization, specialized interests, information services, and publications, are given for each agency. See also section E, no.18.

F19* A directory of information resources in the United States: social sciences. 1965. Prepared by the National Referral Center for Science and Technology, U.S. Library of Congress. Available from the Superintendent of Documents, U.S. Government Printing Office, Washington, D.C. 20402. 218p. paper, $1.50.

Acquaints the reader with about 600 of the major institutions and agencies providing information on various aspects of social science. Describes the social science-related activities, collections, publications, and specific information services of each listed resource. Also gives full addresses and telephone numbers. See also section E, no.18.

F20* A directory of information resources in the United States: water. 1966. Prepared by the National Referral Center for Science and Technology, U.S. Library of Congress. Available from the Superintendent of Documents, U.S. Government Printing Office, Washington, D.C. 20402. 248p. paper, $1.50.

Another very useful information resource directory giving complete addresses, specialized activities, holdings, information services, and publications for each organization and agency entered. Treats the major libraries, information centers, and other sources offering materials and advice in the field of water, excluding oceanography. Indexed by subject and organization. See also section E, no.18.

F21 Directory of national organizations concerned with land pollution control. 1970. Freed Publisher Co., P.O. Box 1144, FDR Station, New York, N.Y. 10022. 24p. $5.

This reference does not claim to be complete because of the frequent changes in the field of environmental science. It does, however, give addresses, phone numbers, officials, activities, committees, purpose, membership, and publications for the major organizations involved with control of land pollution. Because of the growing emphasis on the subject, this directory is a quite useful source.

F21a Directory of national trade and professional associations of the United States. 1966– . Columbia Books, Inc. Ste-300, 917 15th St. NW, Washington, D.C. 20005. Annual. Latest ed. $12.50; earlier editions still available: 1966 $2.80; 1967 $5.50; 1968 $7.95; 1969 $7.95; 1970 $10.

An alphabetical list of national and international trade and professional associations. Gives address, publications, and other pertinent information for each association listed. Covers subjects of interest to agriculturists from agricultural chemicals to zoological gardens.

F22 The directory of the canning, freezing, preserving industries, 1968–69. 2d ed. 1968. E. Judge & Son, P.O. Box 866, Westminster, Md. 21157. 564p. $25.

An alphabetically arranged list of the companies concerned with canning, freezing, and preserving in the food industry. Gives the address, personnel, associations, brands, factories, can or glass size, and volume for every firm included. Provides a geographical, product, and a brand listing in pts. 2–4 of the directory.

F23 Directory of the forest products industry. 1958– . Miller Freeman Publications,

Inc., 500 Howard St., San Francisco, Calif. 94105. Annual. illus. $30.

Designed as a reference tool for those directly or indirectly involved with the forest products industry and for any others desiring information on North American lumber and wood products manufacturers, wholesalers, and industry suppliers. A directory of forest-oriented organizations and a list of forestry schools are in the miscellaneous section. Updated every year.

F24*   Encyclopedia of associations. 1956– . Gale Research, Book Tower, Detroit, Mich. 48226. 3v. Triennial. v.1 $32.50; v.2 $20; v.3 $25.

The most important, comprehensive guide to general sources available. A standard reference list of associations, societies, and organizations, which supplies the address, acronym, membership, activities, research and educational programs, staff, publications, and meetings for each. Volume 1 is the primary listing of the national organizations, volume 2 is a geographic-executive index, and volume 3 is a supplementary guide to new associations not included in the parent volume. Beginning with volume 3 of the 6th edition (1970), the geographical coverage is expanded to include international associations. In addition to organizations and agencies, this source also contains federal projects, social development programs, committees for action, and other groups.

F24a   Encyclopedia of business information sources. 2d ed. 1970. Gale Research Co., 961 Book Tower, Detroit, Mich. 48226. 2v. $47.50.

A detailed listing of business and business-related subjects with a record of sources concerned with each topic. Arrangement is alphabetical by broad subject, such as agricultural engineering, florists, peach industry, potato industry, etc. Contains information of value to agriculturists although intended for business executives.

F25   Farm chemicals directory. 1969– . Meister Publishing Co., 37841 Euclid Ave., Willoughby, Ohio 44094. Annual. $60.

Formerly a part of the *Farm chemicals handbook*, this directory consists of a geographical list of over 500 manufacturers, dealers, and formulators in the United States, Canada, and Puerto Rico, an alphabetical list by company which gives full addresses, a directory of aerial applicators, and an international section giving companies and plants by country. A useful source but too expensive for most small libraries.

F26   Farm chemicals handbook. 1914– Meister Publishing Co., 37841 Euclid Ave., Willoughby, Ohio 44094. Annual. illus. $17.50.

Many helpful features of this handbook such as, the key word index to farm chemicals and the environment, the directory of associations and control officials, dictionary of plant foods, pesticide dictionary, and the buyers' guide to services, equipment, and supplies, are noteworthy. The listing of associations gives conferences to be held and addresses of officers for each organization.

F27   Farm equipment manufacturers directory. No beginning date available. Farm Power Equipment, 2340 Hampton St., St. Louis, Mo. 63139. Annual. $1.

A specialized directory of about 2,500 manufacturers and distributors of farm equipment and related products. A good reference to companies and organizations that may offer free or inexpensive materials and advice on this subject.

F28   Federal outdoor recreation programs; the nationwide plan for outdoor recreation. 1968. U.S. Bureau of Outdoor Recreation, U.S. Dept. of the Interior, Washington, D.C. 20240. 224p. paper, $1.75.

Reviews national outdoor programs developed by various government departments. Indicates the sponsor of the program, the agency for it, and the address of the headquarters, chairmen, or others to contact for information. An excellent outline of the federal outdoor recreation plan.

F29   Feed industry red book. 1938– . Communications Marketing, Inc., 5100 Edina Industrial Blvd., Edina, Minn. 55435. Annual. illus. $3.

Previously entitled *Feed bag red book*, this annual reference book and buyers' guide is a well-known, highly useful list of feed manufacturers and their addresses. Indexed by product. Directed toward manufacturers of feed for livestock and poultry, but helpful for anyone desiring information on feeds and their production.

F30   Food industry sourcebook for communication. 1970– . National Canner Association, 1133 20th St. NW, Washington, D.C. 20036. Irregular. $20.

In 7 sections: government, education, consumerism, environment, industry, publications, and journalists. These major aspects of the food industry are presented in relation to canners, and pertinent data is given under each. The

Guides to Societies, Organizations

emphasis in every section is on the relevance of the main topic to advertising, education, environmentalism, food, fraud and deceptive practices, packaging, labeling, product safety, testing of products, and weights and measures. Indicates people and places to contact, addresses, curriculum, and publications used under each subject heading.

F31 Fortune directory. Fortune Magazine, 541 N. Fairbanks Ct., Chicago, Ill. 60611. Annual. illus. $2.

A general list of the 500 industrial firms which have the largest net profits, assets, and sales volumes. Also provides a directory of the 100 largest industrial firms in foreign countries, the 50 largest banks, transportation companies, utilities, and life insurance companies. Arranged geographically by state and city.

F32* Foundation directory. 1960– . Russell Sage Foundation, 230 Park Ave., New York, N.Y. 10017. Irregular. $12.

Includes over 6,000 foundations with their addresses, donors, officers and trustees, purposes, activities, gifts, expenditures, grants, and specific limitations. Replaces *American foundations and their fields*. The standard reference for information on U.S. foundation organizations, and an excellent source of endowed institutions that may provide materials and advice on various subjects for inquirers.

F33 Governmental guide. (Regular state eds. and educational ed.). 1932– . Governmental Guide, Drawer 299, Madison, Tenn. 37115. Annual. $5.

These guides are issued in 2 forms: educational editions covering all of the states, and separate editions available for each of only 14 states. They supply addresses for the officials and the major branches of the U.S. government, selected boards, house UN officials, congresses, state officials and department heads, college and university heads, county officers, and mayors. Selected statistics are also reported for the states.

F34 Home and garden supply merchandiser green book. 1950– . Home and Garden Supply Merchandiser, P.O. Box 67, Minneapolis, Minn. 55440. Annual. illus. $5 or included with annual subscription of *Home and garden supply merchandiser* for $10 a year; free to qualified readers.

Spans the field of home and garden supply in its recognition of central parts distributors, garden organizations, product categories, sources of equipment and provisions, the most important garden manufacturers, and brand names.

A highly recommended, inexpensive source of merchandisers who may have educational pamphlets, AV media, and other materials for the student, teacher, and librarian interested in these particular agricultural areas.

F35 Implement and tractor product file. 196?– . Intertec Pub Corp., 1014 Wyandotte St., Kansas City, Mo. 64105. Annual. illus. $1 or included with annual subscription to *Implement and tractor* for $3 a year.

The *Product file* gives complete buying information for the farm supplies and equipment, shop supplies, replacement and component parts, and the red listings which are guides to the advertised products. It is primarily important for the addresses given for all of the companies included in the *Red book*.

F36 Implement and tractor red book. 1917– . Intertec Pub. Corp., 1014 Wyandotte St., Kansas City, Mo. 64105. Annual. illus. $1 or included with annual subscription to *Implement and tractor* for $3 a year.

To be used with the *Product file* printed in the March 31st issue of *Implement and tractor*. The *Red book* lists all kinds of farm equipment and machinery and the manufacturers. Intended for those in the farm machinery industry who are responsible for the research, development, service and marketing, sales, and design of these products.

F37 Marketing information guide. 1954–
Prepared by the Business and Defense Service Administration, U.S. Dept. of Commerce. Available from the Superintendent of Documents, U.S. Government Printing Office, Washington, D.C. 20402. Monthly. Dep. item no.215-A, S.D. no.641.11:(nos.). $4.50 a year; paper, $2.50.

A good bibliographic source for materials on marketing because it contains information not easily found in other places. Entries are annotated. Governmental and nongovernmental books, journal articles, pamphlets, and other items are listed, and the source and availability are indicated for each. Intended to serve not only those involved with buying industrial or consumer products but also those in general business or personnel services.

F38 Modern packaging: encyclopedia issue. 1955– . McGraw-Hill, Inc., 330 W. 42nd St., New York, N.Y. 10036. Annual. illus. Included with annual subscription to *Modern packaging* for $10 a year.

Concerned with the dynamics of modern packaging equipment, techniques, and research.

Sections describe packaging news, planning and development, materials, containers and container components, equipment, methods, sources of supply, and specific manufacturers. Significant for this section because its buyers' directory gives addresses for its listed companies. Thus, a useful guide to potential sources of educational materials as well as to general information on packaging technology.

F38a Modern veterinary practice red book. 1964– . American Veterinary Publications, Inc., 114 N. West Street, Wheaton, Ill. 60187. Annual. illus. $5 or included with annual subscription to *Modern veterinary practice.*

Of special interest to veterinary teachers, students, and librarians seeking educational material in veterinary science from product manufacturers and associations. Manufacturers and suppliers of veterinary equipment are arranged alphabetically.

F39 Park maintenance annual buyer's guide. 1948– . Madisen Publisher Div., P.O. Box 409, Appleton, Wis. 54911. Annual. illus. $1 or included with annual subscription to *Park maintenance* for $4 a year.

A representative list of manufacturers and suppliers of park products. Those companies which service parks, recreation areas, golf courses, and campuses are listed alphabetically with their addresses and descriptions of products produced or distributed.

F40 Pesticide handbook—Entoma. 1951– . College Science Publishers, State College, Pa. 16801. Annual. illus. $5; paper, $3.50.

Designed for those concerned with any phase of pest control. Gives about 600 new pesticide products and a helpful listing of poison control centers. Especially useful for its roster of agricultural chemical leaders and coordinators and their addresses.

F41 Real estate information sources. 1963. Gale Research Company, 1400 Book Tower, Detroit, Mich. 48226. 317p. $11.50.

Covers over 1,000 sources for materials and other information on the real estate business, brokerage, appraisal, modernization, building, counseling, finance, land development, law, associations, periodicals, AV aids, and government agencies. An inexpensive guide to the resources available in every specialized facet of real estate. (Also issued as volume 1 of the Management Information Guide series)

F42 Research centers directory: a guide to university-sponsored and other non-profit research organizations. 1960– . Gale Research Co., 1400 Book Tower, Detroit, Mich. 48226. Irregular. $39.50.

Attempts to supply comprehensive descriptions of about 5,000 university and private research groups. The subjects studied by them, their staffs, budgets, and other details are outlined. Indexed by institution, research unit, personal name, and subject, and revised frequently by supplements. A good reference to sources for information on research in progress and completed, but it may be too advanced for the technical school level.

F43* Scientific and technical societies of the United States. 8th ed. 1968. Printing and Publishing Office, National Academy of Sciences, 2101 Constitution Ave., Washington, D.C. 20418. 221p. $12.50.

Arranged alphabetically by society and indexed by key word and geographic location. Gives addresses, history, purpose, membership, meetings, publications, and professional activities for all societies included in this directory. Trade associations, undergraduate groups, fund-raising organizations, small city and county medical and engineering societies, and non-scientifically-oriented groups are omitted. A well-known guide that has become a standard reference book in most libraries.

F44 Seed trade buyer's guide. [1918?]– . Seed World publishers, 434 S. Wabash Ave., Chicago, Ill. 60605. Annual. illus. Included with subscription to *Seed world* for $5 a year.

Not only lists seed growers, dealers, and distributors, wholesalers, and manufacturers of machinery, but also reports the seed laws by state, statistical tables, officers of associations concerned with seed trade, and their addresses. Useful for anyone interested in this phase of agribusiness.

F45* Specialized science information services in the United States: a directory of selected specialized information services in the physical and biological sciences. 1961. Prepared by the National Science Foundation. Available from the Superintendent of Documents, U.S. Government Printing Office, Washington, D.C. 20402. 528p. $1.75.

Consists of companies offering information services in aerospace, agriculture, pharmacy, physics, and other scientific subjects. Company names are grouped alphabetically under broad headings. Scope, user qualifications, collection, publications, and the particular materials and advice provided for requesters are features noted for each entry. An extremely valuable source because it also describes how the materials and other services may be obtained.

F45a Standard periodical directory. 3d ed. 1970. Oxbridge Publishing Co., Inc., 420 Lexington Ave., New York, N.Y. 10017. $25.

A subject approach to the publications of organizations of all types. Arrangement is alphabetical, with a key word cross index to subjects and an index of periodical titles. Includes pertinent buying information for each periodical listed.

F46 A survey of the environmental science organizations in the U.S.A. 1970– . Ecology Center Press, 1360 Howard St., San Francisco, Calif. 94103. Annual. $5.

Organizations listed include college and university centers; nonprofit organizations and associations; citizens groups; federal government departments and agencies; professional societies and academies; and science-oriented industries. Entries contain such information as director and address, phone, a description of the organization, purpose, and activities.

F47 Telephone directories.

These guides prepared by all cities are one of the best sources of local businesses, associations, and agencies and their addresses. Organizations and firms are also conveniently arranged in the yellow pages by subject.

F48 Thomas' register of American Manufacturers. 1905-06– . Thomas Publisher Co., 461 Eighth Ave., New York, N.Y. 10001. Annual. 10v. illus. $50.

A multivolume catalog of over 100,000 manufacturers, about one million producers, and several thousand chambers of commerce and trade associations. Intended for firms dealing with industrial products and for purchasing agents, but useful for anyone as a general source of businesses and related agencies. Volumes 1 to 6 deal with products and services; volume 7 lists the companies, local offices, and phone numbers; volume 8 indexes the preceding parts; and volumes 9 and 10 consist of a catalog file.

F49 Veterinarians' blue book and therapeutic index. 1953– . The Reuben H. Donnelley Corp., 466 Lexington Ave., New York, N.Y. 10017. Irregular. illus. $8 (approx.).

Designed for veterinary student, teacher, and professional. Gives all kinds of helpful information such as the major manufacturers, product descriptions, therapeutic practices, and lists of feed additives, etc., and where to obtain them.

F50 Who's who in the egg and poultry industries. 1929– . Walt Publisher Co., Sandstone Bldg., Mt. Morris, Ill. 61054. Annual. illus. $15.

A roster of manufacturers, chicken breeders, turkey breeders, marketers, processors, and federal agents. A national and sectional associations list, a directory of refrigerated warehouses, and a where-to-buy-it products catalog are other significant sections in this guide.

PUBLISHERS DIRECTORY

AVI. AVI Publishing Co., Box 670, Westport, Conn. 06880

Academic. Academic Press, Inc., 111 Fifth Ave., New York, N.Y. 10003

Addison-Wesley. Addison-Wesley Publishing Co., Inc., Reading, Mass. 01867

Agr'l Consulting Assoc. Agricultural Consulting Associates, Box 330, Wooster, Ohio 44691

Aldine. Aldine-Atherton, Inc., 529 S. Wabash Ave., Chicago, Ill. 60605

Aldus. Aldus Books, Ltd., 17 Conway St., London, W. 1, England

Allen. J. A. Allen & Co., Ltd., Lower Grosvenor Pl., Buckingham Palace Rd., London, S.W. 1, England

Allen & Unwin. George Allen & Unwin, Ltd., 40 Museum St., London, W.C. 1, England

Allyn. Allyn & Bacon, Inc., 470 Atlantic Ave., Boston, Mass. 02210

Amer. Appraisal. American Appraisal Co., 525 E. Michigan St., Milwaukee, Wis. 53201

Amer. Assn. of Cereal Chem. American Association of Cereal Chemists, 1821 University Ave., St. Paul, Minn. 55104

Amer. Bee J. American Bee Journal, Hamilton, Ill. 62341

Amer. Chem. Soc. American Chemical Society, 1155 16th St. NW, Washington, D.C. 20036

Amer. Elsevier. American Elsevier Publishing Co., Inc., 52 Vanderbilt Ave., New York, N.Y. 10017

Amer. Forestry Assn. American Forestry Association, 919 17th St. NW, Washington, D.C. 20012

Amer. Heritage. American Heritage Press, 1221 Ave. of the Americas, New York, N.Y. 10020

Amer. Meat Inst. American Meat Institute, 59 E. Van Buren St., Chicago, Ill. 60610

Amer. Met. Soc. American Meteorological Society, 45 Beacon St., Boston, Mass. 02108

Amer. Nurseryman. American Nurseryman Publishing Co., 343 S. Dearborn, Chicago, Ill. 60604

Amer. Poultry Assn. American Poultry Association, Publications Dept., c/o E. C. Schultz, E. Fourth St., Crete, Nebr. 68333

Amer. Public Health Assn. American Public Health Association, 1740 Broadway, New York, N.Y. 10019

Amer. Soc. of Agron. American Society of Agronomy, 677 S. Segoe Rd., Madison, Wis. 53711

Amer. Tech. Soc. American Technical Society, 848 E. 58th St., Chicago, Ill. 60637

Amer. Voc. Assoc. American Vocational Association, Inc., 1510 H St. NW, Washington, D.C. 20005

Appleton. Appleton-Century-Crofts, 440 Park Ave. S, New York, N.Y. 10016

Arco. Arco Publishing Co., Inc., 219 Park Ave. S, New York, N.Y. 10003

Arnold. E. J. Arnold & Son, Ltd., Butterley St., Leeds 10, England

Assn. of State Foresters. *See* National Association of State Foresters

Auburn Print. Auburn Printing Co., Auburn, Ala. 36830

Bailliere. Bailliere, Tindall, & Cassell, Ltd., 7-8 Henrietta St., Covent Garden, London, W.C.2, England

Ball. George J. Ball, Inc., 250 Town Rd., West Chicago, Ill. 60185

Balt. Balt Publishers, 3315 South St., West Lafayette, Ind. 47905

Barnes. A. S. Barnes & Co., Forsgate Dr., Cranbury, N.J. 08512

Barnes & Noble. Barnes & Noble, Inc., 105 Fifth Ave., New York, N.Y. 10003

Barrows. M. Barrows & Co., Inc. Distributed by Morrow

Publishers Directory

Basic. Basic Books, Inc., Publishers, 404 Park Ave. S, New York, N.Y. 10016

Beacon. Beacon Press, 25 Beacon St., Boston, Mass. 02108

Benjamin. The Benjamin Co., Inc., 485 Madison Ave., New York, N.Y. 10022

Black. A. & C. Black, Ltd., 4, 5, & 6 Soho Sq., London, W. 1, England

Blackwell Scientific. Blackwell Scientific Publications, Ltd., 5 Alfred St., Oxford, England

Bobbs-Merrill. The Bobbs-Merrill Co., Inc., 4300 W. 62nd St., Indianapolis, Ind. 46268

Books. Books, Inc., 5330 Wisconsin Ave. NW, Washington, D.C. 20015

Bowen & Jenkins. Bowen & Jenkins, P.O. Box 262, Westwood, N.J. 07675

Branden. Branden Press, Inc., 221 Columbus Ave., Boston, Mass. 02116

Branford. Charles T. Branford Co., 28 Union St., Newton Centre, Mass. 02159

Brown. William C. Brown Co., Publishers, 135 S. Locust St., Dubuque, Iowa 52001

Burgess. Burgess Publishing Co., 426 S. 6th St., Minneapolis, Minn. 55415

Business Books. Business Books, Ltd., Mercury House, 103-119 Waterloo Rd., London, S.E.1, England

Butterworth. Butterworth & Co. (Publishers), Ltd., 88 Kingsway, London, W.C. 2, England

Calif. State Bd. of Forestry. California State Board of Forestry, Office of Procurement, Document Section, P.O. Box 1612, Sacramento, Calif. 95807

California. University of California Press, 2223 Fulton St., Berkeley, Calif. 94720

Cambridge. Cambridge University Press, 32 E. 57th St., New York, N.Y. 10022

Chandler. Chandler Publishing Co., 124 Spear St., San Francisco, Calif. 94105

Chapman. Chapman & Hall, Ltd. Distributed by Barnes & Noble

Chemical Pub. Chemical Publishing Co., Inc., 200 Park Ave. S, New York, N.Y. 10003

Chemical Rubber. Chemical Rubber Co., 18901 Cranwood Pkwy., Cleveland, Ohio 44128

Chicago. University of Chicago Press, 5801 Ellis Ave., Chicago, Ill. 60637

Classic Pubns. Classic Publications, Ltd., Recorder House, Church St., London, N. 16, England

College Pr. College Press, College Pl., Wash. 99324

College Science. College Science Publications, Box 798, State College, Pa. 16801

Collins. William Collins Sons & Co., Ltd., 144 Cathedral St., Glasgow, C. 4, Scotland

Colo. Flower Growers Assn. Colorado Flower Growers Association, Inc., 909 Sherman St., Denver, Colo. 80203

Columbia. Columbia University Press, 562 W. 113th St., New York, N.Y. 10025

Comstock. Comstock Publishing Associates, division of Cornell University Press. *See* Cornell

Constable. Constable & Co., Ltd., 10-12 Orange St., London, W.C. 2, England

Cornell. Cornell University Press, 124 Roberts Pl., Ithaca, N.Y. 14850

Coward. Coward-McCann & Geoghegan, Inc., 200 Madison Ave., New York, N.Y. 10016

Cranbrook. Cranbrook Institute of Science, 380 Lone Pine Rd., Bloomfield Hills, Mich. 48013

Crosby Lockwood. Crosby Lockwood & Son, Ltd., 26 Old Brompton Rd., London, S.W. 7, England

Crowell. Thomas Y. Crowell Co., 201 Park Ave. S, New York, N.Y. 10003

Crown. Crown Publishers, Inc., 419 Park Ave. S, New York, N.Y. 10016

Dadant. Dadant & Sons, Inc., Second & Broadway, Hamilton, Ill. 62341

Davis. F. A. Davis Co., 1915 Arch St., Philadelphia, Pa. 19103

Day. John Day Co., Inc., 257 Park Ave. S, New York, N.Y. 10010

Delmar. Delmar Publishers, Inc., Mountainview Ave. or P.O. Box 5087, Albany, N.Y. 12205

Denison. T. S. Denison & Co., Inc., 5100 W. 82nd St., Minneapolis, Minn. 55431

Devin-Adair. The Devin-Adair Co., 1 Park Ave., Old Greenwich, Conn. 06870

Dodd. Dodd, Mead & Co., 79 Madison Ave., New York, N.Y. 10016

Doubleday. Doubleday & Co., Inc., Garden City, N.Y. 11530

Dover. Dover Publications, Inc., 180 Varick St., New York, N.Y. 10014

Dufour. Dufour Editions, Inc., Chester Springs, Pa. 19425

Dutton. E. P. Dutton & Co., Inc., 201 Park Ave. S, New York, N.Y. 10003

Educator Books. Educator Books, Inc., Drawer 32, San Angelo, Tex. 76901

Edwards. Edwards Brothers, 2500 S. State St., Ann Arbor, Mich. 48104

Elsevier. *See* Amer. Elsevier

English Universities. English Universities Press, Ltd., St. Pauls House, Warwick La., London, E.C. 4, England

Estates Gazette. Estates Gazette, Ltd., 151 Wardour St., London, W1V 4BN, England

Exposition. Exposition Press, Inc., 50 Jericho Tnpk., Jericho, N.Y. 11753

Faber. Faber & Faber, Ltd., 24 Russell Sq., London, W.C. 1, England

Farm Qtly. The Farm Quarterly, 22 E. 12th St., Cincinnati, Ohio 45210

Farming Pr. Farming Press, Ltd., Lloyd Chambers, Lloyds Ave., Ipswich, Suffolk, England

Fernhill. Fernhill House Ltd., 303 Park Ave. S, New York, N.Y. 10010

Fishing News. Fishing News (Books), Ltd., Ludgate House, 110 Fleet St., London, E.C. 4, England

Fla. Forest Ser. Florida Forest Service, C. H. Coulter, Collins Bldg., Tallahassee, Fla. 32304

Florida. University of Florida Press, 15 N.W. 15th St., Gainesville, Fla. 32601

Publishers Directory

Food Trade Pr. Food Trade Press, 7 Garrick St., London, W.C. 2, England

Food Trade Review. *See* Food Trade Pr.

Freeman. W. H. Freeman & Co., Publishers, 660 Market St., San Francisco, Calif. 94104

Funk & Wagnalls. Funk & Wagnalls, Inc., 53 E. 77th St., New York, N.Y. 10021

Golden. Golden Press, Inc., 1220 Mound Ave., Racine, Wis. 53404

Goodheart-Willcox. Goodheart-Willcox Co., 123 W. Taft Dr., South Holland, Ill. 60473

Gordon & Breach. Gordon & Breach, Science Publishers, Inc., 150 Fifth Ave., New York, N.Y. 10011

Govt. Print. Off. Government Printing Office, Washington, D.C. 20402

Greene. Stephen Greene Press, Box 1000, Brattleboro, Vt. 05301

Greenwood. Greenwood Press, Inc., 51 Riverside Ave., Westport, Conn. 06880

Greystone. Greystone Corp., 225 Park Ave. S, New York, N.Y. 10003

Grossman. Grossman Publishers, 44 W. 56th St., New York, N.Y. 10019

Hafner. Hafner Publishing Co., Inc., 866 Third Ave., New York, N.Y. 10022

Harcourt. Harcourt Brace Jovanovich, Inc., 757 Third Ave., New York, N.Y. 10017

Harper. Harper & Row, Publishers, 49 E. 33rd St., New York, N.Y. 10016

Harrap. George G. Harrap & Co., Ltd., Publishers, 182 High Holborn, London, W.C. 1, England

Harvard. Harvard University Press, 79 Garden St., Cambridge, Mass. 02138

Hawthorn. Hawthorn Books, Inc., 70 Fifth Ave., New York, N.Y. 10010

Hearthside. Hearthside Press, 445 Northern Blvd., Great Neck, N.Y. 11021

Heath. D. C. Heath & Co., 125 Spring St., Lexington, Mass. 02173

Heinemann Med. William Heinemann Medical Books, Ltd., 23 Bedford Sq., London, W.C. 1, England

Hialeah. Hialeah Guild Publishers, Box 206, Miami, Fla. 33137

Hilger. Adam Hilger, Ltd., 98 St. Pancras Way, Camden Rd., London, N.W. 1, England

Hill. Leonard Hill Books, Morgan-Grampian Books, Ltd., 28 Essex St., London, W.C. 2, England

Hill & Wang. Hill & Wang, Inc., 72 Fifth Ave., New York, N.Y. 10011

Hillary. Hillary House Publishers, 303 Park Ave. S, New York, N.Y. 10010

Holt. Holt, Rinehart, & Winston, Inc., 383 Madison Ave., New York, N.Y. 10017

Houghton Mifflin. Houghton Mifflin Co., 2 Park St., Boston, Mass. 02107

Howell. Howell Book House, Inc., 845 Third Ave., New York, N.Y. 10022

Humanities. Humanities Press, Inc., 303 Park Ave. S, New York, N.Y. 10010

Hutchinson. Hutchinson Publishing Group, Ltd., 178-202 Great Portland St., London, W. 1, England

Iliffe. Iliffe Books, Ltd., 42 Russell Sq., London, W.C. 1, England

Illinois. University of Illinois Press, Urbana, Ill. 61801

Indiana. Indiana University Press, Tenth & Morton Sts., Bloomington, Ind. 47401

International Pubn. Ser. International Publication Service, 303 Park Ave. S, New York, N.Y. 10010

International Textbook. International Textbook Co., Scranton, Pa. 18515

Interscience. Interscience Publishers, Inc. *See* Wiley

Interstate. The Interstate Printers & Publishers, Inc., 19 N. Jackson St., Danville, Ill. 61832

Iowa State. Iowa State University Press, Press Bldg., Ames, Iowa 50010

Irwin. Richard D. Irwin, Inc., 1818 Ridge Rd., Homewood, Ill. 60430

Johns Hopkins. The Johns Hopkins Press, Baltimore, Md. 21218

Judge. Edward E. Judge & Sons, 79 Bond St., Westminster, Md. 21157

Kansas. University Press of Kansas, 66 Watson, Lawrence, Kans. 66044

Knopf. Alfred A. Knopf, Inc., 201 E. 50th St., New York, N.Y. 10022

Lane. Lane Magazine & Book Co., Willow Rd. at Middlefield Rd., Menlo Park, Calif. 94025

Lantern. Lantern Press, Inc., 345 Hussey Rd., Mt. Vernon, N.Y. 10010

Lea & Febiger. Lea & Febiger, 600 S. Washington Sq., Philadelphia, Pa. 19106

Lincoln Electric. Lincoln Electric Co., 22801 St. Claire Ave., Cleveland, Ohio 44117

Lippincott. J. B. Lippincott Co., E. Washington Sq., Philadelphia, Pa. 19105

Little, Brown. Little, Brown & Co., 34 Beacon St., Boston, Mass. 02106

Livingston. Livingston Publishing Co., 18 Hampstead Circle, Wynnewood, Pa. 19096

Livingstone. E. & S. Livingstone, Ltd., 15-17 Teviot Pl., Edinburgh 1, Scotland

Lockwood. Lockwood Publishing Co., Inc., 551 Fifth Ave., New York, N.Y. 10017

Longmans. Longmans, Green & Co., Ltd., 74 Grosvenor St., London, W1XOAS, England

Louisiana State. Louisiana State University Press, Baton Rouge, La. 70803

Lucas. Lucas Brothers Publishers, 909 Lowry St., Columbia, Mo. 65201

MacDonald. MacDonald & Co. (Publishers), Ltd., 49 Poland St., London, W. 1, England

Macfarland. C. Stedman Macfarland, Jr., Publisher, 834 South Ave. W, Westfield, N.J. 07090

McGraw-Hill. McGraw-Hill Book Co., 330 W. 42nd St., New York, N.Y. 10036

Publishers Directory

McKay. David McKay Co., Inc., 750 Third Ave., New York, N.Y. 10017

McKnight & McKnight. McKnight & McKnight Publishing Co., U.S. Rte. 66 at Towanda Ave., Bloomington, Ill. 61701

Maclaren. Maclaren & Sons, Ltd., 7 Grape St., London, W.C. 2, England

Macmillan. The Macmillan Co., 866 Third Ave., New York, N.Y. 10022

Mass. Hort. Soc. Massachusetts Horticultural Society, Horticultural Hall, 300 Massachusetts Ave., Boston, Mass. 02115

Massachusetts. University of Massachusetts Press, Munson Hall, Amherst, Mass. 01002

Meister. Meister Publishing Co., 37841 Euclid Ave., Willoughby, Ohio 44094

Mentor. Mentor Books. See New Amer. Lib.

Merck. Merck & Co., Inc., Rahway, N.J. 07065

Messner. Julian Messner, Inc., 1 W. 39th St., New York, N.Y. 10018

Methuen. Distributed by Barnes & Noble

Michigan. University of Michigan Press, Ann Arbor, Mich. 48106

Michigan State. Michigan State University Press, Box 550, East Lansing, Mich. 48823

Miller. The Miller Publishing Co., P.O. Box 1289, Minneapolis, Minn. 55440

Minnesota. University of Minnesota Press, 2037 University Ave. SE, Minneapolis, Minn. 55455

Miramar. Miramar Publishing Co., 1300 W. 24th St., Los Angeles, Calif. 90007

Mor-Mac. Mor-Mac Publishing Co., Inc., Box 984, Fairborn, Ohio 45324

Morrison. Morrison Publishing Co., 515 Woodlands Dr., Clinton, Iowa 52732

Morrow. William Morrow & Co., Inc., 105 Madison Ave., New York, N.Y. 10016

Mosby. The C. V. Mosby Co., 11830 Westline Industrial Dr., St. Louis, Mo. 63141

Murray. John Murray, 50 Albemarle St., London, W. 1, England

Nat. Assn. of State Foresters. National Association of State Foresters, Colorado State Forest Service, Ft. Collins, Colo. 80521

Nat. Audubon Society. National Audubon Society, Nature Centers Division, 1130 Fifth Ave., New York, N.Y. 10029

Nat. Canners Assn. National Canners Association, 1133 20th St. NW, Washington, D.C. 20036

Nat. Farm & Power Equip. Dealers Assn. National Farm & Power Equipment Dealers Association, 2340 Hampton Ave., St. Louis, Mo. 63139

Nat. Geographic. National Geographic Society, 17th & M Sts. NW, Washington, D.C. 20036

Nat. Hist. Pr. Natural History Press, 277 Park Ave., New York, N.Y. 10017

Nat. Pr. National Press Books, 850 Hansen Way, Palo Alto, Calif. 94304

Nebraska. University of Nebraska Press, Lincoln, Nebr. 68508

New Amer. Lib. The New American Library, Inc., 1301 Ave. of the Americas, New York, N.Y. 10019

New York. New York University Press, Washington Sq., New York, N.Y. 10003

North Carolina. University of North Carolina Press, Box 2288, Chapel Hill, N.C. 27514

Northern Nut Growers' Assn. Northern Nut Growers' Association, 4518 Holston Hills Rd., Knoxville, Tenn. 37914

Norton. W. W. Norton & Co., Inc., 55 Fifth Ave., New York, N.Y. 10003

Oceana. Oceana Publications, Inc., Dobbs Ferry, N.Y. 10522

Oklahoma. University of Oklahoma Press, 1005 Asp Ave., Norman, Okla. 73069

Olsen. Olsen Publishing Co., 1445 N. 5th St., Milwaukee, Wis. 53212

Oreg. Hist. Soc. Oregon Historical Society, 1230 S.W. Park Ave., Portland, Oreg. 97205

Oregon State. Oregon State University Press, 101 Waldo Hall, Corvallis, Oreg. 97331

Oxford. Oxford University Press, Inc., 200 Madison Ave., New York, N.Y. 10016

Oxford Book. Oxford Book Co., Inc., 387 Park Ave. S, New York, N.Y. 10016

PUDOC. PUDOC, Duivendaal 6a, Wageningen, Netherlands

Pantheon. Pantheon Books, Inc., 201 E. 50th St., New York, N.Y. 10022

Pegasus. Pegasus Publishing, 4300 W. 62nd St., Indianapolis, Ind. 46268

Pennsylvania. University of Pennsylvania Press, 3933 Walnut St., Philadelphia, Pa. 19104

Pergamon. Pergamon Press, Inc., Maxwell House, Fairview Park, Elmsford, N.Y. 10523

Philosophical Lib. Philosophical Library, Inc., 15 E. 40th St., New York, N.Y. 10016

Plenum. Plenum Publishing Corp., 227 W. 17th St., New York, N.Y. 10011

Praeger. Frederick A. Praeger, Inc., 111 Fourth Ave., New York, N.Y. 10003

Prakken. Prakken Publications, 416 Long Shore Dr., Ann Arbor, Mich. 48107

Prentice-Hall. Prentice-Hall, Inc., Englewood Cliffs, N.J. 07632

Princeton. Princeton University Press, Princeton, N.J. 08540

Putnam. G. P. Putnam's Sons, 200 Madison Ave., New York, N.Y. 10016

Ralston Purina. Ralston Purina Co., Checkerboard Sq., St. Louis, Mo. 63199

Rand McNally. Rand McNally & Co., 8255 Central Park Ave., Skokie, Ill. (address mail to Box 7600, Chicago, Ill. 60680)

Random. Random House, Inc., 201 E. 50th St., New York, N.Y. 10022

Reed. A. H. & A. W. Reed, 182 Wakefield St., Wellington, New Zealand

Reinhold. See Van Nostrand Reinhold Co.

Rodale. Rodale Books, Inc., 33 E. Minor St., Emmaus, Pa. 18049

Roehrs. Roehrs Co., Book Division, East Rutherford, N.J. 07073

Ronald. The Ronald Press Co., 79 Madison Ave., New York, N.Y. 10016

Publishers Directory

Root. A. I. Root Co., Medina, Ohio 44256
Rosen. Richards Rosen Press, Inc., 29 E. 21st St., New York, N.Y. 10010
Rutgers. Rutgers University Press, 30 College Ave., New Brunswick, N.J. 08903
Rutgers Hort. Pubns. Rutgers University Horticultural Publications, The State University, Nichol Ave., New Brunswick, N.J. 08903

St. Martin's. St. Martin's Press, Inc., 175 Fifth Ave., New York, N.Y. 10010
Sams. Howard W. Sams & Co., Inc., Publishers, 4300 W. 62nd St., Indianapolis, Ind. 46268
San Jacinto. San Jacinto Publishing Co., Box 66254, Houston, Tex. 77006
Saunders. W. B. Saunders Co., W. Washington Sq., Philadelphia, Pa. 19105
Scarecrow. Scarecrow Press, Inc., 52 Liberty St., Box 656, Metuchen, N.J. 08840
Schenkman. Schenkman Publishing Co., Inc., 3 Revere St., Harvard Sq., Cambridge, Mass. 02138
Scott. M. L. Scott & Associates, Publishers, P.O. Box 816, Ithaca, N.Y. 14850
Scott, Foresman. Scott, Foresman & Co., 1900 E. Lake Ave., Glenview, Ill. 60025
Scribner. Charles Scribner's Sons, 597 Fifth Ave., New York, N.Y. 10017
Sheridan. Sheridan House, Inc., 257 Park Ave. S, New York, N.Y. 10010
Shoe String. The Shoe String Press, 995 Sherman Ave., Hamden, Conn. 06514
Silver Burdett. Silver Burdett Co., 250 James St., Morristown, N.J. 07960
Simon & Schuster. Simon & Schuster, Inc., 630 Fifth Ave., New York, N.Y. 10020
Smith. Peter Smith, 6 Lexington Ave., Gloucester, Mass. 01930
Soc. of Amer. Foresters. Society of American Foresters, 1010 16th St. NW, Washington, D.C. 20036
Soil Sci. Soc. of Amer. Soil Science Society of America, 677 S. Segoe Rd., Madison, Wis. 53711
South-Western. South-Western Publishing Co., 5101 Madison Rd., Cincinnati, Ohio 45227
Speller. Robert Speller & Sons Publishers, Inc., 10 E. 23rd St., New York, N.Y. 10010
Springer. Springer Publishing Co., Inc., 200 Park Ave. S, New York, N.Y. 10003
Springer-Verlag. Springer-Verlag New York, Inc., 175 Fifth Ave., New York, N.Y. 10010
Stackpole. Stackpole Books, Cameron & Kelker Sts., Harrisburg, Pa. 17105
Stanford. Stanford University Press, Stanford, Calif. 94305
Stechert-Hafner. Stechert-Hafner Service Agency, 260 Heights Rd., Darien, Conn. 06820
Sterling. Sterling Publishing Co., Inc., 419 Park Ave. S, New York, N.Y. 10016
Stipes. Stipes Publishing Co., 10-12 Chester St., Champaign, Ill. 61820
Sydney. Sydney University Press, Press Bldg., University of Sydney, Sydney N.S.W., Australia

Syracuse. Syracuse University Press, Box 8, University Station, Syracuse, N.Y. 13210

Taplinger. Taplinger Publishing Co., Inc., 200 Park Ave. S, New York, N.Y. 10003
Thomas. Charles C. Thomas, Publisher, 301-27 E. Lawrence Ave., Springfield, Ill. 62703
Thomas Print. Thomas Printing & Publishing Co., 724 Desnoyer St., Kaukauna, Wis. 54130
Thomson. Thomson Publications, P.O. Box 5601, Fresno, Calif. 93704
Time. Time-Life Books, Time & Life Bldg., Rockefeller Center, New York, N.Y. 10020 (trade eds. distributed by Little Brown & Co.; library eds. distributed by Silver Burdett Co.)
Toronto. University of Toronto Press, St. George Campus, Toronto 181, Ontario, Canada
Transatlantic. Transatlantic Arts, Inc., North Village Green, Levittown, N.Y. 11756
Tuttle. Charles E. Tuttle Co., Inc., 28 S. Main St., Rutland, Vt. 05701
Twentieth Century. Twentieth Century Fund, Inc., 41 E. 70th St., New York, N.Y. 10021

U.S. Trotting Assn. United States Trotting Association, 750 Michigan Ave., Columbus, Ohio 43215
Univ. Pub. University Publishing, P.O. Box 856, College Station, Tex. 77840
Universe. Universe Books, 381 Park Ave. S, New York, N.Y. 10016

Van Nostrand. Van Nostrand Reinhold Co., 450 W. 33rd St., New York, N.Y. 10001
Viking. The Viking Press, Inc., 625 Madison Ave., New York, N.Y. 10022

Wadsworth. Wadsworth Publishing Co., Belmont, Calif. 94002
Walck. Henry Z. Walck, Inc., 19 Union Sq. W, New York, N.Y. 10003
Warne. Frederick Warne & Co., Inc., 101 Fifth Ave., New York, N.Y. 10003
Washington. University of Washington Press, Seattle, Wash. 98015
Weidenfeld. George Weidenfeld & Nicolson, Ltd., 5 Winsley St., London, W. 1, England
Welding Engineer Pub. Welding Engineer Publications, Inc., P.O. Box 128, Morton Grove, Ill. 60053
Wildlife Soc. The Wildlife Society, 2900 Wisconsin Ave., Suite S-176, Washington, D.C. 20016
Wiley. John Wiley & Sons, Inc., 605 Third Ave., New York, N.Y. 10016
Williams & Wilkins. The Williams & Wilkins Co., 428 E. Preston St., Baltimore, Md. 21202
Wilson. The H. W. Wilson Co., 950 University Ave., Bronx, N.Y. 10452
Witherby. H. F. & G. Witherby, Ltd., 61/62 Watling St., London, E.C. 4, England
World. The World Publishing Co., 110 E. 59th St., New York, N.Y. 10022
Worth. Worth Publishers, Inc., 70 Fifth Ave., New York, N.Y. 10011

AUTHOR INDEX

All references are to entry numbers.

Abraham, George, A409, A489
Abrams, John T., A833
Abrams, LeRoy, A236
Acker, Duane C., A714
Acosta, Allen J., A998
Adams, J. Edison, A205
Adams, John W., A776
Adams, Ora R., A852
Adriance, Guy W., A410
Afzelius, Björn, A122
Agrios, George N., A320
Ahlgren, Gilbert H., A360
Ahlgren, Henry L., A372
Ainsworth, John H., A582
Aitchison, Gordon J., A28
Aiton, Edward W., A1402
Akehurst, B. C., A361
Albaugh, Reuben, A812
Albers, Henry H., A62
Albrecht, Carl F., A1021
Alder, Henry L., A45
Aldrich, Samuel R., A362, A1105
Alexander, Martin, A1055
Alexander, R. McNeill, A657
Alexopoulos, Constantine, A185, A571
Allard, Robert W., A315
Allen, Durward L., A1312
Allen, Garland E., A29
Allen, Shirley W., A583, A1268
Allison, James B., A1253
Althouse, Paul M., A38
Altschul, Aaron M., A1132, A1161

American Association of Cereal Chemists, A1133
American Kennel Club, A715
American Meat Institute Foundation, A1162
American Poultry Association, A823
American Public Health Association, Inc., A1235
Amerine, Maynard A., A1134, A1163
Amos, Arthur J., A1179
Amos, William H., A170
Andersen, Aage J. C., A1164
Anderson, Arthur L., A716
Anderson, David A., A650
Anderson, Dean A., A146
Anderson, Donald B., A300
Anderson, Edgar, A186
Anderson, Harry W., A321
Anderson, Roger F., A886
Anderson, Sydney, A658
Andrewartha, Herbert G., A659
Andrewes, Christopher H., A139, A853, A887
Andrews, F. S., A432
Andriot, John L., E1, E2
Anglemier, A. F., A1156
Arbuckle, Wendell S., A1165
Archer, Sellers G., A364, A1056
Armstrong, George R., A607
Arsdel, Wallace B. Van, *see* Van Arsdel, Wallace B.

Asdell, Sydney A., A660, A717, A718
Ashby, Wallace, A1008
Asimov, Isaac, A123, A274
Atherton, Henry V., A1242
Atkins, Fred C., A412
Audus, Leslie J., A275
Auerback, Charlotte, A124
Aul, Henry B., A490
Aurand, Leonard W., A1159
Austin, Cedric, A1166
Austin, Charles F., A63
Austin, Colin R., A105
Avery, Thomas E., A584, A960
Ayers, John C., A1221, A1236

Baden-Fuller, A. J., *see* Fuller, A. J. Baden
Bailey, Ethel Z., A414, A415
Bailey, Jackson W., A824, A854
Bailey, Liberty H., A237, A238, A413, A414, A415
Bainer, Roy, A1009
Baker, Frederick S., A585
Baker, Herbert G., A187
Baker, Jeffrey J., A29
Bakuzis, Egolfs V., A586
Ball, Vic, A491
Barbour, Michael G., A233
Bardach, John, A941
Barger, Edgar L., A1010
Barkley, William D., A259
Barnard, Colin, A365

Author Index

Barner, Ralph D., A875
Barnes, Arthur C., A366
Barnes, Charles D., A855
Barnes, Ervin H., A322
Barnett, Horace L., A572
Barney, Charles W., A606
Barrett, John W., A587
Barron, Arthur L., A140
Bartrum, Douglas, A492
Barzun, Jacques, A1
Bates, Marston, A1269
Bates, Robert L., A2
Battan, Louis J., A1112, A1113
Baumback, Clifford M., A72
Baumgardt, John P., A416
Baver, Leonard D., A1057
Bawden, Frederick C., A323
Baxter, Dow V., A324
Beal, George M., A3
Bean, Louis H., A64
Bear, Firman E., A1058, A1059
Beasley, Ann R., A157
Beaton, George H., A1249
Beaver, William C., A106
Becker, Manning H., A1345
Becker, P., A934
Beeson, William M., A719, A784
Behme, Robert L., A493
Bell, Frederick W., A942
Bell, Leonie, A565
Bell, Peter R., A276
Bellairs, Angus d'A., A661, A662
Bender, Arnold E., A1125, A1250
Benedict, Stewart, A21
Beneke, Raymond R., A1338
Benice, Daniel D., A46
Benjamin, Earl W., A1339
Bennett, George W., A943
Bennett, Hugh H., A1060
Bennett, Russell H., A720
Benton, Allen H., A171
Benz, Morris, A494, A495
Berg, Harold W., A1163
Bergen, Werner Von, *see* Von Bergen, Werner
Berger, Kermit C., A1061
Bernal, John D., A107
Berry, James B., A239
Bertrand, Alvin L., A1399
Bertrand, Anson R., A1082
Bessey, Ernest A., A240
Beuscher, Jacob H., A85
Biddle, George H., A825
Biester, Harry E., A856
Billiet, Walter E., A972, A1004, A1005
Binkley, Harold, A86
Binsted, Raymond, A1167, A1173, A1226
Bishop, Charles E., A1340
Bishop, Frank E., A1042
Black, Arthur, A496
Black, Charles A., A1062, A1063
Black, John D., A1270
Black, Michael, A277

Blair, W. Frank, A663
Blake, Claire L., A497
Blaxter, Kenneth, A834
Block, Seymour S., A152
Blood, Douglas C., A857
Blount, William P., A721
Boalch, Donald H., A87
Boddie, George F., A858
Bodemer, Charles W., A664
Boeckh, Everard H., A65
Bogart, Ralph, A722
Bohlen, Joe M., A3
Böhning, Richard H., A300
Bold, Harold C., A185, A188, A278
Bolton, Edward R., A1135
Bonner, David M., A125
Bonner, James F., A279
Boom, Boudewijn K., A241
Borek, Ernest, A126
Borgstrom, Georg, A1126, A1258, A1271
Borror, Donald J., A888, A889
Botsford, Harold E., A832
Boughey, Arthur S., A172
Bowen, Edwin G., A723
Boyce, John G., A60
Boyce, John S., A325
Boyd, Anne M., E3
Bradford, Lawrence A., A1341
Brady, Irene, A724
Brady, Nyle C., A1065
Brake, John R., A1342
Brandly, Paul J., A1168
Brandsberg, George, A88
Brater, Ernest F., A1053
Breder, Charles M., A944
Breed, Robert S., A159
Breimyer, Harold F., A1343
Brewbaker, James L., A127
Brewer, Roy, A1064
Brian, Michael V., A890
Brickbauer, Elwood A., A367
Bridger, G. L., A1085
Briggs, Fred N., A316
Briggs, George E., A280
Briggs, Hilton M., A725
Brightman, Frank H., A242
Brilmayer, Bernice, A498
Brinker, Russell C., A1011
Brison, Fred R., A410
Britt, Kenneth W., A588
Britton, Nathaniel L., A243, A248
Brock, Thomas D., A141
Brockman, Christian Frank, A244, A1330
Bromley, Willard S., A589
Brook, Alan J., A189
Brown, Addison, A248
Brown, Arlen D., A1012
Brown, Harry P., A627
Brown, James Bush, *see* Bush-Brown, James
Brown, Leland, A4
Brown, Lester R., A1259

Brown, Louise Bush, *see* Bush-Brown, Louise
Brown, Nelson C., A590
Brown, Robert H., A1013
Browning, Bertie L., A591, A592
Brune, Elmer J., A974
Bruner, Dorsey W., A866
Brush, Warren D., A246
Bruun, Bertel, A698
Bryant, W. R., A1357
Buban, Peter, A961
Buchanan, Robert E., A142
Buchsbaum, Ralph, A665
Buckett, M., A726
Buckman, Harry O., A1065
Bull, Sleeter, A1169
Bunce, Arthur C., A1066
Bunch, Clarence E., A364
Bundy, Clarence E., A727, A728, A739, A740, A826
Bunting, Brian T., A1067
Burgess, Abraham H., A368
Burgess, G. H. O., A945
Burkhardt, Dietrich, A666
Burnett, John H., A573
Burns, Marca, A729
Bursell, E., A891
Burt, David R., A164
Burton, Ian, A1272
Burton, Maurice, A685
Bush-Brown, James, A417
Bush-Brown, Louise, A417
Butcher, Devereux, A246
Buvat, Roger, A281
Byerly, Theodore C., A730

Cain, Robert F., A1157
Calhoon, Richard P., A66
Callison, Charles H., A1273
Campaigne, Ernest, A31
Campbell, John R., A731
Canham, Allan E., A499
Cantor, Leonard M., A1014
Cantzlaar, George L., A1114
Card, Leslie E., A827
Carefoot, Garnet L., A326
Carew, John, A488
Carleton, R. Milton, A418, A500
Carlson, A. C., A821
Carothers, Zane B., A201
Carpenter, Philip L., A143
Carr, Donald E., A1274
Carrington, Richard, A662
Carroll, William E., A772
Carson, Rachel, A945a, A1275
Carter, Robert C., A962
Carter, Walter, A327
Case, Harold C. M., A1344
Casey, Harold, A838
Casey, James P., A593
Cassard, Daniel W., A835
Castle, Emery H., A1345
Champlin, James R., A1337
Chandler, Robert F., A620

194

Author Index

Chandler, William H., A419, A420
Chang, Jen-Hu, A1115
Chapman, R. F., A892
Chapman, Valentine J., A190
Charm, Stanley E., A1204
Chastain, Elijah Denton, A1346
Chatten, Leonard R., A594
Chelminski, Stephen M., A271
Cherry, Elaine C., A502
Chidamian, Claude, A503
Child, Reginald, A421
Childers, Norman F., A422, A423, A431
Childs, Ernest C., A1068
Chittenden, Frederick J., A428
Christensen, Clyde M., A574, A1237
Christopher, Everett P., A424, A425
Chupp, Charles, A328
Cicero, Donald R., A230
Clagett, Carl O., A38
Clar, C. Raymond, A594
Clark, Colin, A1015, A1347
Clark, D. S., A1233
Clark, Eugene, A81
Clark, Fred E., A81
Clark, Robert B., A245
Clarke, Eustace G. C., A863
Clarke, George L., A173
Clarke, Myra L., A863
Clawson, Marion, A1077, A1331, A1348, A1349, A1400
Clepper, Henry, A595, A1276, A1277
Cloudsley-Thompson, J. L., A667
Clowes, Frederick A. L., A282
Cochran, William G., A61
Cochrane, Willard W., A1260, A1396
Coffman, Franklin A., A369
Coggin, J. K., A1029
Cole, E. J., A609
Cole, Harold H., A732, A733
Coles, Embert H., A859
Collier, John W., A596
Collings, Gilbeart H., A1069
Collingwood, George H., A246
Compton, Henry K., A1350
Comstock, Anna, A893
Comstock, John H., A893
Conn, Eric E., A32
Connell, Joseph H., A686
Conover, Herbert S., A504
Cook, Arthur H., A370
Cook, Glen C., A734, A1016
Cook, Ray L., A1070
Cook, Stanton A., A128
Cooke, George W., A1071
Cooley, Richard A., A1278
Coombs, Charles I., A597
Cooper, Malcolm M., A735
Copley, Michael J., A1217, A1218, A1244

Copson, David A., A1205
Corner, Eldred J. H., A191
Corty, Floyd L., A1399
Couch, Houston B., A328a
Coulter, Merle C., A192
Coursey, Donald G., A371
Courtenay, Philip P., A89
Cox, Henry E., A1137
Crafts, Alden S., A351, A352
Crampton, Earle W., A836
Crampton, John A., A90
Crandall, Lee S., A935, A936
Crissy, William J., A67
Critchfield, Howard J., A1116
Crofton, Harry D., A165
Cronquist, Arthur, A202
Cross, Jennifer, A1245
Crouse, William H., A963
Cruess, William V., A1163, A1170
Cumming, Roderick W., A505
Cunha, Tony J., A736, A737, A837
Cupps, P. T., A733
Curry, Othel J., A47
Curtis, Helena, A108, A144, A225
Cutter, Elizabeth G., A283
Cyphers, Emma H., A506

Dack, Gail M., A1222
Dadant, Camille P., A914
Dadant, James C., A914
Dadant, Maurice G., A914
Dallimore, William, A247
Dalzell, James R., A1017
Darling, Frank F., A1279
Darlington, Arnold, A329
Darlington, Cyril D., A129
Darnell, James E., A153
Darrah, Lawrence B., A1246
Darrow, George McM., A426
Dasmann, Raymond F., A1280, A1281, A1313
Daubenmire, Rexford F., A193, A194
Davidson, John, A458
Davidson, Ralph H., A930
Davies, J. Clarence, A1282
Davis, Herbert S., A945b
Davis, John G., A1171
Davis, Kenneth P., A598, A599
Davis, Richard F., A738
Day, E. A., A1158
Day, John A., A1117
DeBach, Paul, A921
Deibel, R. H., A1225
DeLaubenfels, David J., A109
DeLong, Dwight M., A888
Delorit, Richard J., A372
Demark, Noland L. Van, *see* Van Demark, Noland L.
Denisen, Ervin L., A427
Dent, John B., A838
Der Pyl, Leendert Van, *see* Pyl, Leendert Van der

Dersal, William R. Van, *see* Van Dersal, William R.
Desrosier, Norman W., A1018, A1206, A1261
Devey, James D., A1167
Devlin, Robert M., A284
DeZeeuw, Carl, A627
Dezettell, Louis, A964
Diamond, Solomon, A48
Dickson, James G., A330
Diggins, Ronald V., A727, A728, A739, A740, A826
Dillon, John L., A1351
Dittmer, Howard J., A192
Dixon, Malcolm, A1138
Dodds, Thomas C., A147
Dodge, Bernard O., A931
Dodge, J. Robert, A1008
Dodson, Edward O., A110
Doell, Charles E., A1332
Donahoo, Alvin, A1388
Donahue, Roy L., A1072, A1073
Donahue, Warren, A49
Donald, Hugh P., A775
Dorries, W. L., A1352
Douglas, William O., A1314
Douglass, Robert W., A1333
Doutt, Richard L., A926
Dowdell, Dorothy, A429, A600
Dowdell, Joseph, A429, A600
Doyle, William T., A195
Draper, Alec, A965
Duckworth, Ronald B., A430
Duerr, William A., A601
Dukes, Henry H., A741
Dunne, Howard W., A860
Dwyer, James L., A966
Dyke, Stanley F., A1139
Dykstra, Ralph, A861
Dyne, George M. Van, *see* Van Dyne, George M.

Eames, Arthur, A285
Earle, R. L., A1207
Eastin, R. B., E8
Eastman, Edward R., A6
Eck, Paul, A431
Eckbo, Garrett, A507, A508
Eckert, John E., A915, A919
Eddowes, Maurice, A373
Edelman, Jack, A277
Eden, Alfred, A840
Edgar, C. David, A814
Edgar, Carroll, A967
Edgerton, Claude W., A331
Edlin, Herbert L., A196
Edminster, Talcott W., A1026
Edmond, Joseph B., A432
Edmondson, Walles T., A111
Edwards, Clive A., A894
Efferson, John N., A1353
Ehlers, Victor M., A1019
Ehrenfeld, David W., A1315
Ehrlich, Anne H., A1283
Ehrlich, Paul R., A1283

Author Index

Eisenberg, David S., A1284
Ellison, Arthur J., A968
Eltherington, Lorne G., A855
Elton, Charles S., A668
Emery, David A., A68
Ensminger, M. Eugene, A742–A747
Esau, Katherine, A286
Eshelman, Philip V., A1020
Esmay, Merle L., A669
Evans, Elfed, A332
Evans, Howard E., A895
Evans, Ralph M., A7
Everett, Thomas H., A456
Ewer, R. F., A670
Eyre, S. R., A197

Faegri, Knut, A198
Fahn, Abraham, A287
Fairley, James L., A33
Farrall, Arthur W., A1021, A1208
Farris, Edmond J., A748
Feininger, Andreas, A602
Feirer, John L., A969
Feldmaier, Carl, A509
Felt, Ephraim P., A333
Ferguson, Egbert R., A1238
Fernald, Merritt L., A199
Ferrar, William L., A50
Ferris, Roxana J. S., A236
Findlay, Walter P. K., A603
Finney, David J., A51
Fisher, James, A671, A1316
Fisher, Katherine H., A1257
Flint, Wesley P., A903
Fogg, Gordon E., A288
Fogg, John M., A353
Foote, H. Elliot, A1192
Foote, Richard J., A1393
Forbes, Reginald D., A604
Foreman, Charles F., A791
Foss, Edward W., A1022
Foster, Albert B., A1074
Foster, Edwin M., A1223
Foster, Frank G., A510
Foster, H. Lincoln, A511
Foth, H. D., A1089
Fourt, David L., A784
Foust, Harry L., A749
Fowler, Stewart H., A750, A1354
Fox, M. W., A862
Frandson, R. D., A751
Fraser, Allan, A752
Fraser, Margaret N., A729
Frazee, Irving A., A970, A971, A972
Frazier, John B., A512
Frazier, William C., A1224, A1225
Frear, Donald E. H., A922
Free, John B., A200
Free, Montague, A434, A513
Freeman, Orville L., A605
French, Thomas E., A973
Freund, John E., A52

Frisch, Karl von, A916
Fritsch, Rudolf, A885
Frobisher, Martin, A145
Frome, Michael, A605
Fukuda, Kazuhiko, A514
Fuller, A. J. Baden, A34
Fuller, Harry J., A201
Fuller, Melvin S., A234
Fuqua, Mary E., A1257
Fussell, George E., A1023
Futrell, Gene A., A1386

Gabrielson, Ira N., A1317
Galston, Arthur W., A279, A289
Gardiner, George F., A515
Gardner, Eldon J., A130
Gardner, Victor R., A436
Garner, Frank H., A840
Garner, Reuben J., A863
Garner, Robert J., A437
Garratt, George A., A614
Garrett, Stephen D., A1075
Gay, William I., A864
Gebhardt, Louis P., A146
Genders, Roy, A516
Getty, Robert, A749, A865
Giachino, Joseph W., A974, A975
Gilchrist, Francis G., A672
Giles, G. W., A1029
Gilkey, Helen M., A354
Gillespie, James H., A866
Gillies, Robert R., A147
Gilman, William, A8
Gilmour, Darcy, A896
Gleason, Henry A., A202, A248
Glicksman, Martin, A1172
Goin, Coleman J., A673
Goin, Olive B., A673
Goings, Leslie F., A972
Goldblith, Samuel A., A1209, A1251
Goldstein, Milton, A989
Goldstein, Philip, A131
Good, Ronald, A203
Goodall, Daphne M., A753
Goodnight, Clarence J., A674
Goodnight, Marie L., A674
Goodwin, Raymond W., A1140
Goor, Amihud Y., A606
Goose, Peter G., A1173
Gordon, Malcolm S., A675
Gordon, Morris A., A576
Gough, W. James, A880
Gould, Frank W., A204
Gould, Robert F., A923
Govindjee, A302
Graf, Alfred B., A517, A518
Graff, Henry F., A1
Graff, M. M., A519
Graham, Frank, A1285
Graham, Kennard C., A976, A977
Graham, Kenneth, A897
Graham, Samuel A., A898
Graham-Rack, Barry, A1226
Grant, Leonard J., A946

Grantz, Gerald J., A9
Graves, Paul F., A635
Gray, Asa, A199
Gray, Harold, A1024
Gray, James R., A1355
Gray, Peter, A112, A674
Gray, William D., A575
Greeley, R. Gordon, A754
Green, Melvin M., A133
Greenfield, Ian, A438
Greensill, T. M., A439
Greenwood, Douglas, A978
Greenwood, Peter H., A957
Greulach, Victor A., A205
Griffin, Donald R., A676
Griffin, Roger C., A1216
Griffiths, John F., A178
Grimm, William C., A249, A250, A251
Grist, Donald H., A374
Grout, Roy A., A917
Grove, Alvin R., A209
Gulvin, Harold E., A1051
Gunderson, Frank L., A1238
Gunderson, Helen W., A1238
Gunther, Francis A., A924
Gurham, Fred C., A1025
Guthrie, John A., A607

Haber, Ernest S., A453
Hackett, Donald F., A608
Hadwiger, Don F., A1391
Hafez, E. S. E., A755, A756, A757
Hagan, Robert M., A1026
Hagan, William A., A866
Hagedoorn, Arend L., A758
Hagenstein, William D., A649
Hahn, Emily, A937
Hailman, Jack P., A684
Haines, Peter G., A16
Hainsworth, Marguerite D., A677
Haise, Howard R., A1026
Hale, Murray, A828
Hall, Carl W., A1044, A1174, A1175
Hall, Cecil E., A148
Hall, Isaac F., A1356, A1372
Hallowell, Elliot R., A1220
Halnan, Edward T., A840
Hamilton, J. Roland, A1352
Hamilton, James E., A1357
Hamilton, James R., A1027
Hamilton, William J., A678
Hammond, John, A759
Hammonds, Carsie, A86
Hancock, Robert S., A70
Handley, William M., A979
Hannah, Harold W., A91, A867
Hansen, Henry L., A586
Hansen, Vaughn E., A1031
Hanson, Angus A., A440
Harden Jones, F. R., *see* Jones, F. R. Harden
Hardin, Clifford M., A1262

Author Index

Hardman, H., A609
Hardy, Glenn W., A1076
Harlow, William M., A252, A253, A610, A611
Harrar, Ellwood S., A254, A611
Harrar, Jacob G., A254, A343
Harrigan, W. F., A149
Harris, Anthony G., A1028
Harris, Elwin E., A641
Harris, Lorin E., A836
Harrison, James C., A760
Harrison, Sydney G., A206, A247
Hart, George H., A812
Hartmann, Hudson T., A441
Harwood, Robert F., A899
Hawkes, Alex D., A520
Hay, Robert D., A10
Hay, Roy, A521
Hayes, Herbert K., A317
Hayes, Matthew H., A761, A762, A868
Hayward, Herman E., A290
Hazen, Elizabeth L., A576
Hazleton, Jared E., A942
Heady, Earl O., A1358, A1359, A1363
Heaps, Willard A., A1401
Heath, Gordon W., A894
Heath, J. S., A869
Heath, Oscar V., A291
Hebard, Edna L., A69
Hedges, Trimble R., A1360
Hedrick, Theodore I., A1174
Heid, John L., A1210
Heiser, Charles B., A207
Held, R. Burnell, A1077, A1348
Helfman, Elizabeth S., A1286
Hemp, Paul E., A522, A523
Henderson, Alexander E., A763
Henderson, Harry O., A793
Henderson, Isabelle F., A113
Henderson, James A., A857
Henderson, William D., A113
Henle, Hans, A633
Henry, S. Mark, A174
Henson, Edwin R., A376
Herbert, Frederick W., A1287
Herbert, W. J., A870
Herbst, John, A1361
Herms, William B., A899
Herrick, John, A871
Herschdoerfer, S. M., A1239
Hersom, A. C., A1227
Herzberg, R. J., A1198
Hesling, John J., A925
Hewitt, Oliver H., A1318
Hickling, Charles F., A947
Hide, J. C., A1080
Higbee, Edward, A92, A1362
Hill, Albert F., A208
Hill, Donald D., A403
Hill, John B., A209
Hill, Johnson D., A93
Hillis, W. E., A612
Hilterbrand, Luther R., A613

Hiscox, Gardner D., A11
Hoffman, Glenn L., A948
Hoffman, Katherine B., A35
Holloway, Robert J., A70
Holt, John G., A142
Holt, Oris M., A401
Honma, S., A485
Hoover, Norman K., A94, A524
Hopkins, John A., A1363
Hora, Frederick B., A578
Houck, James P., A1381
House, Homer D., A255
Howell, Ezra L., A1029
Hoyt, Murray, A918
Hughes, Harold D., A375, A376
Hull, George F., A525
Hull, T. E., A53
Hulland, E. D., A1227
Hulme, A. C., A442
Humason, Gretchen L., A679
Hume, H. Harold, A443, A444
Hummel, Charles, A1176
Humphreys, Clifford R., A1306
Hungate, Robert E., A764
Hunsley, Roger E., A719
Hunt, Donnell, A1030
Hunt, George M., A614
Hunt, Robert L., A1364
Hunter, W. D. Russell, see Russell-Hunter, W. D.
Huntsberger, David V., A54
Husch, Bertram, A615
Hussey, Norman W., A925
Hutchins, Ross E., A210, A900
Hutchinson, John, A256
Hutchinson, Sir Joseph B., A1263
Hutt, Frederick B., A680
Hyams, Edward S., A445
Hylander, Clarence J., A211, A257

Idyll, Clarence P., A949
Immer, Forrest R., A317
Ingles, Lloyd G., A681
Inglett, George E., A377
Isely, Duane, A355
Ishimoto, Kiyoko, A526
Ishimoto, Tatsuo, A526
Israelsen, Orson W., A1031
Iverson, Edwin S., A950
Ives, J. Russell, A1365

Jackson, Alan A., A445
Jackson, Albert B., A247
Jackson, Ellen, E4
Jackson, Herbert W. A980
Jackson, Nora, A1288
Jacobs, Morris B., A1141, A1177
Jacobson, Paul, A1044
Jaeger, Edmund C., A682
James, Maurice T., A899
James, Sydney C., A1366
Jamieson, B. G. M., A212
Jane, Frank W., A616
Janick, Jules, A378, A379, A446

Jay, James M., A1228
Jaynes, Richard A., A447
Jefferson, Ted B., A981
Jenkins, Ross W., A723
Jenkins, William R., A334
Jenness, Robert, A1142
Jennings, Joseph B., A841
Jensen, Clayne R., A1334
Jensen, Harald R., A1359
Jensen, Rue, A872
Jensen, Stanley M., A617
Jensen, William A., A292
Jeppson, L. R., A924
Jepsen, Stanley M., A617
Jevons, Frederick R., A36
Joffe, Joyce, A1289
Johansson, Ivar, A765
Johns, Glenn F., A448
Johnson, Alex A., A829
Johnson, Arnold H., A1160
Johnson, George S., A975
Johnson, Glenn L., A1341
Johnson, Herbert W., A71
Johnson, William H., A380
Jones, Arthur, A1404
Jones, Arthur W., A166
Jones, Bruce V., A873
Jones, Douglas W. Kent, see Kent-Jones, Douglas W.
Jones, E. W., A646
Jones, F. R. Harden, A951
Jones, Fred R., A1032
Jones, Henry A., A381
Jones, J. Knox, A658
Jones, John R. E., A952
Jones, Mack M., A1033
Jones, Ruth M., A150
Joslyn, Maynard A., A1143, A1199, A1209, A1210, A1251
Judkins, Henry F., A766
Juergenson, Elwood M., A460, A524, A734, A767, A768, A825, A835
Julin, Richard J., A512
Juniper, Barrie E. A282, A293
Juska, Felix V., A440

Kaplan, Robert M., A67
Karplus, Robert, A37
Karr, Harrison M., A12
Kates, Edgar J., A982
Kates, Robert W., A1272
Kauffeld, Carl, A938
Kaufmann, Henry H., A1237
Kauzmann, Walter, A1284
Kavaler, Lucy, A577
Kay, Desmond H., A151
Kays, Donald J., A769
Kays, John M., A769, A770
Kazmann, Raphael G., A1034
Keast, James D., A788
Keener, Harry A., A766
Keeton, William T., A114
Kelley, Pearce C., A72
Kellogg, Charles E., A1078
Kelly, William C., A474

Author Index

Kendeigh, Samuel C., A683
Kenfield, Warren G., A527
Kennedy, Donald, A120
Kennedy, Robert W., A645
Kenneth, John H., A113
Kent, Norman L., A1178
Kent-Jones, Douglas W., A1179
Kerby, Joe K., A73
Kertesz, Zoltan I., A1144
Kester, Dale E., A441
Ketchum, Richard M., A618
Kevan, D. Keith McE., A1079
Kilgore, Wendell W., A926
Kilgour, Gordon L., A33
Kilvert, B. Cory, A528
King, George H., A382
King, Lawrence J., A356
Kingsbury, John M., A213, A214
Kiplinger, Donald C., A530
Kipps, Michael S., A383
Kirchshofer, Rose, A939
Kirk, Franklyn W., A983
Kirkpatrick, Elwood G., A984
Kiser, James J., A716
Kissam, Philip, A1035
Klaf, A. Albert, A55
Kleijn, H., A241
Klingman, Glenn C., A357, A384
Klopfer, Peter H., A684
Klots, Elsie B., A115
Knetsch, Jack L., A1331
Knight, Fred B., A898
Knight, James W., A1180
Knight, Maxwell, A689
Knott, James E., A449
Knowles, Paulden F., A316
Knudsen, Jens W., A116
Knuti, Leo L., A1080
Koch, Peter, A619
Koehler, William R., A771
Koger, Marvin, A736, A737
Kohls, Richard L., A1367
Kohnke, Helmut, A1082
Kolaga, Walter A., A529
Korpi, Milton, A1080
Kosikowski, Frank V., A1181
Kozlowski, Theodore T., A294, A295, A297
Kraft, Ken, A450
Kraft, Pat, A450
Kramer, Amihud, A1240
Kramer, Paul J., A296, A297
Krausz, Norman G. P., A91
Krebs, Alfred H., A95
Kreitlow, B. W., A1402
Krider, Jake L., A772
Küchler, August W., A215

LaCour, L. F., A129
Laidlaw, Harry H., A919
Lamer, Mirko, A1083
Lamp, Benson J., A380
Lampert, Lincoln M., A1182
Lancaster, John F., A985
Lane-Petter, William, A792, A818

Lange, Morten, A578
Lanham, Urless N., A901, A953
Lapage, Geoffrey, A874
Large, Ernest C., A334a
Lasley, John F., A731, A773
Lasser, Jacob K., A74
Lastrucci, Carlo L., A13
Laurie, Alexander, A530, A531
Lauwerys, Joseph A., A1290
Laverton, Sylvia, A1036
Lawrence, Carl A., A152
Lawrence, George H. M., A258
Lawrence, Nelda R., A14
Lawrie, Ralston A., A1183, A1252
Lawson, Alexander H., A532
Lawyer, Kenneth, A72
Laycock, George, A1319
Lee, Donald G., A813
Lee, Donald L., A167
Lee, Frederick P., A533
Leidy, W. Philip, A15, E5
Leng, Earl R., A362
Leonard, Albert H., A774
Leonard, Justin W., A1268
Leonard, Warren H., A385, A387
Leopold, Aldo C., A298, A1320
Lerner, I. Michael, A775
Lessel, Erwin F., A142
Levie, Albert, A1184
Levitt, Jacob, A299
Liener, Irvin E., A1229
Lincoln Electric Company, A986
Little, Angela C., A1146
Little, Van A., A902
Lock, Arthur, A1212
Loomis, Walter E., A235
Loosli, John K., A842
Lowe, Belle, A1145
Lowry, William P., A1118
Lucas, George B., A335
Lungwitz, Anton, A776
Luria, Salvador E., A153
Luthin, James N., A1037
Lutz, Harold J., A620
Luxton, George E., A534
Lytel, Allan, A56, A987

Macan, Thomas T., A175
MacArthur, Robert H., A686
McCance, Margaret E., A149
McCart, Gerald D., A1104
McClellan, Grant S., A1291
McClung, Robert M., A1321
McCollum, John P., A480
McColly, Howard F., A1038
McCoy, C. E., A905
McCoy, Joseph J., A777
McCulloch, Walter F., A621
McCully, Margaret E., A301
McDaniels, Laurence H., A285
McDonald, Elvin, A451, A535
McFarland, John H., A536
McGee, Roger V., A57
MacGillivray, John H., A452, A1368

McGraw, E. L., A1346
McHarg, Ian R., A1292
McHenry, Earle W., A1249
McIntosh, Douglas H., A1119
Mack, Cornelius, A58
Mack, Jerry, A953a
McKee, Alexander, A954
McKenny, Margaret, A260
Mackenzie, David, A778
MacKey, Donald R., A872
Mackinney, Gordon, A1146
McMichael, Stanley L., A75
McNall, Preston E., A1293
McNickle, L. S., A988
McVickar, John S., A386
McVickar, Malcolm H., A386, A1084, A1085
Madow, Pauline, A1335
Mahlstede, John P., A453
Mallette, M. Frank, A38
Mann, Louis K., A381
Manning, Aubrey, A687
Manwill, Marion C., A779
Marchant, Ronald A., A688
Margolis, Louis, A60
Markham, Jesse W., A1086
Marsh, R. Warren, A1213
Marsh, Warner L., A537
Marshall, Charles E., A1087
Marshall, Norman B., A955
Marth, E. H., A1225
Martin, Alexander C., A259, A1322
Martin, Hubert, A927
Martin, James W., A1038
Martin, John H., A385, A387
Masefield, G. B., A206
Mason, Ralph E., A16, A76
Mastalerz, John W., A538
Mather, Kirtley F., A17
Mathre, Don E., A344
Matthews, Leonard H., A689
Matthews, Richard E. F., A336
Matthiessen, Peter, A1323
Matz, Samuel A., A1147, A1148, A1185, A1186, A1187, A1214
May, J. W., A711
Mayber, A. Poljakoff, see Poljakoff-Mayber, A.
Mayer, A. M., A216
Maynard, Leonard A., A842
Mayr, Ernst, A690
Meade, George P., A1193
Mechanic, Sylvia, E6
Meeuse, Bastiaan J. D., A217
Meglitsch, Paul A., A691
Meisel, Gerald S., A69
Mellanby, Kenneth, A1294
Mellor, John W., A1369
Mendenhall, William, A59
Menninger, Edwin A., A218, A539
Merchant, Ival, A875
Merewin, A. W., A272
Merory, Joseph, A1149

Author Index

Metcalf, Clell L., A903
Metcalf, Robert L., A1300
Meyer, Arthur B., A595, A604
Meyer, Bernard S., A300
Meyer, Hans A., A622
Meyer, William E., A454
Michaelis, Arnd, A133
Michell, Arthur S., A649
Middleton, Robert G., A989
Migaki, George, A1168
Mighell, Ronald L., A97
Millar, Charles E., A1088, A1089
Miller, Howard L., A98
Miller, M. W., A579
Miller, Malcolm E., A780
Mills, Gordon H., A18
Milne, Lorus J., A219, A692
Milne, Margery, A219, A692
Milton, John P., A1279
Minifie, Bernard W., A1188
Mirov, Nicholas T., A623
Mitchell, Roger L., A388
Mittleider, Jacob R., A1264
Mohsenin, Nuri N., A1150
Moore, Arthur D., A19
Moore, John R., A1370
Moore, Stanley B., A540
Morrison, Frank, A843
Morrison, Ivan G., A1012
Morrison, John H., A137
Morse, Harriet K., A541
Mortenson, William P., A367, A1356, A1371, A1372
Mosesson, Gloria R., A781
Mosher, Arthur T., A99
Mountney, George J., A1189
Mrak, E. M., A579
Muckle, T. B., A1028
Muenscher, Walter C., A358
Muller, Walter H., A220
Munro, Hamish N., A1253
Munro, James W., A928
Murphy, Richard C., A454
Murphy, Robert, A1324
Murray, William G., A1373, A1375
Musser, Howard B., A455
Muzik, Thomas J., A359
Myers, Louis M., A20

Naether, Carl A., A782
Nanney, David L., A134
Nash, Andrew J., A624
Nason, Alvin, A117
National Association of State Foresters, A652
National Farm Institute, A1265, A1374
National Geographic Society, A693
Neal, Harry E., A1295
Needham, George H., A154
Nehrling, Arno, A542, A543, A543a

Nehrling, Irene, A542, A543, A543a
Neilsen, Knut Schmidt, *see* Schmidt-Neilsen, Knut
Nelson, Aaron G., A1375
Nelson, Andrew N., A1264
Nelson, Arnold L., A1322
Nelson, John A., A1241
Nelson, Kennard S., A530, A544
Nelson, Lewis B., A1085
Nelson, Robert H., A844
Nelson, Werner L., A1100
Nesheim, Malden C., A827, A847
Netboy, Anthony, A956
Neubauer, Loren W., A1039
Neumann, Alvin L., A783
Neurath, Hans, A1151
Newbigin, Marion I., A118
Newlander, John A., A1242
Nicherson, J. T., A1209
Nicholson, B. E., A242
Nikolaieff, George A., A1296
Noble, Mary, A552
Nolan, Carroll A., A83
Noland, George B., A106
Nordby, Julius E., A719, A784
Nordell, Eskel, A990
Norman, Arthur G., A389
Norman, John R., A957
Northen, Henry T., A221, A222, A545
Northen, Rebecca T., A222, A545, A546
Novák, František A., A223
Nye, Nelson C., A785

O'Brien, Robert, A1123
O'Brien, Terence P., A301
Ochse, J. J., A100
Odum, Eugene P., A176, A177
Odum, Howard T., A177
O'Hara, Frederic J., E7
Oldfield, R. L., A991
Olivier, Henry, A1040
Olivo, C. Thomas, A1213
Olson, Robert L., A1244
Oosting, Henry J., A224
Opdycke, John B., A21
Öpik, Helgi, A310
Oppenheimer, Harold L., A786, A787, A788
Ordish, George, A929
Ortloff, Henry S., A547
Osborn, Fairfield, A1297, A1298
Osmaston, Fitzwalter C., A625

Paarlberg, Donald, A1376
Painter, Reginald H., A390
Pangborn, Rose M., A1134
Panshin, Alex J., A626, A627
Parish, Howard I., A230
Parker, Marvin M., A992
Parker, William H., A877
Parry, John W., A1190
Parson, Ruben L., A1299

Partain, Llovd E., A1337
Patten, Bradley M., A789, A831
Patton, Stuart, A1142
Patton, W. J., A993
Paul, Gordon W., A82
Pauli, Frederich W., A1090
Pauling, Linus, A40
Payn, William H., A829
Peairs, Leonard M., A930
Pearson, David, A1137
Pearson, John E., A47
Pearson, Lorentz C., A391
Pederson, Carlton A., A77
Pegram, Calvin W., A794
Pelczar, Michael J., A155
Penn, Phillip, A1288
Pennak, Robert W., A694, A695
Pereira, H. G., A853
Perry, Enos J., A790
Perry, John, A940
Perry, Tilden W., A845
Peterson, Martin S., A1127
Peterson, Roger T., A260, A671
Peterson, Rudolph F., A392
Petrides, George A., A261
Petter, William Lane, *see* Lane-Petter, William
Petterssen, Sverre, A1120
Pettit, Lincoln C., A696
Pfadt, Robert E., A904
Pfahl, Peter B., A548
Pfander, W. H., A848
Pfeiffer, John, A132
Phaff, Herman J., A579
Phillips, Irving D., A312
Phipps, Lloyd J., A1016, A1041
Pigman, William W., A1152
Pijl, Leendert van der, A198
Pimentel, Richard A., A119
Pinney, John J., A549, A550
Pipe, Ted, A994
Pirone, Pascal P., A457, A931
Pitts, James N., A1300
Plakidas, Antonios G., A337
Platt, Robert B., A178
Platt, Rutherford H., A628, A629
Poehlman, John M., A318
Pohl, Richard W., A262
Poljakoff-Mayber, A., A216
Popp, Henry W., A209
Porter, Arthur R., A791
Porter, Cedric L., A263
Porter, George, A792
Potter, Norman N., A1128
Preston, Richard J., A264
Price, Molly, A551
Prince, Ford S., A393
Prodan, Michail, A630
Promsberger, William J., A1042
Purvis, Judson A., A995
Pye, Orrea F., A1255
Pyke, Magnus, A1254, A1266
Pyle, Robert, A536

Quagliano, James V., A41

Author Index

Quisenberry, Karl S., A394

Rabinowitch, Eugene, A302
Rack, Barry Graham, *see* Graham-Rack, Barry
Rand, Austin L., A697
Rath, Patricia M., A76
Raudabaugh, J. Neil, A3
Raven, Peter, A225
Raymore, Henry B., A547
Rayner, William H., A1043
Read, Clark P., A179
Read, Wilfred H., A925
Reaves, Paul M., A793, A794
Reed, Clarence A., A458
Reed, Frank C., A576
Reed, Gerald, A1215
Rehder, Alfred, A265
Reid, Roger D., A155
Rendel, Jan, A765
Reusch, Glad, A552
Revis, Cecil, A1135
Reynolds, John F., A212
Rice, Cedric B. F., A43
Rice, James E., A832
Rice, Victor A., A795
Rich, Stuart U., A631
Richer, A. Chester, A407
Richey, C. B., A1044
Richter, Herbert P., A1045
Rickett, Harold W., A931
Rieger, Rigomar, A133
Riemann, Hans, A1230
Ries, Victor H., A531
Rietz, Carl A., A1129
Riggs, James L., A78
Rimboi, Nicholas R., A983
Rips, Rae E., E3
Robbins, Chandler S., A698
Robbins, Wilfred W., A233, A352
Robert, Henry M., A22
Robert, Sarah C., A22
Robertson, Everett, A49
Robinson, Christine H., A79
Robinson, O. Preston, A79
Robinson, T. J., A878
Rodale, Jerome I., A433
Roessler, Edward B., A45, A1134
Rogers, Eric M., A42
Rogers, John L., A1247
Rolston, L. H., A905
Romer, Alfred S., A699
Roosevelt, Nicholas, A1301
Root, Amos I., A920
Root, E. R., A920
Roper, Lanning, A459
Rose, C. W., A1107
Rose, Graham J., A395
Rose, Joseph N., A243
Rosen, Donn, A944
Rosenberg, Jerome L., A303
Rosengarten, Frederic, A1191
Rosenstock, Henry M., A1018
Ross, Herbert H., A906
Ross, William K., A400

Rouse, John E., A796
Roy, Ewell P., A1377, A1378, A1379
Rubenstein, Harvey M., A553
Rubey, Harry, A1046
Rue, Leonard L., A700
Rumney, George R., A1121
Rusinoff, Samuel E., A996, A997
Russell, Edward J., A1091
Russell-Hunter, W. D., A958
Ruttan, Vernon W., A1380, A1381
Rydberg, Per A., A266, A267
Rydholm, Sven A., A632

Sabersky, Rolf H., A998
Sacharow, Stanley, A1216
Sadleir, R. M., A701
Sainsbury, David, A797
Salisbury, Glen W., A798
Salle, Anthony J., A156
Sams, Howard W., & Co., A999
Sargent, Charles S., A268
Sarner, Harvey, A879
Sartorius, Peter, A633
Sass, John E., A304
Sauchelli, Vincent, A1092, A1093, A1108
Saunders, John W., A702, A703
Scarborough, Clarence C., A462
Schaible, Philip J., A846
Schaller, Friedrich, A1094
Scharf, John H., A1231
Scheer, Arnold H., A460
Scheer, Bradley T., A704
Scher, Sheldon, A781
Scheraga, Harold A., A1153
Schery, Robert W., A226, A461
Schmeckebier, Laurence F., E8
Schmidt, Milton O., A1043
Schmidt-Nielsen, Knut, A705
Schmitt, Marshall L., A961
Schneider, George W., A462
Schonberg, James S., A1382
Schorger, Arlie W., A1325
Schuler, Stanley, A463, A464
Schultz, Harold W., A1154, A1155, A1156, A1157, A1158
Schultz, Theodore W., A1383
Schwab, Glenn O., A1047
Schwarte, Louis H., A856
Scott, John P., A706
Scott, Milton L., A847
Scott, Thomas C., A799
Sears, Paul B., A180, A181
Seiden, Rudolph, A848, A880, A932
Self, H. L., A871
Self, Margaret C., A800
Semenow, Robert W., A80
Semple, Arthur T., A396
Seneviratna, P., A881
Senn, T. L., A432
Seymour, Whitney N., A1336
Sharpe, Grant W., A583
Sharvelle, Eric G., A338

Shaw, Frank R., A915
Shaw, Herbert K. A., A273
Shaw, J. A., A1028
Shedd, C. K., A1008
Shelford, Victor E., A182
Shelton, John S., A23
Shepherd, Geoffrey S., A1384, A1385, A1386
Sherf, Arden F., A328
Shields, Phyllis J., A554
Shippen, John M., A1053
Shirlaw, Douglas W. G., A1109
Shirley, Hardy L., A634, A635
Shoemaker, James S., A465, A466
Shomon, Joseph J., A1326
Short, Douglas J., A801
Shurtleff, Malcolm C., A339
Sidney, Howard, A101
Siegmund, O. H., A876
Sigel, Mola M., A157
Silvan, James C., A802
Simmonds, Norman W., A467
Simmons, Norman O., A849
Simonds, John O., A555
Simpson, George G., A803
Sims, John A., A791
Singer, Rolf, A468
Singleton, W. Ralph, A1049
Sinnhuber, R. O., A1158
Sinnott, Edmund W., A227
Sire, Marcel, A305
Sivetz, Michael, A1192
Slack, Archie V., A1095, A1096
Slade, Samuel L., A60
Slater, Sir William, A1267
Sloane, T. O'Conor, A11
Smith, Alexander H., A269
Smith, Alice L., A158
Smith, Arthur D., A807
Smith, Carroll N., A907
Smith, Clodus R., A1337
Smith, David C., A317
Smith, David M., A636
Smith, Donald B., A1130
Smith, Frank E., A1302
Smith, Frederick E., A708
Smith, Geoffrey Wandesforde, *see* Wandesforde-Smith, Geoffrey
Smith, Guy-Harold, A1303
Smith, Harris P., A1050
Smith, Kenneth M., A340, A908
Smith, Ora, A397, A1197
Smith, Robert L., A183
Smith, Thomas L., A1403
Smith, Vearl R., A804
Smith, William H., A341
Smythe, Reginald H., A805
Snapp, Roscoe R., A783
Snedecor, George W., A61
Snodgrass, Milton M., A1387
Snowden, Obed, A1388
Sorenson, Vernon L., A1389
Soulsby, E. J. L., A882
Spedding, C. R. W., A806

Author Index

Spencer, Guilford L., A1193
Spicer, Edward D., A1006
Spielman, Patrick E., A608
Sporne, K. R., A306
Sprague, Howard B., A342, A469
Sprent, J. F., A168
Sprott, Edgar R., A326
Spurr, Stephen H., A637, A638, A639
Stakman, Elvin C., A343
Stallings, James H., A1097, A1098
Stamm, Alfred J., A640, A641
Stamp, John T., A752
Stanier, Roger Y., A160
Stanley, Wendell M., A161
Stansby, Maurice E., A1194, A1195
Stecher, Paul G., A39
Steel, Ernest W., A1019
Steere, Norman V., A5
Steeves, Taylor A., A235
Steffek, Edwin F., A470, A556
Stehli, George J., A162
Stephenson, George E., A1000
Stephenson, William A., A228
Stern, Herbert, A134
Stevens, Robert A., A1368
Steward, Frederick C., A307
Stewart, William D. P., A308
Stiles, Walter, A309
Stock, Ralph, A43
Stocking, C. Ralph, A233
Stoddard, Charles H., A642, A643, A1348
Stoddart, Laurence A., A807
Stone, Archie A., A1051, A1390
Storer, Tracy I., A707
Storm, Donald F., A867
Stout, Thomas T., A1397
Stowell, Jerald P., A557
Strandberg, Carl H., A1001
Street, Herbert E., A310
Streit, Fred, A644
Strelis, I., A645
Strobel, Gary A., A344
Stuermann, Walter E., A93
Stumbo, Charles R., A1232
Stumpf, Paul K., A32
Sudworth, George B., A270
Sultan, William J., A1196
Suter, Robert C., A398
Sutherland, Sidney S., A24
Sutter, Anne B., A558
Swain, Ralph B., A909
Swan, Lester A., A910
Swanson, Carl P., A135
Sweeney, Robert M., A712
Sweet, Walter C., A2
Sweetman, Harvey L., A933
Swidler, David T., A808
Symonds, George W., A271, A272
Synge, Patrick M., A428, A521, A559

Talbert, Thomas J., A472

Talbot, Ross B., A1391
Talburt, William F., A1197
Tavernor, W. D., A883
Taylor, Clara M., A1255
Taylor, Donald P., A334
Taylor, James A., A1122
Taylor, Kenneth E., A1168
Taylor, Miller L., A1404
Taylor, Norman, A473, A560, A561
Taylor, Warren C., A1011
Telfer, William, A120
Templeton, George S., A809
Teskey, Benjamin J. E., A466
Thatcher, F. S., A1233
Thom, A. S., A1119
Thomas, Robert J., A735
Thompson, Homer C., A474
Thompson, J. L. Cloudsley, see Cloudsley-Thompson, J. L.
Thompson, Louis M., A1099
Thompson, Phillip D., A1123
Thomsen, Frederick L., A1392, A1393
Thorne, Gerald, A169
Thorner, M. E., A1198
Thornton, Horace, A1243
Thorstensen, Thomas C., A26
Threlkeld, John L., A563
Ticquet, C. E., A475
Tiffany, Lewis H., A229
Tisdale, Samuel L., A1100
Titus, Harry W., A850
Topel, David G., A1394
Torrence, Andrew P., A1402
Torrey, John G., A311
Tortora, Gerald J., A230
Tousley, Rayburn D., A81
Toussaint, William D., A1340
Town, Harold C., A1002
Townsend, Gilbert, A1017
Tressler, Donald K., A1127, A1199, A1217
Triebold, Howard, A1159
Trimberger, George W., A810
Trippensee, Reuben E., A1327
Troup, Robert S., A646
Trout, G. Malcolm, A1175, A1244
Tsoumis, George, A647
Tuite, John F., A345
Tukey, Harold B., A476
Turk, Lloyd M., A1089
Turner, John C., A1048
Turner, Rufus P., A1003
Tutt, John F., A868
Twigg, Bernard A., A1240

Udall, Stewart L., A1304
Underwood, Eric J., A851, A1256
Uphof, Johannes C., A231
Urquhart, Frederick A., A911
Usher, George, A232
Usinger, Robert L., A707

Valens, Evans G., A161
Vallarino, L. M., A41
Van Arsdel, Wallace B., A1217, A1218, A1244
Van Demark, Noland L., A798
Van der Pijl, Leendert, see Pijl, Leendert van der
Van Dersal, William R., A1328
Van Dyne, George M., A1305
Van Wijk, W. R., see Wijk, W. R. Van
Vanderford, Harvey B., A1101
Vardaman, James M., A648
Veatch, Jethro O., A1306
Vengris, Jonas, A477
Venk, Ernest A., A1004, A1005, A1006
Vierck, Charles J., A973
Villee, Claude A., A708
Von Bergen, Werner, A811
Von Frisch, Karl, see Frisch, Karl von
Voorhis, Horace J., A1395
Voykin, Paul N., A478

Wackerman, Albert E., A649
Waggoner, Paul E., A1124
Wagner, Philip M., A479
Wagnon, Kenneth A., A812
Waksman, Selman A., A1102
Walden, Howard T., A399
Waldo, Arley D., A1381
Walker, Ernest P., A709
Walker, Harry B., A1039
Walker, John C., A346
Walker, Warren F., A708
Wall, Joseph S., A400
Wallace, George J., A710
Wallace, H. R., A347
Wallace, Luther T., A1387
Wallace, Thomas, A348
Wallis, Michael, A206
Wallwork, John A., A1103
Walsh, Richard G., A1370
Walter, John A., A18
Walters, Arthur H., A1130
Walters, Charles G., A82
Walton, Earnest V., A401
Wanderstock, Jeremiah J., A1129
Wandesforde-Smith, Geoffrey, A1278
Ward, R. C., A1052
Ware, George W., A480
Wareing, P. F., A312
Warnick, A. C., A736, A737
Wasley, G. D., A711
Waterson, A. P., A884
Watt, Kenneth E. F., A1307
Watterson, Ray L., A712
Way, Marjorie S., A564
Way, Robert F., A813
Weaver, Howard E., A650
Webb, Byron H., A1160, A1200
Webb, Edwin C., A1138
Webster, Helen N., A481

Author Index

Webster, John, A580
Weddle, A. E., A562
Weeks, William R., A974, A975
Weier, Thomas E., A233
Weiser, Harry H., A1234
Weiss, Theodore J., A1201
Weisz, Paul B., A121, A234
Welch, Charles D., A1104
Wells, James S., A482
Wells, Walter, A27
Wendt, Herbert, A713
Wenzl, Hermann F. J., A651
Werner, William E., A171
Westcott, Cynthia, A349, A912
Westhues, Melchior, A885
Wetzel, Guy F., A1007
Weyl, Peter K., A959
Wheeler, Bryan E. J., A350
Wheeler, William A., A402, A403
White, Emil, A44
White, Geoffrey W., A163
White, Gilbert F., A1308
White, Philip R., A136
White, Richard E., A889
Whittaker, Robert H., A184
Whitten, Jamie L., A1309
Whittier, Earle O., A1200
Whyte, Robert O., A404
Widner, Ralph R., A652
Wigglesworth, Vincent G., A913
Wijk, W. R. van, A1110
Wilber, Charles G., A1310
Wilcox, Walter W., A1396
Wilde, Sergius A., A653
Williams, Donald B., A1344
Williams, Kenneth A., A1135
Williams, Percy N., A1164
Williams, Sheldon W., A1248
Williams, Stephen, A814
Williams, Watkin, A319
Williams, Willard F., A1397
Williamson, Joseph, A483
Willis, John C., A273
Wilsie, Carroll P., A405
Wilson, Brayton F., A313
Wilson, Carl L., A235
Wilson, Charles M., A314
Wilson, Eva D., A1257
Wilson, George B., A137
Wilson, George F., A934
Wilson, Harold K., A406, A407
Wilson, Helen V. P., A565
Wilson, Katherine S., A227
Wilson, Leonard L., A1398
Winburne, John N., A103
Winchester, Albert M., A138
Wing, James M., A815
Wingate, John W., A83
Winkler, Albert J., A484
Winters, Laurence M., A816
Wiseman, Robert F., A817
Wisler, Chester O., A1053
Wittwer, Sylvan H., A485
Wolf, Frederick A., A581
Wolf, Frederick T., A581
Wolfe, Thomas K., A383
Woodcock, Christopher L. F., A276
Woodin, Ralph J., A98
Woodnott, Dorothy P., A801
Woodroof, Jasper G., A408, A486, A487
Woollen, Anthony, A1131
Woolrich, Willis R., A1219, A1220
Worden, A. N., A818
Work, Paul, A488
Worrell, Albert C., A654
Worthen, Edmund L., A1105
Wright, Dale, A1405
Wright, James C., A1311
Wright, Milton D., A77
Wrigley, Gordon, A104
Wrolstad, R. E., A1157
Wyman, Donald, A566, A567, A568
Wynkoop, Sally, E13

Yapp, William W., A819
Yeager, Joseph H., A1346
Yeates, Neil T. M., A820
Young, Robert J., A847
Youngberg, Chester T., A655
Youtz, H. G., A821

Zacher, Robert V., A84
Zaremba, Joseph, A656
Zeiss, George H., A79
Zeuner, Frederick E., A822
Ziegler, Percival T., A1202
Zim, Herbert S., A698, A1322
Zimmerman, Josef D., A1054
Zion, Robert L., A569
Ziswiler, Vinzenz, A1329
Zopf, Paul E., A1403
Zucker, Isabel, A570
Zweig, Gunter, A1111

TITLE INDEX

All references are to entry numbers.

ABC and XYZ of bee culture, A920
ABC's of electric motors and generators, A987
ABC's of electrical soldering, A964
ACS laboratory guide to instruments, equipment, and chemicals, F1
AGDEX, A98
Abnormal behavior in animals, A862
Abridged reader's guide to periodical literature, B1
Adaptation of domestic animals, A756
Administration of vocational education, rules and regulations, E14
Advance of the fungi, A334a
Advances in environmental science, A1300
Advertising techniques and management, A84
Aerial applicator, B20
Aerial discovery manual, A1001
Agri-business buyers reference, F2
Agri finance, B21
Agricultural chemicals, B22
Agricultural conservation program: handbook, E33
Agricultural economics reports, E34

Agricultural economics research, E35
Agricultural education magazine, B23
Agricultural engineering, A1021, B24
Agricultural engineers' handbook, A1044
Agricultural equipment technology, E15
Agricultural finance, A1375
Agricultural finance review, E36
Agricultural, forestry, and oceanographic technicians, A101
Agricultural genetics, A127
Agricultural labor and its effective use, A1368
Agricultural market analysis, A1389
Agricultural marketing, A1392, E37
Agricultural meteorology, A1124
Agricultural outlook digest, E38
Agricultural physics, A1107
Agricultural policy in an affluent society, A1381
Agricultural price analysis, A1384
Agricultural prices, A1393, E39
Agricultural research, E40
Agricultural science review, E41
Agricultural situation, E42
Agricultural statistics, E43
Agriculture decisions, E44

Agriculture, economics, and growth, A1387
Agriculture handbook, E45
Agriculture in our lives, A95
Agriculture information bulletin, E46
Agriculture teachers directory and handbook, F3
Agronomy journal, B25
Aids to veterinary nursing, A869
Algae, A190, A229
Algae and fungi, A185
All about begonias, A498
All about house plants, A513
All about rock gardens and plants, A529
All about small gas engines, A995
All about thorobred horse racing, A808
Almanac of the canning, freezing, preserving industries, A1203, F4
American agriculture, A92, A97
American bee journal, B26
American beef producer, B27
American cooperatives, A1395
American dairy review, B28
American farm policy, A1376
American fish farmer and world aquaculture news, B29
American forestry—six decades of growth, A595
American forests, B30

203

Title Index

American fruit grower, B31
American fruit grower buyer's directory, F5
American grass book, A364
American horticultural magazine, B32
American journal of botany, B33
American mammals, A678
American nurseryman, B34
American potato yearbook, A363
American tomato yearbook, A411
American vegetable grower, B35
American vegetable grower buyer's directory, F6
American wildlife and plants, A1322
America's endangered wildlife, A1319
America's garden book, A417
America's horses and ponies, A724
America's natural resources, A1273
Analysis of response in crop and livestock production, A1351
Analytical methods for pesticides, plant growth regulators, and food additives, A1111
Anatomy and physiology of farm animals, A751
Anatomy of domestic animals, A749
Anatomy of the cell, A122
Anatomy of the dog, A780
Anatomy of the horse, A813
Animal anaesthesia, A885
Animal behavior, A706
Animal breeding, A758, A816
Animal cell culture methods, A711
Animal conflict and adaptation, A667
Animal ecology, A668, A683
Animal function, A675
Animal gardens, A937
Animal genetics, A680
Animal health and housing, A797
Animal mechanics, A657
Animal morphogenesis, A702
Animal nursing, A873
Animal nutrition, A842
Animal nutrition and health, B36
Animal nutrition and veterinary dietetics, A833
Animal physiology, A704, A705
Animal sanitation and disease control, A861
Animal science, A742
Animal science and industry, A714
Animal species and evolution, A690
Animal tissue techniques, A679
Animals without backbones, A665
Annotated list of selected United States government publications available to depository libraries, E6
Applied animal nutrition, A836
Applied science and technology index, B2
Appraisal digest, B37
Appraisal journal, B38
Approved practices in beautifying the home grounds, A524
Approved practices in beef cattle production, A767
Approved practices in crop production, A367
Approved practices in farm management, A1372
Approved practices in feeds and feeding, A835
Approved practices in fruit production, A460
Approved practices in pasture management, A386
Approved practices in poultry production, A825
Approved practices in sheep production, A768
Approved practices in soil conservation, A1074
Approved practices in swine production, A734
Aquatic productivity, A958
Arithmetic refresher for practical men, A55
Art and science of horseshoeing, A754
Art index, B3
Art of forecasting, A64
Art of home landscaping, A507
Art of shaping shrubs, trees, and other plants, A526
Artificial insemination of farm animals, A790
Atlantic salmon, A956
Atlas and manual of plant pathology, A322
Atlas for applied veterinary anatomy, A865
Attack on starvation, A1261
Audubon, B38a
Automatic machine tools, A1002
Automatic vending—merchandising, catering, A1247
Automotive brakes and power transmission systems, A971
Automotive diagnosis and tune-up, A1007
Automotive electrical systems, A970
Automotive engines, maintenance and repair, A1004
Automotive fuel and ignition systems, A972
Automotive fundamentals, A1005
Automotive maintenance and troubleshooting, A1006
Automotive mechanics, A963
Avian diseases, B39
Azalea book, A533

Bacteriology illustrated, A147
Bakery technology and engineering, A1214
Ball red book, A491
Balsam fir, A586
Bamboos, A532
Bananas, A467
Bargaining power for farmers, A1374
Barley and malt, A370
Basic animal husbandry, A770
Basic electricity, A1003
Basic electricity and an introduction to electronics, A999
Basic electricity for electronics, A989
Basic farm machinery, A1048
Basic mathematics for business analysis, A47
Basic microscopic technics, A150
Basic technology, A1171
Bedding plants, A538
Beef cattle, A783
Beef cattle production, A812
Beef cattle science, A743
Beef production in the South, A750
Beekeeping, A915
Behavior of domestic animals, A755
Beneficial insects, A910
Bergey's manual of determinative bacteriology, A159
Better crops with plant food, B40
Better fruit. Better vegetables, B41
Bibliography of agriculture, B4
Big farmer, B42
Bimonthly list of publications and motion pictures, C1, E47
Biochemical approach to life, A36
Biochemistry, B43
Biochemistry of fruits and their products, A442
Biochemistry of plants and animals, A38
Biological abstracts, B5
Biological and agricultural index, B6
Biological aspects of water pollution, A1310
Biological conservation, A1315
Biological control of insect pests and weeds, A921
Biological methods in crop pest control, A929
Biological techniques, A116
Biology, A108
Biology of cells, A134

Title Index

Biology of higher cryptogams, A195
Biology of organisms, A120
Biology of plant parasitic nematodes, A347
Biology of plants, A225
Biology of populations, A686
Bioscience, B44
Bird migration, A676
Birds of North America, A698
Blueberry culture, A431
Boeckh building valuation manual, A65
Bonsai, A557
Bonsai for Americans, A525
Bonsai, sakei, and bonkei, A493
Book of cacti and other succulents, A503
Book of landscape design, A547
Book of spices, A1191
Book of the domestic rabbit, A782
Book of trees, A249
Booklist, D1
Botanical microtechnique, A304
Botany, A209, A220, A227, A233, A235
Breeding and improvement of farm animals, A795
Breeding field crops, A318
Breeding laboratory animals, A781
Building and equipping the farm shop, A1029
Bulletin of environmental contamination and toxicology, B45
Business management of a small animal practice, A879
Business periodicals index, B7
Byproducts from milk, A1200

C & MS, E48
CCC price support program data, E49
CDC veterinary public health notes, E50
CR crops research (series), E51
CRC critical reviews in environmental control, B46
CRC critical reviews in food technology, B47
CRC handbook of chemistry and physics, A30
CRC handbook of food additives, A1136
CRC handbook of laboratory safety, A5
Cactaceae, A243
Camellia book, A563
Cane sugar handbook, A1193
Canned foods, A1227
Canner/packer, B48
Canner/packer yearbook, F7
Carbohydrates, A1152

Carbonated beverages, A1177
Care and breeding of laboratory animals, A748
Care and feeding of trees, A454
Care and training of the trotter and pacer, A760
Careers in agribusiness and industry, A1390
Careers in conservation, A1276
Careers in horticultural sciences, A429
Careers in natural resource conservation, A1287
Carnations, A501
Catalog of federal assistance programs, E52, F8
Catalogue of FAO publications, 1945–1968, C2
Catfish farming handbook, A953a
Cattle fertility and sterility, A717
Cattleman, B49
Cell, A132, A135
Census of agriculture, E53
Census of commercial fisheries, E54
Census of irrigation, E55
Census of population, E56
Century of service, E16
Cereal chemistry, B50
Cereal crops, A385
Cereal laboratory methods, A1133
Cereal science, A1185
Cereal science today, B51
Cereal technology, A1186
Cheese, A1171
Cheese and fermented milk foods, A1181
Chem sources, F9
Chemical abstracts, B8
Chemical additives in food, A1140
Chemical analysis of foods, A1137
Chemical analysis of foods and food products, A1141
Chemical and biological hazards in foods, A1221
Chemical background for the biological sciences, A44
Chemical control of plant diseases, A338
Chemical processing of wood, A641
Chemical technology of wood, A651
Chemistry, A41
Chemistry and mode of action of herbicides, A351
Chemistry and physiology of flavors, A1154
Chemistry and technology of fertilizers, A1092, A1095
Chemistry and testing of dairy products, A1242

Chemistry for the applied sciences, A35
Chemistry of the soil, A1059
Chemistry of the vitamins, A1139
Chemistry of wood, A592
Chinchillas, A723
Chocolate, cocoa, and confectionery, A1188
Chromatographic methods, A43
Chrysanthemum book, A505
Citrus fruits, A443
City and county directories, F10
Clearinghouse announcements in science and technology: agriculture and food, E57
Climate and agriculture, A1115
Climates of the states, E58
Climatic atlas of the United States, E17
Climatological data, national summary, E59
Climatology and the world's climates, A1121
Cloud physics and cloud seeding, A1112
Coconuts, A421, A487
Code of life, A126
Coffee processing technology, A1192
Cold and freezer storage manual, A1220
College chemistry, A40
Collegiate dictionary of zoology, A694
Color dictionary of flowers and plants for home and garden, A521
Color filmstrips and slide sets of the United States Department of Agriculture, D2
Color of foods, A1146
Combined 16mm film catalog, D3
Coming water famine, A1311
Commercial fertilizer yearbook, F11
Commercial fertilizers, A1069
Commercial fisheries review, E60
Commercial flower forcing, A530
Commercial fruit and vegetable products, A1170
Communicating facts and ideas in business, A4
Communications in business, A27
Communities and ecosystems, A184
Comparative vertebrate anatomy, A673
Compleat manager, A68
Compleat rancher, A720
Complete book of cat health and care, A777
Complete book of gardening under lights, A535

205

Title Index

Complete book of greenhouse gardening, A545
Complete book of the quarter-horse, A785
Complete dog book, A715
Complete guide to bulbs, A559
Complete horseshoeing guide, A817
Compost science, B52
Concepts of forest entomology, A897
Concrete block construction for home and farm, A1017
Congress and the environment, A1278
Conservation, A1289, A1301
Conservation bulletin, E61
Conservation directory, F12
Conservation of natural resources, A1303
Conservation research reports, E62
Conservation yearbook, E63
Conserving American resources, A1299
Conserving natural resources, A1268
Construction and maintenance for farm and home, A1022
Consumer behavior, A82
Contamination analysis and control, A966
Contract farming U.S.A., A1377
Control of growth and differentiation in plants, A312
Control of soil fertility, A1071
Control of the ovarian cycle in the sheep, A878
Cookie and cracker technology, A1187
Cooperative digest and farm power, B53
Cooperative occupational education and work experience in the curriculum, A16
Cooperatives today and tomorrow, A1378
Corn: culture, processing, products, A377
Corsage craft, A552
Costs and returns (FCR 1-), E64
County agents directory, F13
County and city data book, E65
Courage to change, A398
Cowboy arithmetic, A786
Cowboy economics, A787
Cowboy litigation, A788
Creative selling, A71
Crop adaptation and distribution, A405
Crop and livestock reports, E66
Crop growth and culture, A388
Crop production, A372, A376
Crop production (CrPr2-2), E67

Crop production and environment, A404
Crop production in the South, A384
Crop protection, A395
Crop science, B54
Crop technology, A373
Crops and soils, B55
Crossbreeding beef cattle, A736
Cultivated conifers in North America, A414
Cultivation of animal and plant cells, A136
Culture and diseases of game fishes, A945b
Current agricultural serials, A87
Cyanagrams, B56
Cytology, A137

Dairy cattle feeding and management, A793
Dairy cattle in American agriculture, A791
Dairy cattle judging and selection, A819
Dairy cattle judging techniques, A810
Dairy cattle management, A815
Dairy herd management, B57
Dairy industries catalog, F14
Dairy microbiology, A1223
Dairy production, A739
Dancing bees, A916
Deadly harvest, A213
Death of the sweet waters, A1274
Deciduous orchards, A419
Defense against famine, A1096
Desert wildlife, A682
Design and depth in flower arrangement, A506
Design with nature, A1292
Destructive and useful insects, A903
Development in flowering plants, A311
Diagnosis of mineral deficiencies in plants by visual symptoms, A348
Diagnostic methods in veterinary medicine, A858
Dictionary of agricultural and allied terminology, A103
Dictionary of biological terms, A113
Dictionary of botany, A232
Dictionary of economic plants, A231
Dictionary of gardening, A428
Dictionary of natural resources and their principal use, A1288
Dictionary of nutrition and food technology, A1125
Dictionary of the flowering plants and ferns, A273

Diesel and high-compression gas engines, A982
Dietetic foods, A1250
Different kind of country, A1280
Directory of communicators in agriculture, F15
Directory of information resources in the United States: federal government, E18a, F16
Directory of information resources in the United States: general toxicology, E18b, F17
Directory of information resources in the United States: physical sciences, biological sciences, engineering, E18c, F18
Directory of information resources in the United States: social sciences, E18d, F19
Directory of information resources in the United States: water, E18e, F20
Directory of national organizations concerned with land pollution control, F21
Directory of national trade and professional associations of the United States, F21a
Directory of schools offering technical education programs under Title III of the George Barder Act and the Vocational Education Act of 1963, E19
Directory of the canning, freezing, preserving industries, F22
Directory of the forest products industry, F23
Disease in forest plantations, A324
Diseases and pests of ornamental plants, A931
Diseases of feedlot cattle, A872
Diseases of field crops, A330
Diseases of fruit crops, A321
Diseases of poultry, A856
Diseases of poultry (including cage birds), A881
Diseases of swine, A860
Diseases of tobacco, A335
Diseases of turfgrass, A328a
Disinfection, sterilization, and preservation, A152
Diversity of green plants, A276
Doane's agricultural report, B58
Dog breeding, A718
Domestic rabbit production, A809
Down to earth, B59
Drainage engineering, A1037
Drovers journal, B60

Title Index

Drug dosage in laboratory animals, A855
Drying of milk and milk products, A1174
Dwarfed fruit trees, A476

EFLA bulletin, D4
Early embryology of the chick, A831
Earth beneath us, A17
Easy gardening with drought-resistant plants, A542
Ecology, A176
Ecology and field biology, A183
Ecology and resource management, A1307
Ecology of North America, A182
Ecology of populations, A172
Ecology of reproduction in wild and domestic animals, A701
Ecology of soil animals, A1103
Economic botany, A208
Economic crises in world agriculture, A1383
Economic demand for irrigated acreage, A1380
Economics for modern agriculture, A1352
Economics of agricultural development, A1369
Economics of American agriculture, A1396
Economics of American forestry, A654
Economics of irrigation, A1015
Economics of outdoor recreation, A1331
Economics of soil conservation, A1066
Economics of the American lumber industry, A656
Economics of the livestock meat industry, A1397
Ecosystem concept in natural resource management, A1305
Education index, B9
Educational motion pictures, D5
Educational motion pictures for school and community, D6
Educators' grade guide to free teaching aids, C3, D7
Educators' guide to free films, D8
Educators' guide to free filmstrips, D9
Educators' guide to free science materials, D10
Egg industry, B61
Electrical machines, A965
Electricity in horticulture, A499
Electricity on the farm, B62
Elementary organic chemistry, A31
Elementary, secondary, college, educational materials catalog, D11

Elementary surveying, A1011, A1043
Elements of biological science, A114
Elements of biology, A121
Elements of ecology, A173
Elements of park and recreation administration, A1332
Elements of soil conservation, A1060
Elements of statistical inference, A54
Embryology of the pig, A789
Encyclopaedia Britannica Films, Inc. catalog, D12
Encyclopedia of associations, F24
Encyclopedia of business information sources, F24a
Encyclopedia of organic gardening, A433
Encyclopedia of the biological sciences, A112
Energy metabolism of ruminants, A834
Engineering for dairy and food products, A1208
Engineering index, B10
Entomological news, B63
Entomology, A899
Environmental conservation, A1281
Environmental measurement and interpretation, A178
Enzymes, A1138
Enzymes in food processing, A1215
Essential forms for flower arrangement, A564
Essentials of biological chemistry, A33
Essentials of forestry practice, A642
Essentials of marketing management, A73
Essentials of meteorology, A1119
Essentials of modern biology, A117
Essentials of statistics for scientists and technologists, A58
Ethology of mammals, A670
Evaluation of fertility in the bull and the boar, A871
Evergreen orchards, A420
Evolution, A110
Exotic plants illustrated, A517
Exotica 3, A518
Experience programs for learning vocations in agriculture, A86
Experimental cookery from the chemical and physical standpoint, A1145
Experiments in animal behavior, A677
Exploring agribusiness, A1379

Extension service review, E68
Extinct and vanishing animals, A1329
Fabric and mineral analysis of soils, A1064
Factors affecting calf crop, A737
Famine on the wind, A326
Fantastic trees, A218
Farm and personal finance, A1342
Farm and power equipment retailers handbook, A1398
Farm animals, A759
Farm appraisal and valuation, A1373
Farm building design, A1039
Farm business management, A1345, A1346
Farm chemical directory, F25
Farm chemicals, B64
Farm chemicals handbook, A1106, F26
Farm crop production technology, field and forage crop, and fruit and vine production options, E20
Farm electrification, A1013
Farm equipment manufacturers directory, F27
Farm film guide, D13
Farm gas engines and tractors, A1032
Farm index, E69
Farm journal, B65
Farm labor, E70
Farm labor (La 1.), E71
Farm labor fact book, E21
Farm machinery, A1028
Farm machinery and equipment, A1050
Farm management, A1361
Farm management analysis, A1341
Farm management decisions, A1360
Farm management economics, A1359
Farm management handbook, A1356
Farm management in the South, A1364
Farm mechanics text and handbook, A1016
Farm policy, A1385
Farm power and machinery management, A1030
Farm quarterly, B66
Farm real estate market developments, E72
Farm records and accounting, A1363
Farm service buildings, A1024
Farm soils, A1105
Farm supplier, B67

Title Index

Farm tractor maintenance, A1012
Farm welding, A992
Farmers and a hungry world, A1265
Farmers' bulletin, E73
Farmer's digest, B68
Farming techniques from prehistoric to modern times, A1023
Farming the edge of the sea, A950
Farming the sea, A954
Farms and farmers in an urban age, A1362
Federal outdoor recreation programs, F28
Feed additive compendium, A839
Feed formulations handbook, A845
Feed industry, B69
Feed industry red book, F29
Feed management, B70
Feed milling and associated subjects, A849
Feeding, digestion, and assimilation in animals, A841
Feeds and feeding, A843
Feedstuffs, B71
Fertilization, A105
Fertilizer industry, A1086
Fertilizer nitrogen, A1093
Fertilizer solutions, B72
Fertilizer technology and usage, A1085
Fertilizer trends, E74
Field biology and ecology, A171
Field guide to the insects of America north of Mexico, A889
Field guide to trees and shrubs, A261
Field guide to wildflowers of northeastern and northcentral North America, A260
Fifty laboratory exercises for vocational ornamental horticulture students, A522
Fifty years of farm management, A1344
Film evaluation guide, D14
Film news, D15
Films of the U.S. Department of Agriculture, D16
Fire control notes, E75
First lessons in beekeeping, A914
Fish and river pollution, A952
Fish culture, A947
Fish handling and processing, A945
Fish migration, A951
Fish oils, A1194
Fishery leaflet, E76
Fishes, A953
Flora of the prairies and plains of central North America, A266

Flora of the Rocky Mountains and adjacent plains, A267
Floriculture, A531
Flower and plant production in the greenhouse, A544
Flower growing in the North, A534
Flowering shrubs, A570
Flowering trees, A245
Flowering vines of the world, A539
Flowers: free form—interpretive design, A494
Flowers: geometric form, A495
Flowers in the winter garden, A519
Fluid flow, A998
Fluorescent light gardening, A502
Food and society, A1254
Food beverage service handbook, A1198
Food-borne infections and intoxications, A1230
Food composition and analysis, A1159
Food dehydration, A1218
Food engineering, B73
Food enzymes, A1154
Food flavorings, A1149
Food for everyone, A1264
Food industries manual, A1131
Food industry sourcebook for communication, F30
Food marketing, A1246
Food microbiology, A1224
Food oils and their uses, A1201
Food packaging, A1216
Food poisoning, A1222
Food processing technology, E22
Food science, A1128
Food standards and definitions in the United States, A1238
Food technology, B74
Food technology the world over, A1127
Food texture, A1147
Forage and pasture crops, A402
Forage crops, A360
Forages, A375
Forest and shade tree entomology, A886
Forest and the sea, A1269
Forest biometrics, A630
Forest ecology, A637
Forest fire: control and use, A598
Forest history, B75
Forest inventory, A638
Forest management, A599, A622
Forest measurements, A584
Forest mensuration and statistics, A615
Forest ownership for pleasure and profit, A635
Forest pathology, A325

Forest pest leaflet, E77
Forest products, A626
Forest products journal, B76
Forest recreation, A1333
Forest service films, available on loan for educational purposes to schools, civic groups, churches, television, D17
Forest-soil relationships in North America, A655
Forest soils, A620, A653
Forest technology, E23
Forest tree planting in arid zones, A606
Forest trees of the Pacific slope, A270
Forestry and economic development, A633
Forestry and its career opportunities, A634
Forestry handbook, A604
Forests and forestry in the American states, A652
Forging and forming metals, A996
Fortune directory, F31
Foundation directory, F32
Foundations of nutrition, A1255
Foundations of technical mathematics, A49
Fragrant year, A565
Free and inexpensive learning materials, C4, D18
Freezing preservation of foods, A1217
Fresh water biology, A111
Freshwater ecology, A175
Fresh-water invertebrates of the United States, A695
From sea to shining sea, a report on the American environment—our natural heritage, E24
Fruit and vegetable juice processing technology, A1199
Fruit and vegetables, A430
Fruit growing, A462
Fruit key and twig key to trees and shrubs, A252
Fruit varieties and horticultural digest, B77
Fruits for the home garden, A450
Fundamental principles of bacteriology, A156
Fundamentals of applied entomology, A904
Fundamentals of dairy chemistry, A1160
Fundamentals of ecology, A177
Fundamentals of electricity, A976
Fundamentals of engineering drawing, A973
Fundamentals of food engineering, A1204

Title Index

Fundamentals of food processing operations, A1210
Fundamentals of forestry economics, A601
Fundamentals of horticulture, A432
Fundamentals of manufacturing processes and materials, A967
Fundamentals of microbiology, A145
Fundamentals of mycology, A573
Fundamentals of selling, A83
Fundamentals of soil science, A1089
Fungi, A581
Future environments of North America, A1279

Game management, A1320
Garden ideas A to Z, A451
Garden journal, B78
Gardener's almanac of timely hints for home gardeners, A435
Gardener's bug book, A912
Gardener's fern book, A510
Gardening from the ground up, A463
Gardening in the East, A464
Gardening in the lower South, A444
Gardening in the shade, A541
General and applied entomology, A902
General biology, A106
General climatology, A1116
General metals, A969
General physics, A28
General virology, A153
General viticulture, A484
General zoology, A674, A707, A708
Generalized electric machines, A968
Genetic code, A123
Genetical principles and plant breeding, A319
Genetics, A138
Genetics and animal breeding, A765
Genetics in the atomic age, A124
Genetics is easy, A131
Genetics of livestock improvement, A773
Genetics of the dog, A729
Genus pinus, A623
Geography of plants and animals, A109
Geography of soil, A1067
Geography of the flowering plants, A203
Geology, A2
Geology illustrated, A23
Germination of seeds, A216

Getting agriculture moving, A99
Gleanings in bee culture, B79
Glory of the tree, A241
Glossary of genetics and cytogenetics, A133
Goat husbandry, A778
Golf superintendent, B80
Government publications and their use, E8
Government reference books, E13
Governmental guide, F33
Grafter's handbook, A437
Grain and feed journals, B81
Grain crops, A406
Grain, feed, seed, and farm supply technology, E25
Grain storage, A1237
Grain trade, A1382
Grass systematics, A204
Grasses and grasslands, A365
Grassland farming in the humid Northeast, A393
Grassland improvement, A396
Grassland seeds, A403
Gray's manual of botany, A199
Great American forest, A628
Green thumb book of fruit and vegetable gardening, A409
Green thumb book of indoor gardening, A489
Greenhouse gardening, A515
Greenhouse gardening for fun, A497
Greenhouse tomatoes, A485
Ground cover plants, A566
Grounds maintenance handbook, A504
Growing better wool, A763
Growing fruit and vegetable crops, A472
Growing tree, A313
Guide to American English, A20
Guide to flower arranging in 10 easy lessons, A554
Guide to garden flowers, A561
Guide to garden shrubs and trees, A560
Guide to mushrooms and toadstools, A578
Guide to popular U.S. government publications, E1
Guide to site and environmental planning, A553
Guide to southern trees, A254
Guide to the selection, combination and cooking of foods, A1129
Guide to U.S. government serials and periodicals, E2
Guide to understanding the United States Department of Agriculture, E9
Gum technology in the food industry, A1172

Hagan's infectious diseases of domestic animals, A866
Handbook for vegetable growers, A449
Handbook of agricultural occupations, A94
Handbook of algebraic and trigonometric functions, A56
Handbook of chemistry and physics, A30
Handbook of Coniferae and Ginkgoaceae, A247
Handbook of feed stuffs: production, formulation, medication, A848
Handbook of laboratory safety, A5
Handbook of North American nut trees, A447
Handbook of pulp and paper technology, A588
Handbook of refrigerating engineering, A1219
Handling of chromosomes, A129
Harper's English grammar, A21
Harvest of the sea, A941
Harvesting the clouds, A1113
Harvesting timber crops, A649
Health and disease in farm animals, A877
Henley's twentieth century book of formulas, processes, and trade secrets, A11
Herbs, A481
Heredity, A125
Herms's medical entomology, A899
High timber, A597
Historical abstracts, B11
History of domesticated animals, A822
History of fishes, A957
Hive and the honeybee, A917
Hoard's dairyman, B82
Hog farm management, B83
Home and garden bulletin, E78
Home and garden supply merchandiser, B84
Home and garden supply merchandiser green book, F34
Home book of taxidermy and tanning, A9
Home economics research reports, E79
Home garden and flower grower, B85
Home orchid growing, A546
Hops, A368
Horse, A769
Horse: structure and movement, A805
Horseman's encyclopedia, A800
Horses, A803
Horses and horsemanship, A744
Horses of the world, A753

Title Index

Horticultural pests, detection, and control, A934
Horticultural science, A446
Horticulture, B86
Hortus second, A415
How plants get their names, A237
How to control plant diseases in home and garden, A339
How to grow roses, A536
How to know the grasses, A262
How to organize and operate a small business, A72
How to plan modern home grounds, A490
How to prune almost everything, A416
How to read shop drawings, with special reference to welding symbols, A986
How to run a small business, A74
How to shoe a horse, A779
How to speak and write for rural audiences, A6
Hunger signs in crops, A342
Hungry planet, A1258
Hydrology, A1053
Hygiene in food manufacture and handling, A1226

I.A.T. manual of laboratory animal practice and techniques, A801
Ice cream, A1165
Identification of North American pulpwoods and pulp trees, A645
Illustrated flora of the northeastern United States and adjacent Canada, A248
Illustrated flora of the Pacific States: Washington, Oregon, and California, A236
Illustrated genera of imperfect fungi, A572
Implement and tractor, B87
Implement and tractor product file, F35
Implement and tractor red book, F36
Improved nut trees of North America and how to grow them, A458
Improvement of livestock, A722
Index Bergeyana, A142
Index medicus, B12
Index to 8mm motion cartridges, D19
Index to overhead transparencies, D20
Index to 16mm educational films, D21
Index to 35mm educational filmstrips, D22

Individual freedom and the economic organization of agriculture, A1343
Industrial and agricultural education, D23
Industrial fishery technology, A1195
Industrial safety handbook, A979
Industrial wastewater control, A1025
Informal gardening, A528
Ingenious kingdom, A222
Insect colonization and mass production, A907
Insect guide, A909
Insect pests of farm, garden, and orchard, A930
Insect pests of livestock, poultry, and pets, and their control, A932
Insect pollination of crops, A200
Insect resistance in crop plants, A390
Insect virology, A908
Insects, A900, A901
Insects in relation to plant disease, A327
Insects: structure and function, A892
Inside wood: masterpiece of nature, A610
Instrumentation, A983
Intensive livestock farming, A721
Interpretation of aerial photographs, A960
Introducing the insect, A911
Introduction to agricultural economic analysis, A1340
Introduction to agricultural engineering, A1038
Introduction to American forestry, A583
Introduction to animal behavior, A684, A687
Introduction to animal virology, A884
Introduction to applied entomology, A905
Introduction to color, A7
Introduction to computers and data processing, A46
Introduction to computing, A53
Introduction to electric circuits, A980
Introduction to electrical circuit analysis, A962
Introduction to electron microscopy, A148
Introduction to entomology, A893
Introduction to feeding farm livestock, A844
Introduction to forestry, A613
Introduction to fungi, A581

Introduction to insect physiology, A891
Introduction to livestock husbandry, A726
Introduction to livestock production, including dairy and poultry, A732
Introduction to meteorology, A1120
Introduction to microscopy, A163
Introduction to ornithology, A710
Introduction to parasitology, A166
Introduction to plant anatomy, A285
Introduction to plant breeding, A316
Introduction to plant diseases, A350
Introduction to plant physiology, A299, A300
Introduction to plant taxonomy, A258
Introduction to probability and statistics, A45, A59
Introduction to soil microbiology, A1055
Introduction to soil science in the Southeast, A1104
Introduction to soils and plant growth, A1073
Introduction to statistical science in agriculture, A51
Introduction to the physical basis of soil water phenomena, A1068
Introduction to the study of animal populations, A659
Introduction to the study of insects, A888
Introduction to thermal processing of foods, A1209
Introductory animal science, A716
Introductory food science, A1130
Introductory horticulture, A424
Introductory mycology, A571
Introductory physics, A37
Introductory plant science, A221
Introductory soils, A1061
Introductory zoology, A696
Invention, discovery, and creativity, A19
Invertebrate zoology, A691
Iowa veterinarian, B88
Iris book, A551
Irrigation, A1036, A1054
Irrigation and climate, A1040
Irrigation of agricultural lands, A1026
Irrigation principles and practices, A1031

Japanese stone gardens, A514

Title Index

Journal of agricultural and food chemistry, B89
Journal of animal science, B90
Journal of dairy science, B95
Journal of economic entomology, B96
Journal of food science, B97
Journal of forestry, B98
Journal of milk and food technology, B99
Journal of nutrition, B100
Journal of range management, B101
Journal of soil and water conservation, B102
Journal of the American Society for Horticultural Science, B91
Journal of the American Society of Farm Managers and Rural Appraisers, B92
Journal of the American Veterinary Medical Association, B93
Journal of the Association of Official Analytical Chemists, B94
Journal of wildlife management, B103
Judging dairy products, A1241
Judging livestock, dairy cattle, poultry, and crops, A821

Key to the families of flowering plants of the world, A256
Kiplinger agricultural letter, B104
Knowing your trees, A246
Koehler method of guard dog training, A771

Laboratory identification of pathogenic fungi simplified, A576
Laboratory manual for food canners and processors, A1211
Laboratory manual for food microbiology, A1225
Laboratory methods in microbiology, A149
Laboratory studies of chick, pig, and frog embryos, A712
Lameness in horses, A852
Land economics, B105
Land for the future, A1349
Lands beyond the forest, A180
Landscape architecture, A555, B106
Landscape industry, B107
Landscape sketching, A496
Landscape vocabulary, A537
Larousse encyclopedia of animal life, A685
Law and court decisions on agriculture, A91

Law and the farmer, A85
Law for the veterinarian and livestock owner, A867
Lawn book, A461
Lawns, A477
Lawns and ground covers, A483
Laws relating to agriculture, May 29, 1884– , E80
Laws relating to forestry, game conservation, flood control, and related subjects, March 1, 1911– , E81
Laws relating to vocational education and agricultural extension work, May 8, 1941– , E82
Leadership and dynamic group action, A3
Leadership for action in rural communities, A1402
Leaflet (USDA), E83
Learning directory, D24
Library lists, E84
Library of Congress catalog: motion pictures and filmstrips, D25
Life of fishes, A955
Life of insects, A913
Life of plants, A191
Life of the green plant, A289
Life of the pond, A170
Life of yeasts, A579
Life on a little-known planet, A895
Lilies, A509
Linear programming and animal nutrition, A838
Lipids and their oxidation, A1158
List of available publications of the United States Department of Agriculture, C5, E10
List of publications. Canada Department of Agriculture, C6
List of published soil surveys, E85
Lives of wasps and bees, A887
Livestock and livestock products, A730
Livestock and meat economy of the United States, A1365
Livestock and poultry production, A727
Livestock health encyclopedia, A880
Livestock judging and evaluation, A719
Livestock judging handbook, A784
Living landscape, A181
Living plant, A189
Logging, A590
Lost wild America, A1321

Macaroni products, A1176
Machines for power farming, A1051

McMichael's appraising manual, A75
Macmillan wild flower book, A257
Mammalian protein metabolism, A1253
Mammals of the Pacific states, A681
Mammals of the world, A709
Man and food, A1266
Man and land in the United States, A1348
Man must eat, A1267
Management and conservation of biological resources, A1270
Management of artificial lakes and ponds, A943
Management of forests, A625
Management of wild animals in captivity, A935
Management's self-inflicted wounds, A63
Managing southern soils, A1101
Managing the farm business, A1338
Man's impact on nature, A1290
Manual of cultivated plants most commonly grown in the continental United States and Canada, A238
Manual of cultivated trees and shrubs hardy in North America, exclusive of the subtropical and warmer temperate regions, A265
Manual of meat inspection procedures of the Department of Agriculture, E86
Manual of southern forestry, A650
Manual of the trees of North America, A268
Manufacturing processes, A997
Margarine, A1164
Market news reports, E87
Market structure of the agricultural industries, A1370
Marketing and distribution, A76
Marketing bulletin, E88
Marketing farm products, A1386
Marketing in a changing environment, A70
Marketing information guide, F37
Marketing of agricultural products, A1367
Marketing of forest products, A631
Marketing of livestock and meat, A1354
Marketing poultry products, A1339
Marketing research reports, E89
Mathematical reviews, B13
Mathematics for science, A50

Title Index

Mathematics for technical and vocational schools, A60
Mathematics in agriculture, A57
Matter, energy, and life, A29
Meat for the table, A1169
Meat handbook, A1184
Meat hygiene, A1168
Meat management, B108
Meat science, A1183
Meat we eat, A1202
Mechanical power transmission, A978
Mechanics in agriculture, A1041
Merck index, A39
Merck poultry serviceman's manual, A830
Merck veterinary manual, A876
Metabolism of insects, A896
Metal abstracts, B14
Metallurgy of welding, brazing, and soldering, A985
Meteorological and geoastrophysical abstracts, B15
Methods in food analysis, A1143
Methods of animal experimentation, A864
Methods of plant breeding, A317
Methods of soil analysis, A1063
Methods of wood chemistry, A591
Microbial world, A160
Microbiology, A143, A146, A155
Microorganisms in foods, A1233
Microscope, A154
Microscope and how to use it, A162
Microwave heating in freeze drying, electronic ovens, and other applications, A1205
Microwaves, A34
Midwest farm handbook, A96
Midwest farm planning manual, A1366
Milestones in nutrition, A1251
Milk pasteurization, A1175
Milk production and processing, A766
Mineral nutrition of livestock, A851
Miscellaneous publication (USDA), E90
Modern aspects of animal production, A820
Modern breeds of livestock, A725
Modern cereal chemistry, A1179
Modern corn production, A362
Modern dairy cattle management, A738
Modern dairy products, A1182
Modern developments in animal breeding, A775
Modern embryology, A664
Modern farm buildings, A1008
Modern farm power, A1042
Modern food microbiology, A1228

Modern fruit science, A422
Modern hydrology, A1034
Modern insecticides and world food production, A924
Modern marketing of farm products, A1371
Modern mink management, A774
Modern packaging: encyclopedia issue, F38
Modern researcher, A1
Modern veterinary practice, B109
Modern veterinary practice red book, F38a
Modern wood technology, A608
Modes of reproduction in fishes, A944
Molds and man, A574
Monthly catalog of United States government publications, C7
Monthly checklist of state publications, C8
Morphology and taxonomy of fungi, A240
Morphology of gymnosperms, A306
Morphology of plants, A278
Mountain Plains Educational Media Council film catalog, D26
Movement of water in plants, A280
Municipal and rural sanitation, A1019
Mushroom growing today, A412
Mushroom hunter's field guide, A269
Mushrooms and truffles, A468
Mushrooms, molds, and miracles, A577

Nasco agricultural sciences technical catalog, D27
National audio tape catalog, D28
National Farmers Union, A90
National forests of America, A605
National hog farmer, B110
National provisioner, B111
National union catalog, E91
National wildlife, B112
National wool grower, B113
Native inheritance: the story of corn in America, A399
Natural geography of plants, A202
Natural history, A119
Natural history of viruses, A139
Nature and properties of soils, A1065
Nature of animals, A692
Nature's guardians: your career in conservation, A1295
Nematodes, A165
New approach to design principles, A558
New Britton and Brown illustrated flora of the northeastern United States and adjacent Canada, A248
New field book of freshwater life, A115
New illustrated encyclopedia of gardening, A456
New serial titles, E92
News for farmers cooperatives, E93
Nightshades: the paradox plants, A207
Nitrogen fixation in plants, A308
North American trees, A264
Northern logger and timber processer, B114
Notes for breeders of common laboratory animals, A792
Now you're talking, A12
Nuclear radiation in food and agriculture, A1049
Nursery business, B115
Nutrition, A1249
Nutrition and disease in experimental animals, A883
Nutrition of fruit crops, A423
Nutrition of the chicken, A847

Oats and oat improvement, A369
Obedience and security training for dogs, A799
Occupational outlook handbook, E94
Occupational outlook quarterly, E95
Oceanography, A959
Oils, fats, and fatty foods, A1135
On gardens and gardening, A459
1001 questions answered about trees, A629
Onions and their allies, A381
Operating a garden center, A549
Orchard and fruit garden, A445
Orchids, their botany and culture, A520
Organic gardening and farming, B116
Organic pesticides in the environment, A923
Organic way to plant protection, A448
Organization and competition in the Midwest dairy industries, A1248
Origin of life, A107
Origins of American conservation, A1277
Ornamental horticulture as a vocation, A540
Ornamental horticulture source units for vocational teachers, A523
Ornamental horticulture technology, E26
Ornamental waterfowl, A829

Title Index

Ornithology, A697
Our crowded planet, A1298
Our natural resources, A1293
Our plundered planet, A1297
Our public lands, E96
Our soils and their management, A1072
Our wildlife legacy, A1312
Outdoor recreation action, E97
Outdoor recreation in America, A1334
Outdoor recreation in research, E98
Outline of infectious diseases of domestic animals, A875
Outlines of biochemistry, A32
Outlines of plant pathology, A344
Over 2,000 free publications; yours for the asking, C9, E7
Overcoming world hunger, A1262
Oxford book of flowerless plants, A242
Oxford book of food plants, A206

PA [Program aid], E99
Packer, B117
Paper-making practice, A609
Paper quality control, A644
Paper, the fifth wonder, A582
Parasites of North American freshwater fishes, A948
Parasitism, A168
Parasitism and symbiology, A179
Park maintenance, B118
Park maintenance annual buyer's guide, F39
Parks and recreation, B119
Parks for America, a survey of park and related resources in the fifty states, E27
Pastures for the South, A382
Patterns and principles of animal development, A703
Patterns of mammalian reproduction, A660
Peanuts, A408
Pectic substances, A1144
People of rural America, E28
Perfect lawn the easy way, A478
Personnel management and supervision, A66
Pest control, A926, B120
Pesticide handbook—entoma, A922, F40
Pesticide monitoring journal, E100
Pesticides and pollution, A1294
Pests of protected cultivation: the biology and control of glasshouse and mushroom pests, A925
Pests of stored products, A928
Photogrammetry and photointerpretation, A639

Photosynthesis, A274, A288, A302, A303
Physical chemistry and mineralogy of soils, A1087
Physical properties of plant and animal materials, A1150
Physics for the inquiring mind, A42
Physics of plant environment, A1110
Physiological aspects of photosynthesis, A291
Physiology and reproduction and artificial insemination of cattle, A798
Physiology of domestic animals, A741
Physiology of flowering plants, A310
Physiology of lactation, A804
Physiology of nematodes, A167
Physiology of trees, A297
Phytopathology, B121
Pictorial encyclopedia of plants and flowers, A223
Pictorial guide to the mammals of North America, A700
Picture book of annuals, A543
Picture book of perennials, A543a
Planned beef production, A814
Plant agriculture, A378
Plant anatomy, A283, A286, A287
Plant and animal geography, A118
Plant and soil water relationships, A296
Plant cell, A292
Plant cells, A281, A282
Plant communities, A193
Plant disease handbook, A349
Plant disease reporter, E101
Plant diseases and their chemical control, A332
Plant form and function, A230
Plant galls and gall makers, A333
Plant growth, A277
Plant growth and development, A298
Plant growth substances, A275
Plant kingdom, A188
Plant life, A219
Plant nematology, A334
Plant pathological methods: fungi and bacteria, A345
Plant pathology, A320, A346
Plant physiology, A284
Plant propagation, A441, A453
Plant propagation practices, A482
Plant pruning in pictures, A434
Plant science, A379
Plant structure and development, A301
Plant virology, A336
Plant viruses, A340

Plant viruses and virus diseases, A323
Plant world, A201
Plantation agriculture, A89
Plants, A205
Plants and civilization, A187
Plants and environment, A194
Plants and gardens, B122
Plants and man, A196
Plants at work, A307
Plants for man, A226
Plants, man, and life, A186
Plants without leaves, A210
Platyhelminthes and parasitism, A164
Pocket encyclopedia of plant galls in colour, A329
Points of the horse, A761
Poisonous plants of the United States and Canada, A214
Policy directions for U.S. agriculture, A1400
Policy process in American agriculture, A1391
Politics of conservation, A1302
Politics of pollution, A1282
Popular guide to government publications, A15, E5
Population and food supply, A1263
Population growth and land use, A1347
Population, resources, environment, A1283
Pork industry, A1394
Potato processing, A1197
Potatoes, A397
Poultry digest, B123
Poultry: feeds and nutrition, A846
Poultry inspector's handbook, E102
Poultry meat, B124
Poultry production, A826, A827
Poultry products technology, A1189
Poultry science, B125
Poultry tribune, B126
Poultryman's manual, A824
Practical baking, A1196
Practical canning, A1212
Practical course in agricultural chemistry, A1109
Practical dictionary of electricity and electronics, A991
Practical electrical wiring, A1045
Practical food microbiology and technology, A1234
Practical horticulture, A466
Practical leather technology, A26
Practical poultry management, A832
Practice of silviculture, A636
Price lists of government publications, E11

Title Index

Principles and practice of feeding farm animals, A840
Principles, equipment, and systems for corn harvesting, A380
Principles of agricultural entomology, A894
Principles of agronomy, A391
Principles of animal environment, A669
Principles of dairy chemistry, A1142
Principles of farm machinery, A1009
Principles of farm management, A1353
Principles of field crop production, A387
Principles of food science, A1126
Principles of forest entomology, A898
Principles of forest fire management, A594
Principles of genetics, A130
Principles of horticultural production, A436
Principles of horticulture, A427
Principles of hydrology, A1052
Principles of inductive rural sociology, A1403
Principles of management, A62
Principles of marketing, A81
Principles of microbial ecology, A141
Principles of microbiology, A158
Principles of nematology, A169
Principles of nutrition, A1257
Principles of plant breeding, A315
Principles of plant pathology, A343
Principles of plant physiology, A279
Principles of pollination ecology, A198
Principles of real estate law, A69
Principles of refrigeration, A1213
Principles of sensory evaluation of food, A1134
Principles of silviculture, A585
Processed plant protein foodstuffs, A1161
Producing farm crops, A407
Producing vegetable crops, A480
Production of field crops, A383
Production research reports, E103
Production systems, A78
Profitable farm management, A1357
Profitable farm marketing, A1388
Profitable sheep farming, A735
Profitable soil management, A1080
Profitable southern crops, A401

Propagation of horticultural plants, A410
Protecting our environment, A1291
Protein structure, A1153
Proteins and their reactions, A1156
Proteins as human food, A1252
Proteins: composition, structure, and function, A1151
Proteins: their chemistry and politics, A1132
Pruning manual, A425, A470
Psychological abstracts, B16
Public affairs information service bulletin, B17
Publications from commercial, industrial, and organizational sources, C10
Publications lists of state experiment stations, cooperative extension services, and state boards of agriculture, C11
Pulp and paper, A593
Pulping processes, A632
Pulpwood production, A589

Quality and stability of frozen foods, A1244
Quality control for managers and engineers, A984
Quality control for the food industry, A1240
Quality control in the food industry, A1239
Quarterly review of biology, B127
Queen rearing, A919
Questions and answers on real estate, A80
Quiet crisis, A1304

Radiation technology in food, agriculture, and biology, A1018
Raising laboratory animals, A802
Ranch economics, A1355
Range management, A807
Reader's guide to periodical literature, B18
Readings in resource management and conservation, A1272
Real estate information sources, F41
Recent developments and research in fisheries economics, A942
Recent mammals of the world, A658
Recognition of structural pests and their damage, A933
Recognizing flowering wild plants, A250
Recognizing native shrubs, A251
Recommended methods for the microbiological examination of foods, A1231

Recreation in America, A1335
Recreational use of wild land, A1330
Regional silviculture of the United States, A587
Relation of fungi to human affairs, A575
Reproduction, heredity, and sexuality, A128
Reproduction in domestic animals, A733
Reproduction in farm animals, A757
Reptiles, A661
Research centers directory, F42
Retail florist business, A548
Rice, A374
Rivers and watersheds in America's future, A1286
Robert's rules of order, A22
Rock gardening, A511
Roots in the soil, A93
Roots: miracles below, A314
Roots of the farm problem, A1358
Rose, A516
Rumen and its microbes, A764
Rural development newsletter, E104
Rural land tenure in the United States, A1399
Rural life and urbanized society, A1404
Rural recreation for profit, A1337
Rural sociology, B128

SCS national engineering handbook, E105
Safety of foods, A1236
Salesmanship, A67
Salesmanship: principles and methods, A77
Sanitation handbook of consumer protection programs, E106
Science and practice of welding, A993
Science citation index, B19
Science in agriculture, B129
Science materials catalog, D29
Science news, B130
Science of animals that serve mankind, A731
Science of botany, A234
Science of meat and meat products, A1162
Science of weather, A1117
Science of wine, A1166
Science: U.S.A., A8
Scientific American, B131
Scientific and technical societies of the United States, F43
Scientific approach, A13
Scientific feeding of chickens, A850

Title Index

Scientific principles of crop protection, A927
Sea against hunger, A949
Sea around us, A945a
Seaweed in agriculture and horticulture, A228
Secret life of the forest, A618
Secrets of plant life, A305
Seed identification manual, A259
Seed trade buyer's guide, F44
Seed world, B132
Seeds of change, A1259
Seedsmen's digest, B133
Selected films and filmstrips on food and nutrition, D30
Selected United States government publications, C12, E12
Senses of animals, A689
Service and regulatory announcements (SRA-C & MS), E107
Sex life of the animals, A713
Sheep and wool science, A745
Sheep husbandry and diseases, A752
Sheep production, A740
Sheep production and grazing management, A806
Shopwork on the farm, A1033
Shrub identification book, A272
Shrubs and vines for American gardens, A567
Signals in the animal world, A666
Silent spring, A1275
Silvicultural systems, A646
Simplified hydraulics, A988
Since silent spring, A1285
Situation and outlook reports, E108
Small fruit culture, A465
Small gas engines, A1000
Small gasoline engines training manual, A994
Small private forest in the United States, A643
Small urban spaces, A1336
Snakes: the keeper and the kept, A938
Social insect populations, A890
Soil and water conservation engineering, A1047
Soil animals, A1079, A1094
Soil conditions and plant growth, A1091
Soil conservation, A1056, A1082, A1097, E109
Soil conservation in perspective, A1077
Soil fertility, A1088, A1090
Soil fertility and fertilizers, A1100
Soil fungi and soil fertility, A1075
Soil management for conservation and production, A1070
Soil microbiology, A1102
Soil physics, A1057, A1081
Soil-plant relationships, A1062
Soil Science Society of America Proceedings, B134
Soil survey investigations reports, E110
Soil survey reports, E111
Soil testing and plant analysis, A1076
Soil use and improvement, A1098
Soils, A1073
Soils and soil fertility, A1099
Soils in relationship to crop growth, A1058
Soils that support us, A1078
Sorghum production and utilization, A400
Soup manufacture, A1167
South Dakota conservation digest, B135
Southern dairy farming, A794
Soybean, A389
Special English: agriculture, 1,2,3, A102
Specialized science information services in the United States, F45
Spices, A1190
Stable management and exercise, A762
Standard cyclopedia of horticulture, A413
Standard methods for the examination of dairy products, A1235
Standard periodical directory, F45a
Standards of perfection, A823
Starch industry, A1180
State films on agriculture, D31
Statistical abstract of the United States, E112
Statistical bulletins, E113
Statistical methods, A61
Statistical summary, E114
Statistical techniques in forestry, A624
Statistics, A52
Stockman's handbook, A746
Storehouse and stockyard management, A1350
Story of pollination, A217
Story of the plant kingdom, A192
Strategies of American water management, A1308
Strawberry, A426
Strawberry diseases, A337
Structure and properties of water, A1284
Structure of economic plants, A290
Structure of wood, A616
Study of plant communities, A224
Style manual, E29
Subject guide to major United States government publications, E4
Successful farming, B136
Successful gardening without soil, A475
Successful retail salesmanship, A79
Sugar cane, A366
Sugarcane and its diseases, A331
Summaries of federal milk marketing orders, E115
Sunset western garden book, A471
Supermarket trap, A1245
Supplemental irrigation for eastern United States, A1046
Survey of embryology, A672
Survey of the environmental science organizations in the U.S.A., F46
Surveying practice, A1035
Swine feeding and nutrition, A837
Swine production, A728, A772
Swine science, A747
Symbiosis, A174
Symposium on foods, A1155, A1156, A1158

Taxonomy of flowering plants, A263
Taylor's encyclopedia of gardening, horticulture, and landscape design, A473
Technical bulletin, E116
Technical equipment report, E117
Technical writing, A18
Technician education yearbook, A25
Techniques for electron microscopy, A151
Techniques for plant electron microscopy, A293
Techniques of landscape architecture, A562
Technology in winemaking, A1163
Technology of cereals with special reference to wheat, A1178
Technology of food preservation, A1206
Telephone directories, F47
Textbook of dendrology, A611
Textbook of entomology, A906
Textbook of horseshoeing for horseshoers and veterinarians, A776
Textbook of meat inspection, A1243

Title Index

Textbook of veterinary clinical parasitology, A882
Textbook of wood technology, A627
That we may live, A1309
Thermobacteriology in food processing, A1232
They harvest despair; the migrant farm worker, A1405
Thomas' register of American manufacturers, F48
Timber pests and diseases, A603
Tobacco, A361
Tomato paste, puree, juice, and powder, A1173
Too many: a study of the earth's biological limitations, A1271
Toxic constituents of plant foodstuffs, A1229
Trace elements in agriculture, A1108
Trace elements in human and animal nutrition, A1256
Trace elements in plants, A309
Tractors and crawlers, A1020
Tractors and their power units, A1010
Tree farm business management, A648
Tree farms, A600
Tree growth, A294
Tree identification book, A271
Tree maintenance, A457
Tree nuts, A486
Tree pathology, A341
Tree planters' notes, E118
Trees, A602
Trees and forests, A617
Trees for American gardens, A568
Trees for architecture and the landscape, A569
Trees magazine, B137
Trees of North America, A244
Trees of the eastern and central United States and Canada, A253
Trends in parks and recreation, B138
Tropical agriculture, A104
Tropical and subtropical agriculture, A100
Tropical gardening, A439
Tropical plant types, A212
Turf culture, A438
Turf-grass times, B139
Turf management, A455
Turf management handbook, A469
Turfgrass science, A440
Turkey stockman, A828
Turtox biological supplies catalog, D32
Two sides in the NFO's battle, A88

UFAW handbook on the care and management of laboratory animals, A818
Understanding and servicing fractional horsepower motors, A977
Understanding electricity and electronics, A961
Unit operations in food processing, A1207
USDA summary of registered agricultural pesticide chemical uses, E30
U.S. government films, D33
United States government publications, E3
United States standards for grades of (commodity), E119
Urban landscape design, A508
Using commercial fertilizers, A1084
Using electricity on the farm, A1027
Using the microscope, A140
Utilization research report, E120

Vegetable crop management, B140
Vegetable crops, A474
Vegetable diseases and their control, A328
Vegetable production, A452
Vegetable production and marketing, A488
Vegetables for today's gardens, A418
Vegetation and soils, A197
Vegetation mapping, A215
Vertebrate story, A699
Vertebrates of the United States, A663
Veterinarians' blue book and therapeutic index, F49
Veterinary clinical pathology, A859
Veterinary handbook for cattlemen, A854
Veterinary immunology, A870
Veterinary medicine, A857
Veterinary medicine/small animal clinician, B141
Veterinary notes for horse owners, A868
Veterinary parasitology, A874
Veterinary toxicology, A863
Viruses, A144
Viruses and the nature of life, A161
Viruses, cells, and hosts, A157
Viruses of vertebrates, A853
Vocational Education Act of 1963, E31
Vocational education, agricultural education (OE81,000-81,999), E121

Vocational education, miscellaneous (OE80,000-80,999), E122
Vocational education: the bridge between man and his work, E32

Wallaces farmer, B142
Wandering workers, A1401
Ward's catalog: biology and the earth sciences, D34
Water and water use terminology, A1306
Water crisis, A1296
Water in foods, A1148
Water in the garden, A492
Water metabolism in plants, A295
Water spectrum, E123
Water treatment for industrial and other uses, A990
Weather, A1123
Weather and agriculture, A1122
Weather and life, A1118
Weed biology and control, A359
Weed control, A352, A357
Weed identification and control in the north central states, A355
Weed science, B143
Weeds, A358
Weeds of lawn and garden, A353
Weeds of the Pacific Northwest, A354
Weeds of the world: biology and control, A356
Weeds, trees, and turf, B144
Weekly weather and crop bulletin, national summary, E124
Welding encyclopedia, A981
Welding skills and practices, A974
Welding technology, A975
Western forest industry: an economic outlook, A607
Western forest trees, A239
Wheat, A392
Wheat and wheat improvement, A394
When you preside, A24
Where animals live, A688
Who's who in the egg and poultry industries, F50
Wild animals of North America, A693
Wild flowers, A255
Wild flowers and how to grow them, A556
Wild gardener in the wild landscape, A527
Wilderness bill of rights, A1314
Wildlife biology, A1313
Wildlife conservation, A1317
Wildlife for America, A1328

Title Index

Wildlife habitat improvement, A1326
Wildlife in America, A1323
Wildlife in danger, A1316
Wildlife management, A1327
Wildlife sanctuaries, A1324
Wild turkey, A1325
Wild turkey and its management, A1318
Wine-grower's guide, A479
Wondrous world of fishes, A946
Wood and cellulose science, A640
Wood and wood products, B145
Wood as raw material, A647
Wood extractives and their significance to the pulp and paper industries, A612
Wood finishing, A596
Wood machinery processes, A619
Wood preservation, A614
Woods words, A621
Wool handbook, A811
Working reference of livestock regulatory establishments, stations, and officials, E125
World cattle, A796
World fertilizer economy, A1083
World food problem, A1260
World geography of irrigation, A1014
World of bees, A918
World of birds, A671
World of plant life, A211
World of probability, A48
World of reptiles, A662
World of zoos, A939
World wood, B146
World's a zoo, A940
Writing communications in business, A14
Written communications for business administrators, A10

Yams, A371
Yearbook of agriculture, E126
Your future in landscape architecture, A512
Your future in the nursery business, A550
Your guide to the weather, A1114
Your lawn: how to make it and keep it, A500

Zoo man's notebook, A936

SUBJECT INDEX

Cross references are used extensively to simplify finding materials in related areas. All references are to entry numbers.

Abstracting and indexing services, B1–B19
Adaptation (biology), A667, A756
Administrative communication, *see* Communication in management
Advertising, A84
Aerial photography, *see* Photography, aerial
Agribusiness, *see* Agriculture, economic aspects
Agricultural
 aviation, B20
 botany, *see* Botany, economic
 chemicals, A1106, A1111, B22, B59, B64, F25, F26; *see also* Fertilizers and manures; Growth regulating substances; Pesticides; Trace elements; etc.
 chemistry, A1109, B56, B89, B94
 conservation, E33
 contracts, A1377
 credit, A1375, B21, E36
 development, A99
 economics, *see* Agriculture, economic aspects
 education, B23, E121; *see also* Vocational education
 engineering, A1016, A1021, A1022, A1033, A1038, A1041, A1044, B24, E105
 dictionaries of, A1021
 tables of, A1021
 equipment and supplies, A1023, B67, B84, B87, E25, E117, F27, F34, F35
 extension work, E68, E82, F13
 industries, B71, E25; *see also* Agriculture, economic aspects of
 insurance, E36
 laborers, A1362, A1401, A1405, E21, E70, E71
 literature indexing, A98
 machinery, A1009, A1028, A1030, A1032, A1033, A1048, A1050, A1051, A1398, B87, E15, F27, F35, F36
 marketing, *see* Farm produce, marketing of
 meteorology, A1115, A1122, A1124; *see also* Crops and climate
 physics, A1107, A1110; *see also* Atomic energy in agriculture; Crops and climate; Soil physics
 policy, A93, A1376, A1381, A1385, A1391, A1400
 prices, A1384, A1393, E39, E49; *see also* Farm produce marketing; Produce trade
 research, *see* Research

societies, A88, A90, A1374, A1378, A1395
Agriculture, A92, A95, A96, A99, A103, A196, A378, A379, A383, A388, A391, A1298, A1364, A1366, A1368, B25, B42, B44, B55, B58, B65, B66, B68, B136, B142, E41, E45, E46, E69, E73, E83, E90, E99, E116, E126; *see also* specific subjects
 as a profession, *see* Agriculture, vocational guidance
 bibliographies of, A87, B4, B6, E10, E47, E57, E84, E91, E92
 cooperative, B53, E93
 dictionaries of, A103
 economic aspects of, A97, A1066, A1259, A1340, A1343, A1352, A1358, A1362, A1369, A1374, A1377, A1379, A1381, A1383, A1385, A1387, A1390, A1393, A1396, A1400, A1404, B21, B58, B104, B105, E34, E35, E38, E42, E45, E64, E66, E103, E108, E121
 history of, A93, A1023
 laws and legislation, *see* Farm law and legislation
 in the Middle West, A96, A1366

218

Subject Index

in the southern states, A1364
statistics, A51, A61, A1351,
E43, E53, E54, E55, E67,
E90, E113, E114
study and teaching of, B129,
E25, F3
tables of, A1366
taxation, E36
vocational guidance, A25, A86,
A94, A101, A1390; see
also under specific
subjects
Agronomy, see Agriculture; Crop
production; Soils; etc.
Air pollution, B26
Airplanes in agriculture, see
Agricultural aviation
Algae, A185, A190, A229; see also
Seaweed
Anaesthesia in veterinary surgery,
A885
Angiosperms, A256, A263, A311
Animal
breeding, A127; see also Stock
and stockbreeding
communication, A666
diseases, see Veterinary
medicine
ecology, A118, A659, A667–
A669, A675, A683,
A1103, A1313; see also
Adaptation (biology);
Soil fauna; Wildlife
management
food habits, A1322
genetics, A680, A773, A775,
A816
geography, A109, A118
habitations, A688
habits and behavior, A670,
A677, A682, A684, A687,
A693, A706, A713, A756,
A862
industry, A714, A716, A730,
A732, A788, A867,
A1189, B60, B90, B124;
see also Livestock marketing; Poultry industry;
Stock and stockbreeding
law and legislation of, A788,
A867
taxation of, A788
movements, A657
nutrition, see Feeding; Nutrition
physiology, A657, A675, A704,
A705, A741
population, A659, A668, A683,
A686, A1307
Animals
history of, A822
rare, see Rare animals
Annuals (plants), A543
Appraisal, see Valuation
Arid regions, A606

Aromatic plants, A565
Artificial insemination, A790,
A798
Artificial light gardening, A502,
A535
Associations, institutions, etc.,
F21a, F24, F24a, F42, F45a
Atomic energy in agriculture,
A1049
Audiovisual materials, C1,
D1–D34
Automatic machinery, A1102
Automobiles, A963, A972, A1005
maintenance and repair of,
A963, A970, A971,
A1004, A1006, A1007
Azalea, A533

Bacteriology, A142, A147, A156,
A159; see also under
specific subjects
Baking, A1187, A1196, A1214
Bamboo, A532
Banana, A467
Banks and banking, see Agriculture, economic aspects of
Barley, A370
Bee culture, A915, A917, A920
queen rearing, A919
Beef cattle, A736, A743, A750,
A767, A783, A812, A814,
B27
Bees, A887, A914–A920, B26,
B79
diseases of, A917
Begonias, A498
Beverages, A1177, A1198; see
also specific kinds
Biochemistry, see Biological
chemistry
Bioclimatology, A178, A1118; see
also Crops and climate
Biological
chemistry, A29, A32, A33, A36,
A38, A44, B43
control of insects, A448, A910,
A921, A1275
control of pests, A929
control of weeds, A921
specimens, collection and
preservation of, A116
Biology, A106, A108, A112–A114,
A117, A120, A121, B44,
B127; see also specific
subjects
dictionaries of, A113
encyclopedias of, A112
fieldwork, A116, A171, A183
Biometeorology, see Bioclimatology
Birds, see Ornithology
Blueberries, A431
Bonkei, A493
Bonsai, A493, A525, A557
Botanical chemistry, A288; see

also Wood chemistry
Botany, A186, A188, A189, A191,
A192, A201, A205, A209,
A211, A212, A219–A223,
A225, A227, A230, A232–
A235, A305, A391, A415,
A1322, B33; see also specific subjects
classification, see Plant classification
dictionaries of, A232, A415
economic, A186, A187, A196,
A206, A208, A226, A231,
A290, A404
dictionaries of, A231
methodology of, A345
in the Middle West, A266
nomenclature, A237, A273
in the northwestern states,
A248
in the Pacific states, A236
in the Rocky Mountains, A267
taxonomy, see Plant classification
Brazing, A985
Building estimates, A65
Bulbs, A559
Business management, A72, A74;
see also specific subjects

Cactus, A243, A503
Camellia, A563
Candy, A1188
Carbohydrates, A1152, A1157
Carbonated beverages, A1177
Careers, see Occupations; Agriculture, vocational guidance;
etc.
Carnations, A501
Catfishes, A953a
Cats, A777
Cattle, A798, A1355, B49; see also
Beef cattle; Grazing; Livestock judging; Stock and
stockbreeding; Stock
ranges; Veterinary
medicine
anatomy of, see Veterinary
anatomy
breeds and breeding of, A717,
A736, A737, A786,
A796
diseases of, A854, A872
law and legislation, see Animal
industry, law and
legislation
marketing of, A786, A787
physiology of, A834
taxation, A788
Cells, see Cytology; Plant cells
and tissues; etc.
Cereal
crops, see Grain; Corn; Sorghum; Wheat; etc.
products, A1178, A1179, A1186;

Subject Index

Cereal *(continued)*
    see also Macaroni products
Cheese, A1171, A1181
Chemical apparatus, F1
Chemicals, suppliers of, F9
Chemistry, A29, A30, A35, A40, A41, A44; *see also* Formulas; Agricultural chemistry; Botanical chemistry; etc.
    analytic, A1111, A1141; *see also* under specific subjects
    organic, A31; *see also* Food analysis; Carbohydrates; Proteins; etc.
Chickens, *see* Poultry
Chinchillas, A723
Chocolate, A1188
Chromatography, A43
Chromosomes, A129
Chrysanthemums, A505
Cities and towns, planning of, A508, A1336
Citrus fruits, A443
Climatology, A1116, A1121, E17, E58, E59, E126; *see also* Bioclimatology; Meteorology; Weather control
    agricultural, *see* Crops and climate
Climbing plants, A261, A272, A539, A567
Cloud physics, A1112
Cloud seeding, *see* Rainmaking; Weather control
Cocoa, *see* Chocolate
Coconuts, A421, A487
Coffee, A1192
Color, A7
Communicable diseases in animals, A866, A875; *see also* Veterinary hygiene; Veterinary immunology
Communication in management, A4, A10, A14
Communicators in agriculture, F15
Community leadership, A1402
Compost, B52; *see also* Fertilizers and manures
Computers, A46, A53
Concrete construction, A1017
Coniferous trees, A247, A306, A414, A586; *see also* specific kinds
Conservation
    as a profession, *see* Natural resources, vocational guidance
    of natural resources, A92, A605, A1270, A1281, A1289–A1291, A1301, A1305, B38a, B135, E61–E63, F21; *see also* Soil conservation; Water conservation; Wildlife conservation
    directories of, F46
Consumers, A82, E126
Contamination (technology), A966; *see also* Food contamination; Pollution; Sterilization
Cookery, A1129, A1145, A1191; *see also* specific subjects
Cookies, A1187
Cooperative
    agriculture, E53, E93
    education, A16
    societies, A1378, A1395
Corn, A362, A377, A399
    harvesting of, A380
    products industry, A377
Corporations, F31
Correspondence, *see* Letters
Corsages, A552
Creativity, A19
Crop
    dusting, *see* Agricultural aviation
    ecology, A405
    production, A367, A372, A387, A404, E20, E67, E126
    zones, A405
Crops, *see* Field crops; etc.
    and climate, A404, A405, A1115, E124; *see also* Agricultural physics; Agricultural meteorology
Cryptogams, A195, A210, A242, A266; *see also* Algae; Ferns; Fungi
Cytology, A122, A129, A132, A134–A137; *see also* Plant cells and tissues; Histology

Dairy
    bacteriology, A1223
    cattle, A738, A791, A793, A810, A815, A819, B57
    industry and trade, A738, B28, B82
    products, A1174, A1181, A1182, A1200, A1241, A1248, B28, F14
    analysis and examination of, A1160, A1235, A1241, A1242
Dairying, A739, A766, A794, A1208, B28, B57, B82, B95; *see also* Milk; Veterinary hygiene
Data processing, A46, A53
Deficiency diseases in plants, A342, A348; *see also* Trace elements
Dendrology, *see* Trees
Desert fauna, A682
Dictionaries of
    agricultural engineering, A1021
    agriculture, A103
    biology, A113
    botany, A232, A415
        economic, A231
    electronics, A991
    food, A1125
    gardening, A415, A428
    genetics, A133
    lakes, A1306
    landscape architecture, A537
    lumbering, A621
    natural resources, A1288
    nutrition, A1125
    plants, ornamental, A521
    reservoirs, A1306
    water supply, A1306
    zoology, A694
Diesel motor, A982
Dietetic food, A1250
Disinfection and disinfectants, A152; *see also* Sterilization
Distributive education, A16
Dog breeds and breeding, A715, A718, A729
Dogs, A715; *see also* Police dogs; Watchdogs
    anatomy of, A780
    training of, A771, A799
Domestic animals, A730, A731, A733, A741, A756, A867, B36; *see also* Stock and stockbreeding; Cattle; Swine; etc.
    anatomy of, *see* Veterinary anatomy
    behavior of, A755; *see also* Animal habits and behavior
    diseases of, *see* Veterinary medicine
    history of, A822
    parasites of, A874, A882, A932
    physiology of, *see* Veterinary physiology
Drainage, A1037, A1068
Drugs, A39
    dosage of, A855
Drying apparatus for food, A1170; *see also* Food canning and preserving; Freeze-drying
Dwarf fruit trees, A476
Dynamos, A987

Ecology, A109, A118, A141, A172, A173, A176–A178, A180–A184, A1115, A1269, A1281, A1305, A1307; *see also* Conservation of natural resources; Animal ecology; Animal population; Bioclimatology; Forest ecology; Fresh-water

Subject Index

ecology; Plant ecology; etc.
Economic
 assistance, domestic, E52, F8
 botany, *see* Botany, economic
 entomology, *see* Insects
Ecosystems, A184
Eggs, A1339, B61, B126, F50
Electric
 circuits, A962, A980
 engineering, A961, A976, A991, A1003
 machinery, A965, A968, A987
 measurements, A976
 motors, A965, A977, A987, A1042
 wiring, A1045
Electricity, A976, A989, A999, A1003
 effect on plants, A499
 in agriculture, A1013, A1027, B24, B53, B62
Electromagnetic waves, A34, A1205
Electronics, A961, A989, A991, A999
 dictionaries of, A991
Embryology, A664, A672, A703, A712, A789, A831
Encyclopedias of
 biology, A112
 gardening, A413, A417, A456, A471, A473
 organic gardening, A433
 vegetable gardening, A433
 veterinary medicine, A880
 zoology, A685
Endowments, F32
Engineering instruments, A983
English language grammar, A20, A21, A102
Entomology, A886, A888, A893, A897, A898, A902–A906, A913, B63; *see also* Insects
Environmental
 engineering, directories of, F46
 measurement, A178
 policy, A1278, A1282, A1292, B38a; E24; *see also* Conservation of natural resources; Human ecology; Man, influence on nature; Pollution
Enzymes, A370, A1138, A1154, A1215
Epidemics in animals, *see* Communicable diseases in animals
Erosion, A1060, A1298; *see also* Soil conservation; Soils
Evergreens, *see* Coniferous trees
Evolution, A107, A110, A690, A699, A1290
Experimental animals, *see* Laboratory animals

Extinct animals, A1321, A1329; *see also* Paleontology; Rare animals

Farm
 accounting, A1363
 animals, *see* Domestic animals; Stock and stockbreeding
 buildings, A398, A669, A762, A797, A1008, A1017, A1024, A1039
 equipment, *see* Agricultural equipment and supplies
 laborers, *see* Agricultural laborers
 law and legislation, A85, A91, E44, E80, E81, E82, E107
 machinery, *see* Agricultural machinery
 management, A1338, A1341, A1344–A1346, A1353, A1356, A1357, A1359–A1361, A1363, A1364, A1372, B42, B92, B136
 mechanics, *see* Agricultural engineering
 organizations, *see* Agricultural societies; name of organization
 produce, A1393, E120; *see also* Field crops; Food industry and trade; Produce trade
  grading of, A821, E119
  marketing of, A488, A1370, A1371, A1386, A1388, A1392, E37, E48, E87–E89, E126; *see also* specific commodities
  storage of, A928, A1217; *see also* specific produce
 shops, *see* Workshops
 supplies, *see* Agricultural equipment and supplies
 workers, *see* Agricultural laborers
Farmers' Educational and Cooperative Union of America, A90
Farming, *see* Agriculture
Farms, valuation of, A1373
Fats, *see* Oils and fats
Fauna, *see* Forest fauna; freshwater fauna; Zoology
Feed mills, A849
Feeding, A833, A835, A836, A838-A845, A848
 mathematical models for, A838
Feeds, A228, A833, A835, A836, A839, A840, A843–A845, A848, A849, A851, B67, B69–B71, B81, F29; *see also* Forage plants; Poultry, feeding and feeds; etc.

Ferns, A510; *see also* Cryptogams
Fertilization (biology), A105
Fertilizer industry, A1083, A1085, A1086, B72, E74
Fertilizers and manures, A228, A1069, A1071, A1084, A1085, A1092, A1093, A1095, A1096, A1100, A1105, B22, B40, B52, B64, B72, E74, F11
Fibers, A645; *see also* Wool
Field
 biology, *see* Biology field work
 crops, A318, A367, A372, A373, A376, A383, A384, A387, A391, A401, A407, A1264, B54, B55, E51; *see also* Forage plants; Grain; Horticulture; Corn; Wheat; etc.
  diseases and pests of, A330, A390
  in the southern states, A384, A401
Files and filing (documents), A98
Finance, A1342
Fir, *see* Coniferous trees
Fish
 culture, A943, A945b, A947, A953a, B29
 ponds, A943
Fisheries, A949, A956, A958, A1195, B29, E54, E60, E76
 economic aspects of, A942
 in New England, A942
 products of, A945, A1194, A1195
Fishes, A944, A946, A953, A955, A957
 diseases and pests, A945b, A948
 migration of, A951
 physiology of, A952
Fishing, A943, A1323
Flatworms, A164
Flavor, A1149, A1155
Floriculture, A451, A491, A519, A530, A531, A534, A538, A544, B85, B86; *see also* Annuals (plants); Bulbs; House plants; Perennials; Plant-breeding; specific kinds of flowers
Flower
 arrangement, A494, A495, A501, A506, A554, A558, A564; *see also* Corsages
 forcing, *see* Forcing (plants)
Flowering trees and shrubs, A245, A570
Flowers, A223, A256, A521, A561, A565; *see also* Floriculture; Wildflowers; specific kinds
 marketing of, A548

Subject Index

Fluid dynamics, A998
Food, A196, A1125, A1126, A1129–A1131, A1261, A1266, B97, B99, E126; *see also* specific subjects
  additives, A1111, A1136, A1140, A1221
  adulteration and inspection of, A1137; *see also* Meat inspection
  analysis of, A1133–A1135, A1137–A1139, A1141, A1143, A1150–A1154, A1157–A1159, A1231, B89
  bacteriology of, *see* Food, microbiology of
  canned, A1227
  canning and preserving of, A1170, A1203, A1209, A1211, A1212, A1216, B48, F4, F7, F22
  color of, A1146
  composition of, A1148, A1229, E45
  contamination of, A1221, A1229, A1230, A1236, B45
  dictionaries of, A1125
  dietetic, A1250
  drying, A1218
  frozen, A1217, A1244, F22
  habits, A1254
  handling, A1226
  industry and trade, A1126–A1128, A1130, A1131, A1161, A1172, A1187, A1204, A207, A1208, A1210, A1245, A1246, A1266, B47, B73, B74, E22, F7, F22, F30; *see also* specific subjects
  quality control of, A1239, A1240
  standards of, A1238
  law and legislation, A1236, A1238
  microbiology of, A1209, A1221, A1224, A1225, A1227, A1228, A1230–A1234
  plants, A206, A269
  poisoning, A1222, A1229
  preservation, A1049, A1128, A1205, A1206, A1216, A1220, A1224, F7; *see also* specific products
  preservatives, A1215
  research, B97
  supply, A949, A1252, A1258–A1267, A1271, A1383; *see also* specific products
  texture, A1147
Forage plants, A360, A375, A398, A402, A403, A848; *see also* specific kinds

Forcing (plants), A530; *see also* Greenhouse management
Forecasting, A64
Forest
  ecology, A618, A637
  fauna, A618
  fires, A594, A598, E75
  industries, *see* Wood-using industries
  insects, A886, A897, A898, E77
  management, A585, A599, A601, A604, A622, A625, A635, A642, A646, A648, A650, A652, B146; *see also* Tree farms
  policy, A601
  products, A626, A631, B76, B98; *see also* Lumber trade; Wood; Wood pulp; Wood-using industries
  recreation, A605, A1333
  reserves, A605, A1330
  soils, A620, A653, A655
  surveys and surveying, A639
Forestry
  law and legislation, A652, E81
  schools and education, E23
  vocational guidance, A101, A634
Forests and forestry, A583, A585, A587, A597, A600, A604–A606, A613, A615, A617, A618, A628, A630, A633, A634, A636, A642, A643, A646, A652, A655, B30, B75, B98, E117, E118
  economic aspects of, A601, A648, A654
  history of, A595, B75
  law and legislation, A652
  mensuration, A584, A615, A624, A638
  on the Pacific Coast, A270
  in the southern states, A650
  statistics of, A630
  in the West, A607
Forging, A996; *see also* Welding
Formulas, A11, A39
Foundations (endowment), F32
Freeze-drying, A1205
Freshwater
  biology, A111, A115, A170, A958
  ecology, A170, A175
  fauna, A695
Frozen food, A1217, A1244, F22
Fruit, A206, A430, A442, B32, B77; *see also* specific kinds and names
  culture, A409, A419, A422, A423, A445, A450, A460, A462, A465, A472, B31, B34, B41, B77, E78, F5; *see also* Dwarf fruit

trees; Grafting; Nurseries (horticulture); Plant propagation; Pruning; Viticulture; etc.
  diseases and pests of, A321; *see also* Plant diseases
  forcing of, *see* Forcing (plants)
  juices, A1199, A1217
  trade, B34, F5
  tropical, *see* Tropical fruit
Fungi, A185, A240, A334a, A572, A574–A577, A579, A580, A581; *see also* Bacteriology; Mushrooms; Mycology; Soil microorganisms; Truffles
  pathogenic, A325, A576; *see also* Plant diseases
Fungicides, A152, A332, A338, B22; *see also* Agricultural chemicals
Fungus diseases of plants, A334a, A345, A1237

Game
  and game birds, A1320, A1327, B39; *see also* Hunting; Wildlife management; Wild turkeys; etc.
  preserves, A1324
  protection, A1320, B103, E81; *see also* Wildlife conservation
Garden pools, A492
Gardening, A409, A435, A439, A459, A463, A464, A471, A522, A524, A531, A534, A541, A542, A565, B78, B85, B116, B122, E78; *see also* specific subjects
  dictionaries of, A415
  encyclopedias of, A413, A417, A428, A433, A456, A471, A473
  equipment and supplies, *see* Agricultural equipment and supplies
  in the northeastern states, A435
  in the southern states, A444
  in the tropics, A439
  in the West, A471
Gardens, miniature, A493
Gas and oil engines, A982, A994, A995, A1000
Generators, A987
Genetics, A123, A124, A126–A128, A130, A131, A133, A138, A317, A319; *see also* Adaptation (biology); Animal genetics; Biology; Chromosomes; Evolution; Heredity
  dictionaries of, A133
Geography, A92
Geology, A2, A17, A23

222

Subject Index

Germination, A216
Ginkgo, A247
Goats, A745, A778
Government
 information services, E18a–e, F16–20
 publications, guides to, A15, E1–E13
Grafting, A437; *see also* Plant propagation
Grain, A385, A406, A1133, A1178, A1185, A1186, B50, B51; *see also* specific plants
 analysis and chemistry, A1179
 products, *see* Cereal products
 storage, A1237
 trade, A1382, B81, F2
Grapes and grape culture, A479, A484; *see also* Wine and wine making; Viticulture
Grasses, A204, A262, A364, A365, A393, A396, A438, A455, B80, B139, B144, E126; *see also* Forage plants; Grazing; Pastures
 diseases and pests of, A328a, A440, A469
 identification of, A204, A262, E90
 in the southern states, A382
Grassland farming, *see* Pastures; Stock ranges
Grazing, A806, A807; *see also* Forage plants; Pastures; Stock ranges
Greenhouse gardening and management, A485, A497, A499, A502, A515, A530, A535, A544, A545
 diseases and pests of plants, A925
Ground
 cover plants, A272, A566
 water, A1068
Grounds maintenance, A504
Growth (plants), A277, A294, A298, A311, A312, A313, A388, A404, A1073, A1091
Growth regulating substances, A275
Gums and resins, A1172
Gymnosperms, A247, A306, A414, A586; *see also* individual kinds

Herbage, *see* Grasses
Herbicides, A351, A357, A359, A395, B143
Herbs, A206, A481
Heredity, A125, A138; *see also* Evolution; Genetics
Histology, A136, A150, A679, A711; *see also* Cytology; Microscope and microscopy; Plant anatomy

Hogs, *see* Swine
Honey, A917
Hops, A368
Horse
 breeds, A753, A761, A785; *see also* names of specific breeds
 racing, A808
Horses, A724, A744, A761, A762, A769, A800, A803; *see also* specific breeds
 anatomy of, A805, A813, A852; *see also* Veterinary anatomy
 diseases of, A852, A868; *see also* Veterinary medicine
Horseshoeing, A754, A776, A779, A817, A852
Horticulture, A424, A427, A428, A432, A436, A446, A466, A1264, B32, B86, B91; *see also* specific subjects
 study and teaching of, A523, E26
 vocational guidance, A429, A540, A550
House plants, A489, A513, A517, A518; *see also* Artificial light gardening
Human ecology, A1280, A1283, A1292; *see also* Environmental policy; Man, influence on nature; Population
Hunting, A1323
Hydraulic control, A988
Hydrology, A1001, A1034, A1052; *see also* headings beginning with Water
Hydroponics, A475
Hygiene, public, A1019; *see also* Disinfection and disinfectants; Food adulteration and inspection; Meat inspection; Pollution; Water supply; etc.

Ice cream industry, A1165
Immunity, A168
Indexing and abstracting services, B1–B19
Industrial
 safety, A5, A979
 wastes, A1025
 Information services, E18a–E18e, F16–F20, F21a, F24a, F45, F45a
Input-output analysis, A838
Insect
 galls, A329, A333
 societies, A887, A890
Insecticides, A924, A932, B120
Insects, A390, A886, A888, A893, A900–A904, A906, A911, A912, A928, A930, A932,
 B63, B96, E126; *see also* Entomology; Forest insects; Wasps; etc.
 as carriers of disease, A327, A899, A908
 behavior of, A895
 control of, A448, A894, A905, A910, A921, A1275
 cultures and culture media, A907
 identification of, A889, A909
 physiology of, A891, A892, A896, A913
Institutions and societies, F24, F42
Instructional materials, *see* Audiovisual materials
Insurance, agricultural, E36
Internal combustion engines, *see* Gas and oil engines
Inventions, A19
Inventory control, A1350
Invertebrates, A665, A691, A695
Iris (plant), A551
Irrigation, A1014, A1015, A1026, A1031, A1036, A1040, A1046, A1054, A1380, E55

Laboratories, safety measures of, A5
Laboratory animals, A748, A781, A792, A801, A802, A818, A855, A864, A883
Lactation, A804
Lakes, dictionaries of, A1306
Land, A1347–A1349, B105, E96, E116, E126; *see also* Agriculture; Cities and towns, planning of; Real property; Regional planning; etc.
 drainage, *see* Drainage
 surveying, *see* Surveying
 tenure, A1399; *see also* Farm management
 valuation of, A1373
Landscape
 architecture, A508, A537, A553, A555, A562, A569, A1336, B106, B107
 terminology of, A537
 vocational guidance, A512, A540
 drawing, A496
 gardening, A438, A490, A507, A524, A527, A528, A547, B106, B107; *see also* Grounds maintenance
Law and legislation, A69, A80, A85, A91, A652, A788, A867, A879, A1236, A1238, A1278, A1314, E14, E31, E44, E80–E82
Lawns, A328a, A438, A455, A461, A469, A477, A478, A483,

Subject Index

Lawns *(continued)*
A500; *see also* Grasses; Ground cover plants
Leadership, A3, A24, A1402
Leather, A9, A26
Legumes, A382
Leisure, A1335; *see also* Recreation
Letters, A4, A10, A14, A27
Lilies, A509
Linear programming, A838
Lipids, A1158
Livestock, *see* Domestic animals; Stock and stockbreeding
  housing, A669, A797; *see also* Farm buildings; Stables
  judging, A719, A784, A810, A819, A821
  marketing, A1365, B60
Log scaling, *see* Forests and forestry, mensuration
Lumber trade, A656, F23
Lumbering, A590, A649, B114, B146, F23
  terminology of, A621

Macaroni products, A1176
Machine shop
  practice, A986, A997, A1016, A1033, A1041
  tools, A1002, A1029
Malt, A370
Mammals, A658, A678, A693, A700, A709, A935, A936; *see also* specific subjects
  bibliographies of, A709
  of the Pacific Northwest, A681
Man, influence on nature, A1279, A1290, A1292, B46; *see also* Environmental policy; Pollution
Management, A62, A68; *see also* specific types of management
Manufacturers, F48
Manures, *see* Fertilizers and manures
Mapping, A215
Margarine, A1164
Marine
  biology, A945a, A958, A1310
  resources, A941, A949, A950, A954; *see also* Fisheries; Fishery products
Marketing, A70, A76, A81, A1246 A1365, A1370, B60, B117, E89, F37; *see also* specific commodities, Consumers; Distributive education; Retail trade
  management, A73
Mathematics, A47, A49, A50, A55–A57, A60; *see also* Probabilities
Meat, A820, A1162, A1169, A1183, A1184, A1202
  industry and trade, A1162, A1184, A1354, A1365, A1397, B108, B111, E125; *see also* specific subjects
  inspection, A1168, A1243, E86
Mechanical drawing, A973, A986
Medical entomology, *see* Insects as carriers of disease
Meetings, A3, A22, A24
Metabolism, A32, A834, A896; *see also* Mineral metabolism; Nutrition; Plants, nutrition of; Protein metabolism; etc.
Metalwork, A967, A969, A996, A997, A1022; *see also* Brazing; Solder and soldering; Welding
  machinery for, A967
Meteorology, A1114, A1117, A1119, A1120, A1123, E59; *see also* Climatology; Weather control
Microbiology, A141, A143, A145, A146, A149, A155, A158, A160, A1223; *see also* Algae; Bacteriology; Fungi; Mycology; Veterinary microbiology; Virology; etc.
Microscope and microscopy, A140, A148, A150, A151, A154, A162, A163, A293, A304; *see also* Histology; Stains and staining (microscopy)
Microwaves, A34, A1205
Migrant labor, A1401, A1405
Milk, A766, A804, A1182, A1200, B28, B99
  bacteriology of, A1223
  composition of, A1142
  dried, A1174
  marketing of, A1248, B28, E115
  pasteurization of, A1175
  plants, A1174, F14
  trade, A1248, B28
Mineral metabolism, A851
Minks, A774
Morphogenesis, A312, A702, A703; *see also* Plant morphology
Mushroom culture, A412, A468
Mushrooms, A468, A577, A578; *see also* Fungi
  diseases and pests of, A925
  edible, A269
  poisonous, A269
Mycology, A240, A571–A581

National Farmers Organization, A88
National parks and reserves, A1330, E27; *see also* Wilderness areas
Natural history, A119, A180, B38a
Natural resources, A92, A181, A1268, A1272, A1273, A1277, A1279, A1281, A1288, A1293, A1299, A1302–A1304, B30, B112; *see also* specific resources
  dictionaries of, A1288
  law and legislation, A1278, A1314
  vocational guidance, A1276, A1287, A1295
Nematode diseases of plants, A169, A334; *see also* Plant diseases
Nematodes, A165, A167, A334, A347
Nightshades, A207
Nitrogen
  fertilizers, A1093
  fixation, A308
Nonvascular plants, *see* Cryptogams
Nurseries (horticulture), A538, A549, B34, B115
  vocational guidance, A550
Nutrition, A833, A841, A842, A851, A1125, A1249–A1251, A1254–A1257, A1261, A1266, B36, B100; *see also* specific subjects
  dictionaries of, A1125
Nuts and nut trees, A447, A458, A486

Oats, A369
Occupations, A25, E94, E95; *see also* specific fields
Oceanography, A941, A945a, A959; *see also* Marine biology; Marine resources
  vocational guidance, A101
Oil engines, *see* Gas and oil engines
  hydraulic machinery, A988
Oils and fats, A1135, A1201
Onions, A381
Orchards, *see* Fruit, culture
Orchids and orchid culture, A520, A546
Organic
  chemistry, *see* Chemistry, organic
  gardening, A433, A448, B116
  encyclopedias of, A433
Ornamental horticulture, A489–A570, B32, B34, B78, B85, B86, B91, B106, B107, B115, B122, B137, B139, B144, E26
Ornithology, A671, A676, A697, A698, A710, A829, B38a
Outdoor recreation, A1333, A1334, B138, E97, E98,

Subject Index

E126; *see also* Forest recreation; Parks; Recreation; Wildlife conservation
economic aspects of, A1331, A1337
Oxidation, A1158

Pacers, A760
Packaging, F38
Paleontology, A699; *see also* Extinct animals
Pamphlets, *see* Publications, inexpensive
Papermaking and trade, A582, A588, A593, A609
quality control of, A644
Parasites
of domestic animals, A874, A882, A932
of fishes, A948
Parasitism, A164, A168, A179
Parasitology, A166; *see also* Veterinary medicine
Parks, A1336, B118, B119, B138, E27, F39; *see also* Landscape gardening; National parks and reserves; Zoological gardens
management of, A1332
Parliamentary practice, A22
Pastures, A364, A365, A386, A393, A396, A806, A807; *see also* Forage plants; Grasses; Grazing
in the southern states, A382
Peanuts, A408
Pectin, A1144
Perennials, A541, A543a
Personnel management, A63, A66; *see also* Communication in management
Pesticides, A395, A922, A927, A1294, B64, B120, E30, E100, F26, F40; *see also* Fungicides; Herbicides; Insecticides
and the environment, A1285
toxicology of, A923, A1275, A1309, B45
Pests and pest control, A921–A934, B120; *see also* Insects; Parasites
Pets, diseases of, A873
Photographic
interpretation, A960, A1001
surveying, A639
Photography, aerial, A215, A639, A960
Photosynthesis, A274, A288, A291, A302, A303
Physics, A28, A30, A37, A42
Phytopathology, *see* Plant diseases
Pine, A414, A623; *see also* Coniferous trees; Trees

Plant
anatomy, A283, A285–A287, A290, A293, A301, A304, A305; *see also* Wood, anatomy
breeding, A315–A319
cells and tissues, A136, A281, A282, A285, A286, A292, A301, A312
chemistry, A288; *see also* Wood chemistry
classification, A199, A236–A273
diseases, A169, A320–A322, A325–A327, A332, A334, A334a, A338, A339, A343–A346, A349, A350, A448, A931, B121, E45, E101, E126; *see also* diseases and pests under specific plants
ecology, A118, A193, A194, A197, A198, A203, A224, A405, A1110; *see also* Forest ecology; Symbiosis
geography, A109, A118, A197, A202, A203
growth, *see* Growth (plants)
morphology, A276, A278, A286, A290, A304, A306, A312
names, *see* Botany, nomenclature
pathology, *see* Plant diseases
physiology, A219, A275, A279, A284, A289, A298–A300, A307, A308, A310, A419, A1049, A1059, A1062; *see also* Fertilization of plants; Germination; Growth (plants); Symbiosis; Trees, physiology
propagation, A410, A437, A441, A453, A482; *see also* Seeds
protection, B20
protoplasm, A281
regulators, A275
viruses, A323, A336, A340
Plant-water relationships, A280, A295, A296
Plantations, A89
Plants, cultivated, A238, B122; *see also* Aromatic plants; Annuals (plants); Floriculture; Flowers; Gardening; Perennials; Tropical plants; etc.
drought resistance of, A542
edible, A206, A269
effect of electricity on, A499
effect of light on, A535; *see also* Artificial light gardening
nutrition of, A423, A1071, A1076, A1108, B89; *see also* Deficiency diseases

in plants; Hydroponics; Trace elements; etc.
ornamental, A491, A518, A521, A531, A538, A545, A560, B32; *see also* individual plants
dictionaries of, A521
reproduction of, A128, A198, A200, A217, A278
water requirements of, A280, A296, A1040
Plastics, A997
Poison control, information services on, E18b, F17
Poisonous plants, A213, A214, A269
Police dogs, A771, A799
Politics and government, directories of, F33
Pollination, A198, A200, A217
Pollution, A1282, A1283, A1294, A1300, A1315, B46; *see also* Man, influence on nature; Pesticides and the environment; Water pollution
Ponies, A724
Population, A1260, A1263, A1283, A1298, A1347, E56; *see also* Animal populations
rural, E28
Pork industry and trade, A1394, B83, B110
Potato products, A1197
Potatoes, A363, A397, A1197
Poultry, A727, A821, A824–A827, A831, A832, A1189, B123–B126, E120; F50; *see also* Turkeys
diseases of, A824, A830, A856, A881, B39
feeding and feeds, A846, A847, A850
grading of, A823
industry, A1189, B124, E102
marketing of, A1339
Power transmission, A978; *see also* Automobiles; Oil hydraulic machinery
Probabilities, A45, A48, A59
Produce trade, A1367, A1370, A1371, A1386, A1389, A1392, B117, E88, F6; *see also* Dairy products; Farm produce; Food supply; Fruit trade; Grain trade; etc.
Production
management, A78
systems, A78
Property valuation, *see* Valuation, etc.
Protein metabolism, A1253
Proteins, A1132, A1151, A1153, A1156, A1161, A1252

Subject Index

Pruning, A416, A425, A434, A470
Public
  health, see Hygiene, public
  speaking, A6, A12
Publications, inexpensive, C1–C12
Pulpwood, see Wood pulp

Quality control, A984; see also under specific subjects
Quarter horse, A785

Rabbits and rabbit breeds, A782, A809
  diseases of, A809
Race horses, A760, A808
Radiation, A1018
Radioisotopes, A1049
Rainmaking, A1112
Ranch life, A720, A787
Rare animals, A1316, A1319, A1321, A1329
Real estate, A69, A80, E72
  business, B37, B38, F41
  law and legislation, A69, A80
  property, A69, A80; see also land tenure
  valuation, A75, A1373, B37, B38, B92
Recreation, A1330, A1335, B30, B118, B119, B135, F28, F39; see also Forest recreation; Outdoor recreation
Reforestation, A324; see also Tree planting
Refrigeration and refrigerating machinery, A1213, A1217, A1219, A1220
Regional planning, A1279
Registers, F33
Reproduction, A660, A701, A717, A733, A757, A798, A878, A944; see also Cytology; Embryology; Fertilization (biology); Genetics; Plants, reproduction; etc.
Reptiles, A661, A662, A938
Research, A1, E40, E79, E103, E116, E120, F42
Reservoirs, dictionaries of, A1306
Resins, see Gums and resins
Retail trade, A79
Rice, A374
Rivers, A1286; see also Erosion; Water pollution
Rock gardens, A511, A514, A529
Root crops, A206; see also specific crops
Roots (botany), A314
Roses, A516, A536
Rumen, A764
Ruminants, A834; see also specific names
Rural
  communities, A1402
  conditions, A1400, A1404
  population, E28
  sociology, A1402, A1403, B128, E104; see also Urbanization

Safety measures, see Industrial safety
Saikei, A493
Salesmen and salesmanship, A67, A71, A77, A79, A83; see also Advertising; Distributive education; Marketing
Sanitation, A1019, E106; see also Disinfection and disinfectants; Pollution; Water purification; Water supply
Science, B130, B131
  history of, A8
  information services, E18c, F18
  societies, F43
Scientific methods, A13, A48
Seaweed, A228; see also Algae
Seed
  industry and trade, A403, B132, B133, F44
  plants, A266; see also Angiosperms; Gymnosperms
  production, A403, B132, B133, F44
Seeds, A216, A259, E45, E90, E126
Seepage, A1037, A1068
Semen, A871
Sense organs, A666
Senses and sensation, A689, A1134
Sex (biology), A713
Sheep, A735, A740, A745, A752, A768, A806, A1355, B113
  anatomy of, see Veterinary anatomy
  breeding of, A878
  diseases of, A752, A763
  physiology of, A834
Shop practice, see Machine shop practice
Shrubs, A251, A261, A272, A541, A560, A567, A570; see also Coniferous trees; Landscape gardening; Topiary work
  identification of, A252, A265, A272
Silos, A398
Silviculture, see Forests and forestry
Small business management, A72
Snakes, A938
Social insects, see Insect societies
Social sciences information services, E18c, F18
Soil
  chemistry, A1059, A1087
  conservation, A1047, A1056, A1060, A1066, A1070, A1072, A1074, A1077, A1080, A1082, A1097, A1098, A1297, B30, B102, E45, E105, E109
  erosion, see Erosion
  fauna, A1079, A1094, A1103; see also Insects; Soil microorganisms
  fertility, A1070, A1075, A1088, A1090, A1099, A1100; see also Organic gardening
  formation, A1067
  microorganisms, A1055, A1075, A1102
  moisture, A1068; see also Drainage; Irrigation; Plant-water relationships
  physics, A1057, A1064, A1068, A1081
  science, A1061, A1065, A1078, A1089, A1101, A1104, B134
  surveys, E45, E85, E110, E111; see also Photographic surveying; Surveying
  types, see Soils, classification of
Soils, A197, A620, A653, A1058, A1061, A1062, A1065, A1070, A1072, A1073, A1078, A1080, A1089, A1091, A1099, A1101, A1104, A1105, B55, B134, E110, E111, E126
  analysis of, A1063, A1064, A1076, A1109
  classification of, A1067, A1078
  of the southern states, A1101
Solder and soldering, A964, A985
Sorghum, A400
Soups, A1167
Soybean, A389
Species, A690
Spices, A1190, A1191
Stables, A762
Stains and staining (microscopy), A679, A711
Starch industry, A1180
Starvation, A1261
Statistical inference, see Probabilities
Statistics, A45, A48, A51, A52, A54, A58, A59, A61, E65, E112; see also under specific subjects
Sterilization, A152, A1232
Stock and stockbreeding, A669, A714, A716, A721, A722, A725–A727, A730–A732, A742, A746, A757–A759, A765, A770, A773, A775, A795, A807, A816, A871, B36, B49, B60, B90; see

226

Subject Index

also Artificial insemination; Cattle breeds and breeding; Sheep breeding; etc.
  in the West, A720
Stock ranges, A720, A806, A807, A1355, B101
Stores or stock room keeping, A1350
Strawberries, A426
  diseases and pests of, A337
Succulent plants, A503; *see also* Cactus
Sugar, manufacture and refining of, A1193
Sugarcane, A331, A366
  diseases and pests of, A331
Supermarkets, A1245
Supervision of employees, A66
Surveying, A1011, A1035, A1043; *see also* Photographic surveying; Soil surveys
Sweet potatoes, A371
Swine, A728, A734, A747, A772, A789, B83, B110
  anatomy of, *see* Veterinary anatomy
  diseases of, A860
  feeding and feeds of, A837
Symbiosis, A174, A179

Tanning, A9, A26
Taxidermy, A9
Teaching aids and devices, *see* Audiovisual materials
Technical education, *see* Vocational education
Technology, B131; *see also* Inventions
  history of, A8
  information services, E18c, F18
Thoroughbred horse, A808
Tillage, A1101
Tobacco, A361
  diseases and pests of, A335
Tomato products, A411, A1173
Tomatoes, A411, A485
Tools, *see* Agricultural equipment and supplies
Topiary work, A526
Town planning, *see* Cities and towns, planning of
Toxicology information services, E18b, F17
Trace elements, A309, A1108, A1256
Tractors, A1010, A1012, A1020, A1032, A1042, F36
Tree
  care, A454, A457
  farms, A600, A648
  planting, A324, A454
  surgery, *see* Tree, care
Trees, A218, A241, A245, A246, A294, A313, A560, A568, A585, A602, A611, A629, B137, B144, E45, E90, E126; *see also* specific subjects
  diseases and pests of, A324, A325, A341, A457, A603, A886; *see also* Forest insects; Fruit, diseases and pests of; Fungicides; Insecticides; Insects; Plant diseases; etc.
  flowering of, *see* Flowering trees and shrubs
  identification of, A239, A244, A247, A249, A252–A254, A261, A264, A265, A268, A270, A271, A560, A611
  ornamental, A245, A526, A569
  of the Pacific Coast, A270
  physiology of, A294, A297, A313
  of the southern states, A254
  of the West, A239
Tropical
  agriculture, A89, A104
  crops, A89, A100; *see also* specific crops
  fruit, A420; *see also* Citrus fruits, Banana; etc.
  plants, A212, A518
Trotters, A760
Truck farming, A480, F7
Truffles, A468
Turf, *see* Lawns
  management, A328a, A455, B80, B139, B144; *see also* Grasses; Lawns
Turkeys, A828; *see also* Wild turkeys

U.S. Department of Agriculture, E9, E10, E16, E90, E126 (1962)
Underdeveloped areas, A633
Urban design, *see* Cities and towns, planning of
Urbanization, A1362, A1404

Valuation, A65, A75, A1373, B37, B38, B92
Vegetable
  culture, B35, B41
  forcing, *see* Forcing (plants)
  gardening, A409, A418, A449, A452, A472, A474, A480, A488, F6, F7; *see also* Gardening
  encyclopedias of, A433
  juices, A1199
Vegetables, A430, A452, B32, B140; *see also* specific vegetables
  diseases and pests of, A328; *see also* Insects; Plant diseases; etc.
  marketing of, B35, B140
Vegetation mapping, A215
Vending machines, A1247
Vertebrates, A663, A699; *see also* specific classes
  anatomy of, A673
Veterinary
  anatomy, A749, A751, A865; *see also* specific animals
  hospitals, law and legislation, A879
  hygiene, A797, A861, A867
  immunology, A870
  medicine, A857, A867, A869, A873, A875–A877, B36, B88, B93, B109, B141, E50, E126, F38a, F49; *see also* Communicable diseases in animals; Domestic animals; Parasites of domestic animals; Cattle, diseases; etc.
  diagnosis of, A858, A882
  encyclopedias of, A880
  law and legislation, A867
  practice of, A879, F38a
  microbiology, A866; *see also* Virus diseases
  parasitology, *see* Parasites of domestic animals, Parasitology
  pathology, A859, A861
  physiology, A751, A757, A820; *see also* specific animals
  toxicology, A863
Vines, *see* Climbing plants
Vineyards, *see* Grapes; Viticulture
Virology, A157
Virus diseases, A853, A884, A908
  of plants, A323, A336, A340, E45
Viruses, A139, A144, A153, A161, A853, A884; *see also* type of virus
Vitamins, A1139, A1251
Viticulture, A479, A484; *see also* Wine and wine making
Vocational
  education, E14, E15, E19, E20, E22, E23, E25, E26, E31, E32, E82, E121, E122
  guidance, A25, E94, E95; *see also* specific fields

Wasps, A887
Watchdogs, A771, A799
Water, A1284, E81, E123, E126
  conservation of, A1047, A1072, B30, B102, E45
  gardens, A492
  information services, E18e, F20
  pollution, A952, A1274, A1291, A1296, A1310, A1311,

Subject Index

Water *(continued)*
    B45; *see also* Industrial wastes
  purification, A990
  resources development, A1286, A1296, A1308, E18e, F20
  supply, A1053, A1148, A1274, A1296, A1311, E18e, F20
    dictionaries of, A1306
    underground, A1068; *see also* Seepage
Waterfowl culture, A829
Watersheds, A1286
Weather control, A1113; *see also* Rainmaking; Water supply
Weed control, A352, A355, A356, A357, A359, A395, B143, B144; *see also* Herbicides
Weeds, A353, A356, A358, B143, B144, E45
  of the Middle West, A355
  of the Pacific Northwest, A354
Welding, A974, A975, A981, A985, A986, A992, A993
Wheat, A392, A394, A1179
Wild turkey, A1318, A1325
Wilderness areas, A1280, A1330
Wildflowers, A255, A257, A260, A556
  identification of, A250, A260
Wildlife
  conservation, A940, A1270, A1275, A1312, A1315–A1317, A1323, A1324, A1327–A1329, B30, B38a, B112, F12; *see also* Rare animals; Forest reserves; Game protection; National parks and reserves; etc.
  law and legislation, A1314
  management, A1313, A1326, A1327, B103; *see also* Animal populations
Wine and wine making, A1163, A1166; *see also* Grapes; Viticulture
Wood, A627, A640, A641, A647, A651, B145
  anatomy of, A610, A616
  chemistry, A591, A592, A612, A632, A640, A651
  deterioration, A603
  finishing, A596
  identification of, A239, A645, E45
  preservation of, A614
Wood pulp, A588, A589, A632, A645, A651
Wood-using industries, A608, A617, A626, B76, B146, F23
  in the West, A607
Woodwork, A608, A619, A1022, B145
Woodworking machinery, A619
Wool, A745, A763, A811, A820
  trade and industry, A811, B113
Workshops, A1029
Worms, intestinal and parasitic, *see* Nematodes; Nematode diseases of plants
Writing, A1, A6, A18, A27, E29

Yams, A371
Yeasts, A579

Zoological gardens, A935, A936, A937, A939, A940
Zoology, A674, A692, A696, A707, A708, A1322, A1323; *see also* specific subjects
  dictionaries of, A694
  ecology, *see* Animal ecology
  encyclopedias of, A685
  experimental, A864
  variation, A690; *see also* Evolution
Zoos, *see* Zoological gardens